1-800 641-3000

ろ1 ⅼⅼ ⅼ· shingtst

Fairf                      ;

D0870923

# THERMODYNAMICS AND STATISTICAL MECHANICS

PETER T. LANDSBERG

DOVER PUBLICATIONS, INC., NEW YORK

Published in Canada by General Publishing Company, Ltd., 30 Lesmill Road, Don Mills, Toronto, Ontario.
Published in the United Kingdom by Constable and Company, Ltd., 3 The Lanchesters, 162–164 Fulham Palace Road, London W6 9ER.

This Dover edition, first published in 1990, is an unabridged and corrected republication of the work originally published in 1978 by Oxford University Press, Oxford, England.

Manufactured in the United States of America
Dover Publications, Inc., 31 East 2nd Street, Mineola, N.Y. 11501

*Library of Congress Cataloging-in-Publication Data*

Landsberg, Peter Theodore.
Thermodynamics and statistical mechanics / Peter T. Landsberg.
p. cm.
"An unabridged and corrected republication of the work originally published in 1978 by Oxford University Press, Oxford, England" — T.p. verso.
Includes bibliographical references and index.
ISBN 0-486-66493-7 (pbk.)
1. Thermodynamics. 2. Statistical mechanics. I. Title.
QC311.L28 1990
536′.7—dc20 90-44302
CIP

# Preface

Our subject is an old one, but I have tried to present it here as alive, modern, and relevant. To achieve this I have occasionally departed from conventional presentations, and I have sometimes used examples not previously incorporated in a text of this kind. Although the book ranges widely, it is suitable for self-study and for undergraduate or first-year post-graduate work: mathematical steps are usually displayed in detail, and there are worked solutions to 120 problems. In addition, for the reader who confines his attention to those sections whose numbers are printed in bold type, the book contains a brief introduction to thermodynamics and statistical mechanics of about 130 pages. Sections outside this *core* deal with topics of current interest *which can be dealt with simply*, asterisked sections being a little more advanced. By this criterion, I had to omit consideration of dense systems, field theory, more details on phase changes, etc. However, I have tried not to shirk any interesting questions of principle.

Savings in space for the core material have been obtained by the following device. The statistical mechanical theory of ideal gases is presented in a unified way, and from a quantum statistical viewpoint, by assuming a simple density-of-states function $N(E) \propto E^s$, where $s$ is a constant. One can then cover, in a single argument, the theory of the system in both the non-relativistic ($s = \frac{1}{2}$) and the relativistic limit ($s = 2$) for both fermions and bosons (page 201), thus saving both effort and space. One finds that statistical mechanics yields the simple equation $pv = gU$ for these systems, where $g = 1/(s+1)$ is a constant (page 185); such systems can also be examined by purely thermodynamic arguments (problem 6.6). The sense in which thermodynamics and statistical mechanics emphasize different facets of physical systems is thus clarified and some theoretical structure is imposed on the many important applications of these ideas.

The new look I have tried to give our subject arose from the following thoughts. (a) The conversion of energy and conservation of energy will be be in the forefront of people's minds for a long time, and our subject is essential for their understanding. Furthermore, since life exists by virtue of a solar energy gradient, an unobtrusive introduction to a few ideas from irreversible thermodynamics has been included in the discussion of these matters. (b) The description of the universe as a whole, and its development in time, furnishes an exciting area in which our subject can be utilized. (c) Black holes obey a form of mechanics, the laws of which resemble those of thermodynamics, and a semi-quantitative introduction is given also to this topic. Whilst not allowing these subjects to distort the presentation (see the table at the end of this Preface), some allowance has been made for them, notably in chapters 13 and 15.

A preliminary form of sections 13.8 to 13.11 was used in lectures which I gave in a course on Solar Energy Conversion at the International Centre for Theoretical Physics, Trieste, and I am grateful to the organisers of this course for this opportunity to try out the material. I am also most grateful,for stimulating discussions,to Dr. S. Canagaratna, Sri Lanka (sections 13.8 to 13.10), and to the following colleagues and collaborators at the University of Southampton: Ray D'Inverno (Summary of key results), Steve Hope (chapter 17), Jim Mallinson (section 13.9) and David Tranah (sections 7.6, 13.6,15.8, and 18.6), and Miss Jo Clyde who helped to correct the manuscript.

For those (the author is one) who are fascinated by the interaction between physics and mathematics, several unusual points of interest are raised. How is thermodynamics related to the theorem that the arithmetic mean of two numbers is less than the geometric mean (section 5.5)? Can an increase in entropy be described in terms of the notion of equivalence classes (section 5.6)? Can the Euler formula concerning the vertices, edges, and faces of a convex polyhedron be related to the Gibbs phase rule (section 7.4)? At the same time,interesting new physics and chemistry have not been neglected, for example by hinting in section 7.8 at connections between reaction rates and a generalized concept of phase transitions.

The question of why the entropy of an adiabatically isolated system increases, and why there is an 'arrow of time', is of wide philosophical interest, but it has received different answers from different experts. One element in the understanding of this matter, namely that the fine-grained entropy is constant in time, is explained in one of the few formal mathematical passages in this book (section 9.5).

Compared with my 1961 book the present work is easier mathematically and ranges more widely. It is therefore *not a second edition*, and I had regretfully to omit some topics which remain of interest in the earlier book (temperature-dependent energy levels and the method of transition probabilities are examples) with the result that this book is of no greater length than the 1961 book. However, the *core* section of this book is much shorter! But then seventeen years after writing a big book an author should have learned enough about his subject to be able to write a shorter one - if he cares to do so!

Southampton University,
England
8th August 1978

P.T.L.

*Applications*† *discussed in this book (usually only in an introductory manner)*

| Gases and radiation | Astrophysics and related topics | Solids | Physical chemistry |
|---|---|---|---|
| 3.3 Velocity of sound. | 3.4 Polytropic fluid. | 3.2, *3.3* Stretching of wires. | 7.4 Phase rule. |
| 6.2, *6.9*, *7.5*, *8.1*, *8.2*, van der Waals gas. | 7.6, 13.6 Negative heat capacity. | 7.9 Ferromagnetic phase transition. | 7.6 Stability of phases. |
| *9.7* Maxwell distribution. | *11.8*, Ch. 12, *12.5* - *12.7* Gas of particles having relativistic speeds. | *7.3*, *7.5* Joule-Thomson cooling. | 7.10 Chemical reactions. |
| 11.7, *12.2* Equipartition. | 13.4, 13.5 Matter and radiation in Universe. | *7.11* Adiabatic demagnetization. | *7.5* Virial coefficients. |
| 13. Radiation. | 13.6 Radiation and black holes. | 10.3-10.6 Negative temperature and spins. | *7.6* Lowering of freezing point. |
| 14.5 Gas heat capacities. | 13.7 Solar constant. | 11.7, *12.2* Equipartition. | 8, 15.6 Mass action law. |
| 16.6 Bose condensation. | *13.4* -*5* Electrons, positrons in equilibrium with radiations. | 13.8 Electroluminescent diode. | 13.8 Photosynthesis efficiency. |
| | 15.7 Collapse to black holes. | 14 Debye theory, phonons. | *14.9* Vapour pressure of Einstein solid. |
| | 15.8 Black hole thermodynamics. | 14.5 Low temperature metal heat capacity; rotons. | *15.6*, *15.7* Adsorption isotherms. |
| | 18.6 General relativity and thermodynamics. | *14.4* Anharmonic oscillator. | 17.2 Electrochemical potential. |
| | 18.8 A maximum temperature? | 15.2-15.6 Metals and semi-conductors. | |
| | | 15.9 Population inversion and lasers. | |
| | | 17 Transport theory. | |
| | | 18.4 Solar cell. | |

†This list is not exhaustive and gives an indication only. Numbers give chapters and sections. Italic numbers give problem numbers.

# Contents

# PART A: THERMODYNAMICS

# 1. Nature and scope of thermodynamics

Thermodynamics is a study which seeks to establish quantitative relationships among macroscopic variables. These variables describe an arbitrary physical system, which is very large compared with atomic dimensions, when this system is in any one of a large number of equilibrium states. Thus, unless the contrary is explicitly stated, as in the study of surface tension phenomena for instance, surface effects can be neglected compared with volume effects. Thermodynamics also seeks to establish relationships between the values of the variables which specify initial and final equilibrium states of a system which is interacting with another system. Some remarks will elucidate this definition.

(*a*) The macroscopic variables include quantities like pressure, volume, molecular concentrations, and the like, which can all be measured by known methods. They may also include electrical and magnetic variables and possibly some specific thermodynamic quantities. The existence and properties of these latter quantities such as temperature, entropy, and chemical potential can be discussed with the foundations of the subject, and it is then convenient to assume that the use of typically thermodynamic quantities has been avoided in the specification of thermodynamic states. We shall here be less abstract, however, and treat temperature and heat as familiar quantities. Typically microscopic information, such as the occupation probabilities of quantum states, are never needed for thermodynamics. It is assumed in our definition that one can normally agree on what constitutes a suitable set of macroscopic variables for a physical system in a given group of equilibrium states.

(*b*) Since the physical system under investigation must undergo certain changes when its thermodynamic properties are studied, one must also agree on the domain of permissible changes. For example, if a gas is confined to a cylinder with an adjustable piston, one may call this a system S. If the piston is moved, one will normally regard this modification as a state of the system S, rather than as a state of an entirely new system T. However, if the piston is removed completely, so that the gas can escape into a larger volume, one may or may not regard the resulting arrangement as belonging to a new system.

(*c*) When a system, together with its domain of permissible changes, has been specified in ordinary language and a set of values for its thermodynamic variables has also been chosen, one can ask another person to set up copies of this system for which these variables have the same values within certain tolerances which must also be specified. One can now ask a third person to attempt to find a macroscopic difference between these systems (other than their location in space and time) by making arbitrary measurements to the agreed accuracy. If he succeeds in finding a macroscopic difference, one includes the value of the variable measured by him in the specification of the first system, so that this value must now also be exhibited by the copies of the system. The third person is now again asked to find a difference. By repeating this procedure a situation is eventually attained such that the systems under consideration have to be regarded as macroscopically similar by any third observer. One then says that one has arrived at a macroscopic specification of the state of the first system.

(*d*) Our definition presumes also that agreement can be obtained on what constitutes an equilibrium state. It used to be said that an equilibrium state is one in which the physical variables of the system are all constant in time. However, we now know that the fluctuations which these variables undergo under normal conditions can in fact never be removed. Therefore it would be better to look for the absence of systematic trends in the time averages of the physical variables involved (over times judged as 'reasonable'). The times required may be very large when the systematic changes which are suspected are very slow, and this has created difficulties in the past, for instance in connection with the third law of thermodynamics, for states which are not true equilibrium states. However, the science of thermodynamics has grown up precisely because it is normally possible to make convenient conventions on what one is to regard as necessary macroscopic variables, as permissible changes of a system, as equilibrium states, and so on.

Another complication must be borne in mind. Suppose one is cooling a certain mass of water, so that one has at one time ice and water in equilibrium, and at another time only ice. It is clear that the variable specifying the amount of water present may be redundant. If the permissible changes of the system include the vaporization of the water and the dissociation of the water molecules, then the variable specifying the number of hydrogen and oxygen molecules present may sometimes be required and sometimes be redundant. The macroscopic variables of real interest to be associated with a physical system will therefore depend on what one might call the approximate state of the system ('ice only' or 'ice and water', for example). Having specified that state, it will always be possible to give a maximum number, $n$ say, of independent macroscopic variables which define the equilibrium state of the system uniquely. It is clear that the nature of these variables and their number depend in general on the approximate state of the system. It will depend in addition on personal preference, since alternative sets of independent variables will in general be available. However, $n$ must satisfy $n \geq 2$. We shall call macroscopic variables which can enter into the specification of an equilibrium state simply thermodynamic variables and the state a thermodynamic state. By regarding the thermodynamic variables of a system as co-ordinate axes of an $n$-dimensional space, one can picture an equilibrium state of a system as represented by a point in the appropriate phase space.

Since the nature of the thermodynamic variables and their number can vary within wide limits, the basic theoretical framework of thermodynamics must be kept very general. This has the advantage of giving the theory a wide range of application, but this is balanced by the drawback that thermodynamic reasoning is in general unsuitable for giving insight into the details of physical processes.

This last observation, together with the remark that thermodynamics leaves microscopic variables on one side, leads to the conclusion that a thermodynamic theory is necessarily incomplete. For any system to which thermodynamics can be applied, a more exhaustive theory should exist which yields insight into the detailed physical processes involved. This deeper theory must also give some account of the nature of equilibrium states, and therefore of fluctuation phenomena, and of course it must lead to the thermodynamics of the system. Such theories are provided by statistical mechanics. Since a statistical-mechanical theory must exist for every system to which thermodynamics can be applied, it follows that the whole of thermodynamics should in fact be deducible from statistical mechanics if the latter is formulated in a sufficiently general way. This presents problems which are still under discussion (see section 9.4).

# 2. Basic concepts; the laws of thermodynamics

## 2.1. EQUATIONS OF STATE; QUASI-STATIC PROCESSES

Taking the main thermodynamic variables such as pressure ($p$), volume ($v$) and temperature ($t$) as intuitively clear, one finds that for a given mass of gas in equilibrium two of these determine the third. Such relations are called *equations of state*:

$p = p(v,t)$ is a *thermal* equation of state ;

$U = U(v,t)$ is a *caloric* equation of state ;

here $U$ is the total internal energy of the gas and will be discussed more formally later (section 2.3). By changing the variables and after each change allowing the system to return to equilibrium before one is estimating the values of the variables, one can plot various states characteristic of the systems in the thermodynamic phase spaces. Curves drawn through these points are such that each point represents an equilibrium state. One can imagine a system being taken slowly through all the states represented on such a curve. This process is called a *quasi-static* process. Typical curves (or quasi-static processes) are isotherms, for which temperature is kept constant, isobars, for which the pressure is kept constant, and isochores, for which the volume is kept constant. It is desirable to represent such families of empirical curves by (approximate) analytical expressions in an *equation of state*. Frequent use has also to be made of processes which involve non-equilibrium states and which are therefore not quasi-static. Such processes are called *non-static*.

## 2.2. ADIABATIC PROCESSES

An *adiabatic process* is one which does not allow the passage to or from the system of heat or radiation. However, one can use pistons, rods, stirrers, and any mechanical appliances in such a process. In this way one can raise the temperature of a system adiabatically by frictional rubbing (a non-static process), and lower its temperature adiabatically by gaseous expansion, which may be quasi-static or non-static. Suppose now that two arbitrary systems 1 and 2, say, are placed in an *adiabatic enclosure*, i.e. they can interact only adiabatically with the outside world. If they can interact thermally with each other, then they will reach thermal equilibrium after a certain time, and this is a restriction on the previously independent set of variables $\{x_1\},\{x_2\}$ of systems 1 and 2 respectively. Thus *thermal equilibrium* is specified by a relation $f_{12}(\{x_1\},\{x_2\}) = 0$. This relation is assumed to be unique in the sense that, if all but one variable is known, $f = 0$ determines the remaining variable uniquely. This remaining variable is in fact a measure of the temperature. We shall assume that each system is in equilibrium specified by only one such temperature function $t$. The generalization to several temperatures is possible and is important if a system itself contains an adiabatic partition of some sort which does not allow the passage of heat through it. Its equilibrium states are then specified only when the temperatures of the various parts of the system are also given.

*Example.* If two ideal gases have pressure, volume, and temperature related by

$$p_i v_i = A_i t_i \quad (i = 1, 2)$$

then thermodynamic equilibrium between them implies

$$\frac{p_1 v_1}{A_1} - \frac{p_2 v_2}{A_2} = 0 .$$

This relation presumes that each gas occupies its own volume $v_i$ only and that the partition separating them is *not* adiabatic, i.e. it allows heat to flow between them, so that in this way thermal equilibrium can be established. In this case the set of variables $\{x_1\}$ consists of $p_1$ and $v_1$, and

$$f_{12}(\{x_1\}, \{x_2\}) = \frac{p_1v_1}{A_1} - \frac{p_2v_2}{A_2} .$$

## 2.3. THE LAWS OF THERMODYNAMICS

The four laws of thermodynamics each assert something about the thermodynamic functions of general physical systems in thermodynamic phase space. The co-ordinates of the phase space **E** represent a set of $n$ independent macroscopic variables of the system, where $n$ is some integer, but they are otherwise arbitrary. What we have done so far assumes that in E there are sets of points β (say) such that any two can be linked by an adiabatic process. For such sets the *zeroth law* asserts the existence of an empirical temperature function for each system. One should discuss the conditions for an empirical temperature scale to be satisfactory, but this would lead too far here. Sometimes the zeroth law is stated as asserting only the transitive property of thermal equilibrium: If two bodies 1,2 are in equilibrium with a third body 3, then they are in equilibrium with each other. One then deduces the existence of an empirical temperature function from three statements, each expressing the constraint on the thermodynamic variables $x$ imposed by the equilibrium between pairs of systems:

$$f_{13}(x_1,x_3) = 0, \quad f_{23}(x_2,x_3) = 0, \quad f_{12}(x_1,x_2) = 0 .$$

One does so by requiring that any two must imply the third. We need not here go into the subtleties of these arguments (see section 3.6). A particularly important system, for which the value of the remaining variable determined by $f_{12}(x_1,x_2) = 0$ is read off a scale, is a thermometer. The length of a mercury column, for example, is taken as a measure of the temperature $t$ of itself and of the system with which it is in thermal equilibrium. Such temperature scales $t$ are called *empirical* to distinguish them from the absolute temperature scale $T$, which will be discussed later. We now state the first law, using the concept of a *closed system*; this is a system whose mass is fixed. This law leads to the existence of another thermodynamic variable, called the internal energy, $U$ say. The law states: The energy supplied to (or furnished by) a closed system in an adiabatic process which links two equilibrium states represented in sets β is independent of the intermediate states. *This formulation allows one to go outside the states represented in the phase space, and away from equilibrium, to achieve this linkage.* As a consequence of the first law a scalar point function $U(x)$ can be defined on the sets β of **E**. If two points $x_1, x_2$ of β are linked adiabatically, the energy difference involved is then $|U(x_1) - U(x_2)|$ and is independent of how the linkage has been achieved. If points $x_1, x_2$ can be linked without heat transfer then linkage should be more readily realizable if heat transfer is allowed. In a general (i.e. not necessarily adiabatic) process let $W(1,2)$ be the mechanical work done on the system in effecting the linkage, while $W_a(1,2)$ is the work required for an adiabatic linkage. Then

$$W_a(1,2) = U(2) - U(1) .$$

Let the saving in mechanical work to effect the linkage in a non-adiabatic process be $Q(1,2)$ where

$$Q(1,2) + W(1,2) = U(2) - U(1) = W_a(1,2) .$$

It is clear that $Q(1,2)$ can then be interpreted as the *heat energy* supplied to the system.

The process may be non-static, i.e. it can proceed via non-equilibrium states. Here $U(1,2) = U(2) - U(1)$ cannot depend on the intermediate states, but $W(1,2)$ can do so, as shown below. It follows that $Q$ is also path-dependent, for only in this way can $U(2) - U(1)$ be path-independent. Thus a better notation for these quantities would be $Q(1,\ldots,2)$, $W(1,\ldots,2)$. If the process is infinitesimal one can write, denoting by $\mathrm{d}W$ the work done by the fluid,

$$\boxed{\mathrm{d}'Q - \mathrm{d}'W = \mathrm{d}U} \qquad (2.1)$$

where d' denotes an *inexact* (or *incomplete*) differential. This means that functions $Q(\mathbf{x}), W(\mathbf{x})$ of position $\mathbf{x}$ in the phase space may not exist. For if they did, $Q(1,2)$ would be $\int_1^2 dQ = Q_2 - Q_1$, and would therefore be path-independent. We now illustrate the path dependence of the work done.

If a fluid expands *quasi-statically* the work done by it on an external piston in an incremental process is $p\mathrm{d}v$ where $p$ is the pressure and $\mathrm{d}v$ the change in volume. This can be seen most simply for a cylindrical shape. If $A$ is the cross-sectional area, $pA$ is the force exerted on the piston confining the fluid. If it is allowed to move out through a distance $\mathrm{d}l$, then the work done is

$$pA\mathrm{d}l = p\mathrm{d}v, \quad \mathrm{d}v \equiv A\mathrm{d}l \ .$$

Suppose the system is changed from state A to state B along path 1 or path 2 as shown in Fig.2.1. Then the work

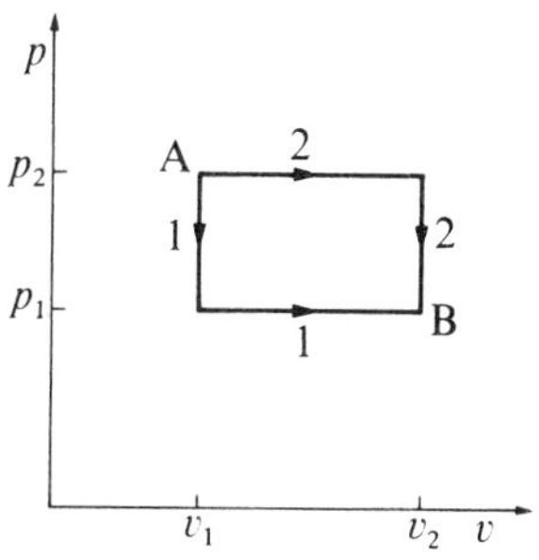


Fig.2.1. Paths 1 and 2 join the states A and B and require different amounts of work to be done.

done on a vertical path is zero since $v$ is constant. Hence

$$W_1 = p_1(v_2 - v_1) \quad \text{and} \quad W_2 = p_2(v_2 - v_1) \ .$$

The work done in linking states A and B quasi-statically thus depends on the path chosen. This path-dependence implies (see Chapter 4) that the increment of work is *inexact*.

In the nineteenth century, and before, much ingenuity was

applied to discovering a machine which, acting cylically, was able to supply work to its surroundings although it was effectively isolated. It can be called a *perpetual motion machine of the first kind.* If each cycle contains a point in a set β of the appropriate thermodynamic phase space, one is involved in a repeated linkage of this state with itself and the same internal energy $U$ would occur again and again. Hence there can be no flow of energy from such a closed system. The perpetual motion engine would therefore violate the first law of thermodynamics. It also violates the *law of the conservation of energy.* This applies also to non-equilibrium situations such as particle collisions, and the two laws overlap, neither being more general than the other.

The *second law* operates only in open, connected subsets of sets β. These will be denoted by γ. The purpose of this restriction is to leave out complications which could arise from boundary points as noted on p. 31. The states of these boundary points are settled by the *third law* of thermodynamics. These two laws are discussed later. However, one can say that the second law asserts, roughly speaking, the existence of additional functions of the state of a system, namely the entropy $S$ and the absolute temperature $T$. The third law then asserts that states at $T = 0$ cannot lie in a set β nor therefore in a set γ; they are in fact *unattainable by any method.*

Here is an alternative summary of the 'laws': The first law says that you cannot get anything for nothing; the second law says that you can get something for nothing, but only at the absolute zero; the third law (section 6.3) implies that you cannot get to the absolute zero. The fourth law (section 7.1) does not fit into this summary. In fact, there cannot really be a fourth law for the following reason. There were three main personalities whose work led to the formulation of the first law: J.R.Mayer, H. von Helmholtz, and J.P. Joule. Two people, Carnot and Clausius, were the main pioneers of the second law, while only one person, Nernst, was involved in the original statement of the third law. Thus nobody can formulate a fourth law. Now the zeroth law does not fit into this scheme, but then everybody knows about that law!

# 3. Thermodynamics of simple systems using the first law only

## 3.1. FLUIDS

The simplest systems are fluids of a single chemical substance which is either gaseous or liquid. Their equilibrium states are specified by two independent variables which can be chosen from an empirical temperature $t$, volume $v$, and pressure $p$. If $d'Q$ is an increment of heat added quasi-statically then one can write for such a quasi-static process

$$\boxed{d'Q = C_v dt + l_v dv = C_p dt + l_p dp = m_v dv + m_p dp} \quad (3.1a)$$

where the coefficients are themselves functions characteristic of the fluid. The heat capacities at constant volume and constant pressure are denoted by $C_v$ and $C_p$ respectively and are defined as the limits as $\Delta t$ tends to zero of ratios of increments

$$C_v = \left\{\lim \frac{\Delta Q}{\Delta t}\right\}_v , \quad C_p = \left\{\lim \frac{\Delta Q}{\Delta t}\right\}_p . \quad (3.1b)$$

The heat required per unit rise of pressure at constant temperature is called the latent heat of pressure increase and is denoted by $l_p$. The latent heat of volume increase is $l_v$. The $m$'s are not independent quantities for we have

$$(C_p - C_v)dt = l_v dv - l_p dp \quad (3.2)$$

so that substitution for $dt$ in equation (3.1$a$) yields

$$d'Q = \frac{C_p}{C_p - C_v}(l_v dv - l_p dp) + l_p dp .$$

It follows that

$$\boxed{m_v = \frac{l_v C_p}{C_p - C_v} , \quad m_p = -\frac{l_p C_v}{C_p - C_v}} \quad (3.3)$$

In the manipulation of partial differentials it is important to observe that if $A,B,C$ be three variables subject to

$$f(A,B,C) = 0 \tag{3.4}$$

then

$$\boxed{\left(\frac{\partial A}{\partial C}\right)_B = \frac{1}{(\partial C/\partial A)_B}} \tag{3.5}$$

and

$$\boxed{\left(\frac{\partial A}{\partial B}\right)_C \left(\frac{\partial B}{\partial C}\right)_A \left(\frac{\partial C}{\partial A}\right)_B = -1 \quad .} \tag{3.6}$$

Equation (3.4) is the abstract statement of an equation of state for simple fluids. The proof of (3.5) and (3.6) proceeds by deducing from (3.4) that

$$\mathrm{d}f = \left(\frac{\partial f}{\partial A}\right)_{B,C} \mathrm{d}A + \left(\frac{\partial f}{\partial B}\right)_{C,A} \mathrm{d}B + \left(\frac{\partial f}{\partial C}\right)_{A,B} \mathrm{d}C = 0 \ . \tag{3.7}$$

For constant $A$ this becomes

$$\left(\frac{\partial f}{\partial B}\right)_{C,A} \left(\frac{\partial B}{\partial C}\right)_A = - \left(\frac{\partial f}{\partial C}\right)_{A,B}$$

that is

$$\left(\frac{\partial B}{\partial C}\right)_A = - \frac{(\partial f/\partial C)_{A,B}}{(\partial f/\partial B)_{C,A}}$$

and similarly

$$\left(\frac{\partial C}{\partial A}\right)_B = - \frac{(\partial f/\partial A)_{B,C}}{(\partial f/\partial C)_{A,B}} \tag{3.8}$$

$$\left(\frac{\partial A}{\partial B}\right)_C = - \frac{(\partial f/\partial B)_{A,C}}{(\partial f/\partial A)_{B,C}} \ .$$

Multiplying these three equations together yields equation (3.6). Interchanging $A$ and $C$ in (3.8) yields (3.5). These results are used, for example, in problem 3.4.

If the first law is involved, one can write (2.1) as

$$\boxed{d'Q = dU + p\,dv} \tag{3.9}$$

where $d'W = p\,dv$ is the mechanical work done *by* the fluid so that only $d'Q - d'W$ is left for the increase in internal energy. As an application we shall consider the Grüneisen ratio which is also important in the thermodynamics of solids

$$\boxed{\Gamma \equiv \frac{\alpha_p v}{K_t C_v},} \tag{3.10}$$

where the coefficient of volume expansion $\alpha_p$ and the compressibility $K_X$ are defined by

$$\boxed{\alpha_p \equiv \frac{1}{v}\left(\frac{\partial v}{\partial t}\right)_p, \qquad K_X \equiv -\frac{1}{v}\left(\frac{\partial v}{\partial p}\right)_X}$$

and $X$ is some thermodynamic constraint for the partial derivative. In (3.10) the isothermal compressibility $K_t$ occurs. We shall now show that $\Gamma$ can be written as

$$\Gamma = \frac{v}{(\partial U/\partial p)_v} \tag{3.11}$$

Using (3.5) and (3.6) with $(A,B,C) \equiv (t,v,p)$

$$\Gamma = -\frac{v(\partial v/\partial t)_p}{C_v(\partial v/\partial p)_t} = \frac{v}{C_v}\left(\frac{\partial p}{\partial t}\right)_v \tag{3.12}$$

but

$$\begin{aligned} d'Q &= C_v dt + l_v dv \\ &= dU + p dv \\ &= \left(\frac{\partial U}{\partial t}\right)_v dt + \left\{\left(\frac{\partial U}{\partial v}\right)_t + p\right\} dv \end{aligned} \tag{3.13}$$

so that

$$\boxed{C_v = \left(\frac{\partial U}{\partial t}\right)_v = \left(\frac{\partial U}{\partial p}\right)_v \left(\frac{\partial p}{\partial t}\right)_v .} \tag{3.14}$$

From (3.12) and (3.14) one obtains (3.11).

Many solids have reasonably constant values of $\Gamma$ lying between values 1 and 3 at moderate temperatures. Examples are NaCl (1.63), Cu (1.96), and Na (1.25). Integration of (3.11) suggests an equation of state of such solids given by

$$U = \frac{1}{\Gamma} pv + f(v)$$

where $f(v)$ is a 'constant' of integration.

Conversely, we shall find that black-body radiation and simple theories of electrons in metals lead to an equation of state $pv = gU$ where $g$ is a constant of order unity. Equation (3.11) enables one to infer at once that such systems have *constant* Grüneisen ratios $\Gamma = g$. Such inferences are typical of thermodynamic arguments. Systems having $pv = gU$ will be called *ideal quantum gases* (see problem 6.6 ).

Equation (3.14) may be given an alternative proof by noting that for any function $f$ of two independent variables (the third being fixed by the equation of state)

$$\mathrm{d}f = \left(\frac{\partial f}{\partial p}\right)_v \mathrm{d}p + \left(\frac{\partial f}{\partial v}\right)_p \mathrm{d}v \tag{3.15}$$

$$= \left(\frac{\partial f}{\partial p}\right)_v \left\{\left(\frac{\partial p}{\partial v}\right)_t \mathrm{d}v + \left(\frac{\partial p}{\partial t}\right)_v \mathrm{d}t\right\} + \left(\frac{\partial f}{\partial v}\right)_p \mathrm{d}v$$

$$= \left(\frac{\partial f}{\partial p}\right)_v \left(\frac{\partial p}{\partial t}\right)_v \mathrm{d}t + \left\{\left(\frac{\partial f}{\partial p}\right)_v \left(\frac{\partial p}{\partial v}\right)_t + \left(\frac{\partial f}{\partial v}\right)_p\right\} \mathrm{d}v$$

but

$$\mathrm{d}f = \left(\frac{\partial f}{\partial t}\right)_v \mathrm{d}t + \left(\frac{\partial f}{\partial v}\right)_t \mathrm{d}v .$$

Comparing coefficients of $\mathrm{d}t$ and $\mathrm{d}v$ ,

$$\left(\frac{\partial f}{\partial t}\right)_v = \left(\frac{\partial f}{\partial p}\right)_v \left(\frac{\partial p}{\partial t}\right)_v$$

$$\left(\frac{\partial f}{\partial v}\right)_t = \left(\frac{\partial f}{\partial p}\right)_v \left(\frac{\partial p}{\partial v}\right)_t + \left(\frac{\partial f}{\partial v}\right)_p .$$

The first of these results justifies (3.14) (with $f = U$). The second shows formally how to differentiate (3.15) with respect to volume under the constraint of constant pressure.

## 3.2. OTHER SIMPLE SYSTEMS

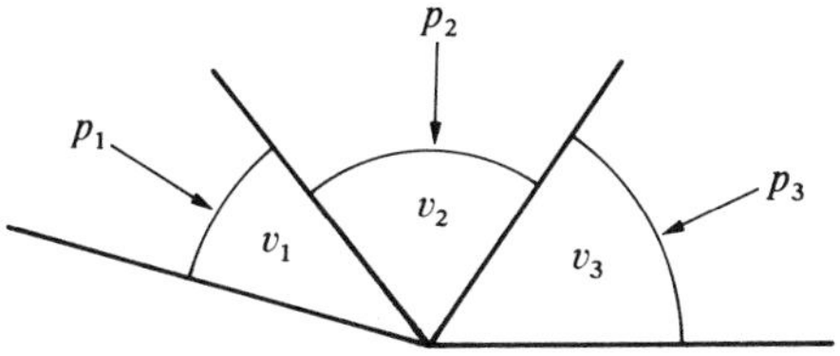


Fig.3.1. A fluid containing two rigid partitions.

A fluid may contain rigid partitions which *prevent* pressure equalization but *allow* temperature equalization. In that case there are $n$-1 'deformation variables' $v_1,\ldots,v_{n-1}$ and one thermal variable such as the temperature or the internal energy. Thus one sees that even for simple fluids thermodynamic space can be $n$-dimensional where $n$ is any integer $n \geq 2$.

More generally, one can think of many physical situations in which work terms of the type $p_s\,\mathrm{d}v_s$ can usefully be introduced. So far we have considered only the compression of fluids and we shall consider below the stretching of a wire. In addition the formalism applies to a magnetic system in which the work term can be $M\mathrm{d}H$ or $H\mathrm{d}M$, where $H$ is the magnetic field and $M$ is the magnetization. The variable $v$ is here replaced by either $H$ or $M$ which, like $v$, can be altered quasi-statically and arbitrarily within wide limits. For surface films $p$ is replaced by the surface tension and $\mathrm{d}v$ by the increment of area. There are in addition a variety of chemical systems. The interpretation of the variables $v_s$ is intended to cover these situations.

Suppose that a simple wire of length $l$ at tension $F$ has length $l_0$ at zero tension. If it is within the elastic limit and at constant temperature $t$, Hooke's law $F = A(l-l_0)$ holds. For a quasi-static change, replacing $p$ by $F$ and $v$ by $-l$,

$$d'Q = dU - F\,dl$$

where the work done *by* the rod on being stretched is $-F\,dl$ (i.e. positive work is done *on* the rod).

The sign of the work term $d'W$ can cause trouble. If (3.9) is written in the form

$$d'Q - p dv = dU \quad \text{or} \quad d'Q + F dl = dU$$

one sees that the difference in sign between $d'W = -pdv$ and $d'W = Fdl$ arises from the fact that $-dv$ and $+dl$ both correspond to work done on the system: to compression of the gas or to stretching of a wire

The linear expansion coefficient (at constant $F$) and Young's modulus at constant $X$ are

$$\alpha \equiv \frac{1}{l}\left(\frac{\partial l}{\partial t}\right)_F \qquad Y_X \equiv \frac{l}{A}\left(\frac{\partial F}{\partial l}\right)_X \tag{3.16}$$

$A$ being the cross-sectional area of the wire. This definition of Young's modulus is a development of the elementary notion

$$Y = \frac{\text{stress}}{\text{strain}} = \frac{F/A}{\delta l/l}\,.$$

As an example note that

$$\left(\frac{\partial t}{\partial F}\right)_l = -\left\{\left(\frac{\partial l}{\partial t}\right)_F\left(\frac{\partial F}{\partial l}\right)_t\right\}^{-1} = -\left\{\alpha l\,.\,Y_t\,\frac{A}{l}\right\}^{-1} = -\frac{1}{Y_t\alpha A}\,.$$

It will be taken for granted that $Y_t > 0$, i.e. that tensile stress produces an elongation. Then the above equation tells one that a wire which expands on heating (i.e. $\alpha > 0$) is cooled on stretching (i.e. $(\partial t/\partial F)_l < 0$). This conclusion would be by no means obvious without the sort of theoretical apparatus provided by thermodynamics.

## 3.3. THE VELOCITY OF SOUND

The velocity of sound is closely connected with the adiabatic compressibility, and its measurement is thus of importance for some ways of estimating the compressibility. In sound waves pressure and density changes are propagated through a fluid, the pressure variations for audible sound being of the order of $10^{-7}$ atm† in air. We shall discuss the case of small disturbances propagated in one direction (the $x$-axis).

Suppose two cross-sections of area $A$ of a fluid are in positions $x+y(x,t)$ and $x+\delta x+y(x+\delta x,t)$ at time $t$ as a result of the disturbance, their undisturbed positions being $x, x+\delta x$. The displacement is specified by the function $y(x,t)$. The thickness has therefore been changed from $\delta x$ to

$$\delta x+y(x+\delta x,t) - y(x,t) = \delta x+\left(\frac{\partial y}{\partial x}\right)_t \delta x$$

$$= \left\{1 + \left(\frac{\partial y}{\partial x}\right)_t\right\}\delta x .$$

It follows that the density has been changed from an undisturbed value $\rho$ to a new value $\rho+\delta\rho$ (say). This new density is obtained from mass conservation $\rho\delta x = (\rho+\delta\rho)\{1 + (\partial y/\partial x)_t\}\delta x$ to be

$$\rho + \delta\rho = \frac{\rho}{1+(\partial y/\partial x)_t}$$

$$\doteqdot \rho\left\{1 - \left(\frac{\partial y}{\partial x}\right)_t\right\} ,$$

i.e.

$$\delta\rho = -\rho\left(\frac{\partial y}{\partial x}\right)_t . \qquad (3.17)$$

† The unit of force in SI units is the Newton: $1\text{N} = 1\ \text{kg m s}^{-2} = 1\ \text{J m}^{-1}$ where the joule ($= 1\ \text{kg m}^2\ \text{s}^{-2}$) is the unit of energy. The unit of pressure is $\text{Nm}^{-2}$. However, SI units also allow the use of the bar as a unit of pressure. This is $10^5\ \text{N m}^{-2}$. The atmosphere ($101{\cdot}325\ \text{N m}^{-2}$) is not an allowed unit of pressure in SI.

Suppose that the pressures on the ends of our typical slab of material are $p+\delta p(x,t)$ and $p+\delta p(x+\delta x,t)$ where $p$ is the uniform pressure in the absence of a disturbance. The force on the slab is

$$A\ \delta p(x,t) - A\ \delta p(x+\delta x,t) = -A\left\{\frac{\partial(\delta p)}{\partial x}\right\}_t \delta x \qquad (3.18)$$

and this is equal to the mass of the slab multiplied by its acceleration

$$A\left\{1 + \left(\frac{\partial y}{\partial x}\right)_t\right\}\delta x.(\rho+\delta\rho).\left(\frac{\partial^2 y}{\partial t^2}\right)_x = A\rho\ \delta x\left(\frac{\partial^2 y}{\partial t^2}\right)_x . \qquad (3.19)$$

The equation of propagation of the disturbance is, by (3.18) and (3.19),

$$\begin{aligned}\left(\frac{\partial^2 y}{\partial t^2}\right)_x &= -\frac{1}{\rho}\left\{\frac{\partial(\delta p)}{\partial x}\right\}_t \\ &= -\frac{1}{\rho}\frac{d(\delta p)}{d(\delta\rho)}\cdot\left\{\frac{\partial(\delta\rho)}{\partial x}\right\}_t \\ &= \frac{d(\delta p)}{d(\delta\rho)}\left(\frac{\partial^2 y}{\partial x^2}\right)_t .\end{aligned}$$

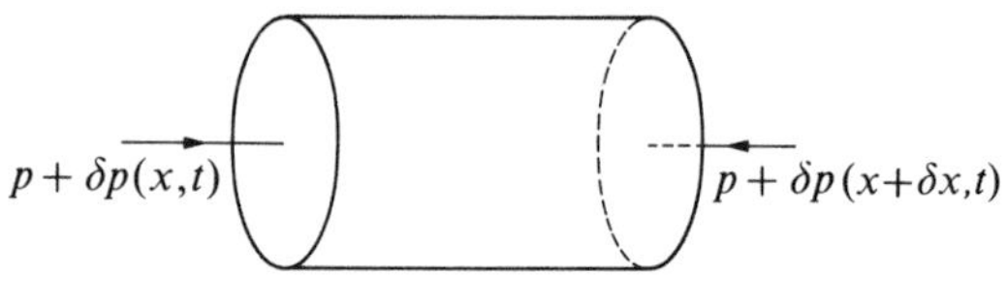


Fig.3.2. An incremental slab of fluid.

In the last step we have used (3.17) and the assumption has been made that the pressure $p+\delta p$ depends on the density $\rho+\delta\rho$ through some definite function which does not depend explicitly on $x$ and $t$. This relation will, however, depend on the way the disturbance is propagated. In many cases the propagation is such that it is slow enough to be considered quasi-static and the frequencies involved are low enough to allow only negligible exchange of heat between the regions of compression and the regions of rarefaction. Such a process is quasi-static adiabatic, and one finds the familiar wave equation for a disturbance having velocity $V$ and an identification of this velocity:

$$\left(\frac{\partial^2 y}{\partial t^2}\right)_x = V_a^2 \left(\frac{\partial^2 y}{\partial x^2}\right)_t \qquad V_a^2 = \left\{\frac{\partial(\delta p)}{\partial(\delta\rho)}\right\}_a = \frac{\partial p}{\partial \rho} . \tag{3.20}$$

In the last equation we have used the fact that

$$p + \delta p = p + \frac{\partial p}{\partial \rho}\,\delta\rho + \frac{1}{2}\,\frac{\partial^2 p}{\partial \rho^2}\,(\delta\rho)^2 + \ldots$$

so that $\partial(\delta p)/\partial(\delta\rho)$ is for small $\delta\rho$ approximately the same as $\partial p/\partial\rho$:

$$\frac{\partial(\delta p)}{\partial(\delta\rho)} = \frac{\partial p}{\partial \rho} + \frac{\partial^2 p}{\partial \rho^2}\,\delta\rho + \frac{1}{2}\,\frac{\partial^3 p}{\partial \rho^3}\,(\delta\rho)^2 + \ldots \sim \frac{\partial p}{\partial \rho} .$$

If the disturbance is such as to allow temperature equilibrium with its surroundings, then one has isothermal instead of adiabatic conditions and

$$V_t^2 = \left(\frac{\partial p}{\partial \rho}\right)_t . \tag{3.21}$$

In a general case a thermodynamic condition '$X$ is a constant' may be imposed. If $m = \rho v$ is the mass of the fluid in volume $v$

$$\left(\frac{\partial p}{\partial v}\right)_X = -\left(\frac{\partial p}{\partial \rho}\right)_X \frac{m}{v^2}$$

the compressibility is

$$K_X = -\frac{1}{v}\left(\frac{\partial v}{\partial p}\right)_X = \frac{1}{v}\cdot\frac{v^2}{m}\left(\frac{\partial \rho}{\partial p}\right)_X = \frac{1}{\rho V_X^2} .$$

The bulk modulus of elasticity is also sometimes used:

$$B_X \equiv \frac{1}{K_X} = V_X^2\rho \ . \tag{3.22}$$

The heat capacity per unit mass is called the specific heat and is denoted by $c_v$ or $c_p$ depending on whether volume or pressure are kept constant. These quantities are $C_v/m$ or $C_p/m$ where $m$ is the mass of material present.

The quantity $c_v$ will now be used to show that the wavelength of a sound wave is long enough to make its propagation adiabatic. Suppose the centres of regions of compression and rarefaction are at a temperature difference $\Delta t$ owing to the rise in temperature on compression and the drop in temperature on rarefaction. They are half a wavelength ($\lambda/2$) apart. The heat energy required to raise the intervening mass through the temperature difference is

$$Q_1 \equiv \frac{\rho A \lambda}{2} c_v \ (\Delta t) \ ,$$

where $\rho$ is the density of the gas and $A$ the cross-sectional area of the material, conceived as a cylinder. If $k$ is the thermal conductivity, the heat conducted in the time $\lambda/2V$ which it takes the wave to cover a distance $\lambda/2$ is

$$Q_2 \equiv kA\left(\frac{\lambda}{2V}\right) \frac{\Delta t}{\lambda/2}$$

where the last factor is an estimate of the temperature gradient. For adiabatic propagation one needs $Q_2 \ll Q_1$, i.e.

$$\lambda \gg \frac{2k}{V\rho c_v}$$

For air the right-hand side is of the order $3 \times 10^{-7}$ which corresponds to the wavelength of ultraviolet light and is much smaller than the wavelength of sound waves. Hence the wave propagation is adiabatic.

### 3.4. HOMOGENEITY OF EXTENSIVITY

Suppose an equilibrium system is placed next to an exact copy of it and the partition between them is withdrawn so as to make the system twice the size. Some variables have their values doubled in this process; they are the *extensive* variables; others have the same value in the original and the doubled system: they are the *intensive* variables. Although this distinction will be introduced more formally in section 7.1, we can observe already here that $t$ and $p$ are clearly intensive, while $v$ and $d'Q$ are extensive. Each side of an equation must have the same character as regards its extensive nature. Hence (3.1a) shows that $C_v$ is extensive (since $d'Q$ is extensive and $dv$ could be zero), $l_v$ is intensive (since $dv$ is extensive and $dt$ could be zero), $C_p$, $l_p$ and $m_p$ are extensive, and $m_v$ is intensive. Note that the identification (3.3) is in agreement with this requirement of *homogeneity of extensivity*.

An equation $pv = at$ can now immediately be rewritten in the form

$$pv = Nbt$$

where $N$ is the number of molecules and is therefore extensive, while $b$ is an intensive constant of dimension energy per degree. The equation for a classical ideal gas

$$pv = Nkt \ ,$$

where $k$ is a universal constant, therefore satisfies homogeneity of extensivity.

Another equation to be used is for an *ideal quantum gas*

$$pv = gU$$

where $g$ is a dimensionless constant. This is also in agreement with this principle.

## *3.5. THE IDENTIFICATION OF $m_v$ AND $m_p$

Another approach to a proof of (3.3) might be as follows. One uses (3.1) replacing $dt$ by

$$dt = \left(\frac{\partial t}{\partial p}\right)_v dp + \left(\frac{\partial t}{\partial v}\right)_p dv \qquad (3.23)$$

to find

$$d'Q = C_v\left(\frac{\partial t}{\partial p}\right)_v dp + \left\{l_v + C_v\left(\frac{\partial t}{\partial v}\right)_p\right\} dv \ .$$

Then

$$m_v = l_v + C_v\left(\frac{\partial t}{\partial v}\right)_p , \quad m_p = C_v\left(\frac{\partial t}{\partial p}\right)_v \ . \tag{3.24}$$

This approach is also correct. However, to prove the equivalence of (3.24) and (3.3) requires additional manipulation.

From (3.2) one sees that

$$\left(\frac{\partial t}{\partial p}\right)_v = -\frac{l_p}{C_p - C_v} , \quad \left(\frac{\partial t}{\partial v}\right)_p = \frac{l_v}{C_p - C_v} \ . \tag{3.25}$$

Substituting (3.25) into (3.24) yields

$$m_v = \frac{l_v(C_p - C_v) + C_v l_v}{C_p - C_v} = \frac{l_v C_p}{C_p - C_v}$$

$$m_p = -C_v \frac{l_p}{C_p - C_v} \ .$$

These are the same equations as were obtained before. The difference between the two proofs is simply a question of manipulation. In the first proof we replaced d$t$ by a differential form in d$p$ and d$v$ in which the coefficients were already identified in terms of $C$'s and $l$'s. In this section we followed the *more general* procedure of going via (3.23). However, this turns out to be longer, since we had still to go back to (3.2).

## *3.6. THE ZEROTH LAW

If $u_K$ denotes the complete set of the thermodynamic variables for a system K, let $F_{12}(u_1,u_2) = 0$ be a relation specifying thermal equilibrium between systems 1 and 2. Then the zeroth law states that any two of the relations

$$F_{12}(u_1,u_2) = 0, F_{23}(u_2,u_3) = 0, F_{31}(u_3,u_1) = 0 \tag{3.26}$$

imply the third (transitivity). It is then inferred that functions $t_j(u_j)$ ($j = 1,2,3$) exist such that thermal equilibrium between systems 1 and 2 may be equivalently expressed by

$$t_1(u_1) = t_2(u_2).$$

Such a result assumes certain uniqueness properties of thermal equilibrium. Sometimes this caution has not been made explicit. In fact the following assumption is needed. Suppose that the values of all but one of the variables $u_1$ are specified. Then for each state of system 2 there must be a unique value of this remaining variable for which the systems 1 and 2 are in equilibrium.

As an illustration of alternative mathematical situations, let $a_{ij}, b_{ij}$ be positive numbers, and let

$$F_{ij}(u_i, u_j) \equiv a_{ij}(\theta_i - \theta_j)^2 + b_{ij}(\phi_i - \phi_j)^2$$

where $(i,j) = (1,2), (2,3), (3,1)$. It is then true that any two of the equilibrium conditions (3.26) imply the third but there are now two 'temperatures', since the equilibrium condition for systems 1 and 2 is equivalent to

$$\theta_1 = \theta_2 \quad \text{and} \quad \phi_1 = \phi_2 .$$

This example has been suggested by the specification of a line $j$ ($j$=1,2,3) by the position of a point on it, together with two angles $\theta_j, \phi_j$ in a Cartesian co-ordinate system. The equilibrium condition $F_{ij} = 0$ is then the condition for parallelism of the lines $i$ and $j$, and transitivity is valid. However, a single 'temperature' function is clearly not adequate; there are *two* such functions $\theta$ and $\phi$. A system with a built-in adiabatic partition can also have two temperatures.

The numbering of the laws of thermodynamics is the result of historical accident. The first and second laws came with the basic completion of thermodynamics in the middle of the nineteenth century. The zeroth laws resulted from work by Carathéodory in 1909, but did not enter the textbooks until rather later. There were then no obvious numbers available, and because of its rather basic nature it was felt that this law has to precede the first law. For this reason this law was called the zeroth law. The third law is associated with Walther Nernst (1906). It can be expressed in various different ways (see section 6.3 and problem 6.7).

PROBLEMS

3.1. Let $(\ )_\mathrm{a}$ indicate a quasi-static adiabatic process and $\gamma \equiv C_p/C_v$ the heat capacity ratio. Show that

$$\gamma = \frac{(\partial p/\partial v)_\mathrm{a}}{(\partial p/\partial v)_t} .$$

Express in terms of $\gamma$ (only)

$$\frac{(\partial v/\partial t)_\mathrm{a}}{(\partial v/\partial t)_p} \quad \text{and} \quad \frac{(\partial p/\partial t)_\mathrm{a}}{(\partial p/\partial t)_v}$$

Hence show that in quasi-static adiabatic processes in fluids for which $pv = At$, $A$ being a constant, the following quantities are constants:

$$pv^\gamma, \quad tv^{\gamma-1}, \quad t^\gamma/p^{\gamma-1} .$$

3.2. Show that $K_t/K_\mathrm{a} = \gamma$.

3.3. A wire of length $l$ is stretched by a force $F$. Show that the ratio of isothermal to adiabatic Young's modulus satisfies

$$\frac{Y_t}{Y_\mathrm{a}} = \frac{C_l}{C_F}$$

where $C_F$, $C_l$ are heat capacities at constant force and constant length respectively.

3.4. Water has $\alpha_p < 0$ for $0°\mathrm{C} < t < 4°\mathrm{C}$ and $\alpha_p > 0$ for $t > 4°\mathrm{C}$. Show that water can be either cooled or heated by adiabatic compression. (A complete proof requires the assumption $K_t > 0$ and the result, which follows from problem 5.2, that $C_p - C_v > 0$.)

3.5. By considering $\ln v$ show that

$$\left(\frac{\partial \alpha_p}{\partial p}\right)_t = -\left(\frac{\partial K_t}{\partial t}\right)_p .$$

3.6. For a polytropic fluid of index $n$ , the relation $pv^n$ = constant holds. Show that the velocity $V$ of sound in such a fluid satisfies

$$V^2 = np/\rho \ .$$

Treating air as a polytropic fluid evaluate its index at atmospheric pressure. ($p$ = 76 cm Hg, density of mercury is 13·6 g cm$^{-3}$, $\rho$ = 0·00129 g cm$^{-3}$ , $V$ = 332 m s$^{-1}$ .)

3.7. Show that, for a temperature range which is small enough to treat $\alpha_p$ as a constant throughout it, the volume of the system is given by

$$v = v_0 \exp\{\alpha_p(t-t_0)\}$$

where $v = v_0$ at an empirical temperature $t_0$ and the pressure is constant throughout. Show also that for small enough $\alpha_p$

$$v \doteqdot v_0 + \alpha_p(t-t_0)v_0 \ .$$

3.8. Two force fields in a plane are in terms of unit vectors given by

$$\mathbf{F}_1 = x^2\hat{\mathbf{X}} + y^2\hat{\mathbf{Y}} \quad \text{and} \quad \mathbf{F}_2 = y^2\hat{\mathbf{X}} + x^2\hat{\mathbf{Y}} \ .$$

Which is conservative and why?

# 4. Exact and inexact differentials

### *4.1. CONDITION FOR EXACTNESS OF d'$F$[†]

The expression

$$\mathrm{d}'F = \sum_{j=1}^{n} Y_j(x_1,\ldots,x_n)\mathrm{d}x_j = \mathbf{S}\cdot\mathrm{d}\mathbf{r}$$

is called a one-form or a Pfaffian form and, with the $n$-dimensional vectors

$$\mathbf{S} \equiv (Y_1,\ldots,Y_n) \quad \text{and} \quad \mathrm{d}\mathbf{r} \equiv (\mathrm{d}x_1,\ldots,\mathrm{d}x_n) \ ,$$

can be written as a scalar product.[††] If d'$F$ is in fact an exact differential it can be written as d$F$, and

$$\mathrm{d}F = \sum_{j=1}^{n} \frac{\partial F}{\partial x_j}\,\mathrm{d}x_j$$

is the same equation. This means that if d'$F$ is exact then $Y_j$ can be written as

$$Y_j = \frac{\partial F}{\partial x_j} \qquad (j=1,\ldots,n) \ .$$

If exactness holds, then the $\frac{1}{2}n(n-1)$ equations

$$\frac{\partial^2 F}{\partial x_j \partial x_k} = \frac{\partial Y_j}{\partial x_k} = \frac{\partial Y_k}{\partial x_j} \qquad (j,k = 1,2,\ldots,n)$$

are satisfied. These are therefore necessary conditions for exactness of d$F$. They also turn out to be sufficient conditions. For three dimensions they can be written as

[†] Readers not interested in mathematics should merely read the summary in section 4.6.

[††] The functions $Y_j$ will be assumed single-valued and at least once differentiable.

$$\frac{\partial Y_1}{\partial x_2} - \frac{\partial Y_2}{\partial x_1} = \frac{\partial Y_2}{\partial x_3} - \frac{\partial Y_3}{\partial x_2} = \frac{\partial Y_3}{\partial x_1} - \frac{\partial Y_1}{\partial x_3} = 0$$

$$\nabla \times \mathbf{S} = 0 \quad \text{or} \quad \text{curl } \mathbf{S} = 0.$$

If the integral of $d'F$ depends on the path between points A and B,

$$P_1 \equiv \underset{\text{path 1}}{\int_A^B} d'F, \qquad P_2 \equiv \underset{\text{path 2}}{\int_A^B} d'F$$

then normally $P_1 \neq P_2$. It follows that the integral from A to B along path 1 and back along path 2 from B to A is

$$\oint d'F = P_1 - P_2 \ (\neq 0, \text{ normally}) \ .$$

When this path dependence is known to occur, then the differential may be inferred to be inexact. Thus one sees that if the line integral of some vector field **S** around any closed curve in a domain is to vanish, then $\mathbf{S}.d\mathbf{r}$ must be exact, that is $\nabla\times\mathbf{S} = 0$ in that domain. Hence **S** must be expressible as the gradient of some scalar function in that domain.

*4.2. CONDITION FOR $d'F = 0$ TO HAVE AN INTEGRATING FACTOR

If $u d'F = 0$ and this is an exact differential, then there exists a solution function $f(x_1,\ldots,x_n)$ such that

$$f(x_1,\ldots,x_n) = c$$

where $c$ is a constant. One then has a whole family of solution curves depending on the parameter $c$.

A *necessary* condition for a solution of

$$d'F = \sum_j Y_j dx_j = 0$$

is that

$$Y_i\left(\frac{\partial Y_k}{\partial x_j} - \frac{\partial Y_j}{\partial x_k}\right) + Y_j\left(\frac{\partial Y_i}{\partial x_k} - \frac{\partial Y_k}{\partial x_i}\right) + Y_k\left(\frac{\partial Y_j}{\partial x_i} - \frac{\partial Y_i}{\partial x_j}\right) = 0$$

($i,j,k=1,2,\ldots,n$, and all different).

To prove this, assume that the solution has the form

$$f(x_1,\ldots,x_n) = c$$

and that $u(x_1,\ldots,x_n)$ is the integrating factor. Then $u\mathrm{d}'F = \mathrm{d}f$ and

$$u(x_1,\ldots,x_n)Y_j = \frac{\partial f}{\partial x_j}$$

where $j$ may have any of the values $1,2,\ldots,n$. Hence for any $i=1,\ldots,n$ and $i \neq j$,

$$\frac{\partial}{\partial x_i}(uY_j) = \frac{\partial^2 f}{\partial x_i \partial x_j} = \frac{\partial}{\partial x_j}(uY_i) .$$

It follows that

$$u\left(\frac{\partial Y_j}{\partial x_i} - \frac{\partial Y_i}{\partial x_j}\right) = Y_i \frac{\partial u}{\partial x_j} - Y_j \frac{\partial u}{\partial x_i} .$$

Similarly

$$u\left(\frac{\partial Y_k}{\partial x_j} - \frac{\partial Y_j}{\partial x_k}\right) = Y_j \frac{\partial u}{\partial x_k} - Y_k \frac{\partial u}{\partial x_j}$$

$$u\left(\frac{\partial Y_i}{\partial x_k} - \frac{\partial Y_k}{\partial x_i}\right) = Y_k \frac{\partial u}{\partial x_i} - Y_i \frac{\partial u}{\partial x_k} .$$

Multiplying these three equations respectively by $Y_k$, $Y_i$ and $Y_j$, and summing, yields the required result.

The *sufficient* condition is the same as the necessary condition, but the proof requires a longer argument which will not be given here.

For $n = 3$, the condition is with $\mathbf{S} \equiv (Y_1,\ldots,Y_n)$

$$\mathbf{S}.\nabla\times\mathbf{S} = 0$$

which is a weakened version of the condition $\nabla\times\mathbf{S} = 0$ for exactness.

## *4.3. THE EQUATION $\mathrm{d}'F = 0$ IN TWO VARIABLES HAS ALWAYS AN INTEGRATING FACTOR

Let the one-form be

$$\mathrm{d}'F = Y_1\mathrm{d}x_1 + Y_2\mathrm{d}x_2 \equiv \mathbf{S}.\mathrm{d}\mathbf{r} .$$

Hence the equation $d'F = 0$ is

$$\frac{dx_2}{dx_1} = -\frac{Y_1}{Y_2} .$$

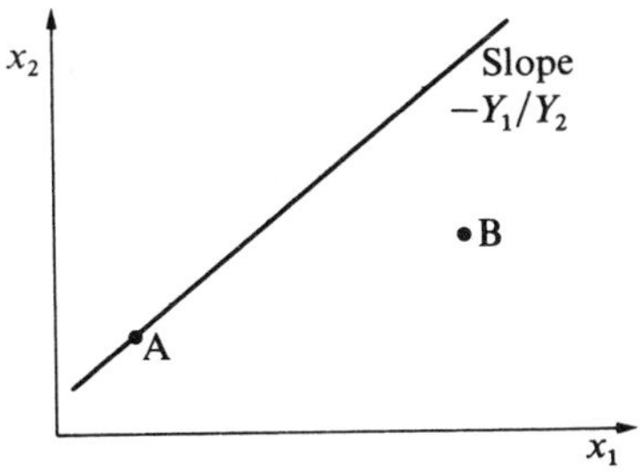


Fig.4.1. A line of slope $-Y_1/Y_2$ through a point A.

This equation must be soluble, for one can draw through any point $(x_1,x_2)$ a line of slope $-Y_1/Y_2$ and this is the slope of the solution curve at this point. From a point A one can go to a neighbouring point B by moving along the line segment of slope $-Y_1(x_{1A},x_{2A})/Y_2(x_{1A},x_{2A})$. Similarly one can go from B to C, etc., thus obtaining an approximate solution curve. The process becomes exact if A,B,C,... lie infinitesimally close to each other. Thus one may assume that the equation $d'F = \mathbf{S}.d\mathbf{r} = 0$ has a solution of the form

$$f(x_1,x_2) = C,$$

that is

$$df = \frac{\partial f}{\partial x_1} dx_1 + \frac{\partial f}{\partial x_2} dx_2 \;\; (\equiv \mathbf{R}.d\mathbf{r}) = 0$$

$C$ being a constant. Since $df = 0$ and $dF = 0$, at any point $P(x_1,x_2)$, an element $d\mathbf{r}$ has perpendicular to it (i) the vector $\mathbf{S}\,(x_1,x_2)$ arising from $d'F$ and (ii) a vector $\mathbf{R}(x_1,x_2)$ arising from $df$. These two vectors must therefore be parallel so that

$$\frac{1}{Y_1}\frac{\partial f}{\partial x_1} = \frac{1}{Y_2}\frac{\partial f}{\partial x_2} \;\; (\equiv g(x_1,x_2)) .$$

It follows that

$$g\,\mathrm{d}'F = g\,Y_1\mathrm{d}x_1 + g\,Y_2\mathrm{d}x_2 = \frac{\partial f}{\partial x_1}\,\mathrm{d}x_1 + \frac{\partial f}{\partial x_2}\,\mathrm{d}x_2 = \mathrm{d}f\ .$$

Hence $\mathrm{d}'F = 0$ has an integrating factor. A rigorous proof of this result tends to be rather longer, and is not needed in the present context.

For $n = 3$ the above argument can fail, since the equation $\mathrm{d}'F = 0$ can now furnish an infinity of slopes at any point and a solution curve cannot be constructed by the simple intuitive procedure indicated above. Note that if $\mathrm{d}'F$ has one integrating factor $g(x_1,\ldots,x_n)$ then it has an infinity of integrating factors.

## *4.4. ANOTHER CONDITION FOR $\mathrm{d}'F = 0$ TO HAVE AN INTEGRATING FACTOR

Suppose a domain in a phase space is completely filled by non-intersecting level surfaces such that for any increment of a line in a surface $\mathrm{d}'F = 0$, while for any increment of a line intersecting different surfaces $\mathrm{d}'F \neq 0$. Then $\mathrm{d}'F = 0$ has an integrating factor. To see this we again write

$$\mathrm{d}'F \equiv \sum_{j=1}^{n} Y_j(x_1,\ldots,x_n)\mathrm{d}x_j \equiv \mathbf{S}.\mathrm{d}\mathbf{r}\ .$$

Suppose also that the family of $(n-1)$-dimensional 'surfaces' is given by $f(x_1,\ldots,x_n) = C$ so that each value of $C$ picks out one level surface. Differentiating,

$$\mathrm{d}f = \sum_j \frac{\partial f}{\partial x_j}\,\mathrm{d}x_j = \mathbf{R}.\mathrm{d}\mathbf{r}\ .$$

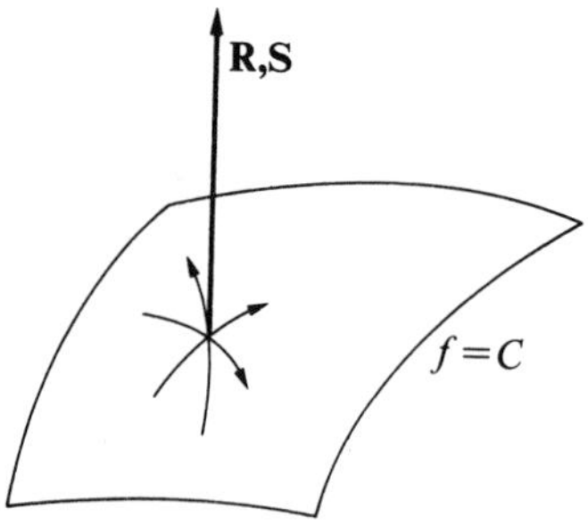


Fig.4.2. The vectors **R** and **S** are normal to the level surface $f = C$.

We now arrive at a situation which is analogous to that in section 4.3. At each point P in phase space *any* element $d\mathbf{r}$ of a line in a surface $f = C$ is orthogonal to two vectors **R** and **S** whose directions are functions of the coordinates of P. It follows that throughout the simultaneous range of definition of the vectors **R**,**S** these are parallel vectors, and they are perpendicular to the element of the surface at P. Hence there exists a scalar function of position such that $g\mathbf{S} = \mathbf{R}$, and it follows that

$$gd'F = g\mathbf{S}.d\mathbf{r} = \mathbf{R}.d\mathbf{r} = df \ .$$

Thus $d'F = 0$ has an integrating factor.

If P is actually on the boundary of a (connected) region in which the adiabatic linkage of points can be satisfactorily defined, there may be difficulties concerning the existence of elements $d\mathbf{r}$ at P. It is, therefore, necessary to suppose that P is an *interior* point of such a region (i.e. it cannot lie on a boundary). Thus P must lie in an open domain, i.e. a domain from which boundary points have been removed.

*4.5. EXAMPLE FOR THE ABSENCE OF AN INTEGRATING FACTOR WHEN $n = 3$

Consider now the equation

$$d'F = x_1 dx_2 + k dx_3$$

where $k$ is a non-zero constant. If $gd'F = df$ then

$$gx_1 \, dx_2 + kg dx_3 = \sum_j \frac{\partial f}{\partial x_j} dx_j \ .$$

Hence

$$\frac{\partial f}{\partial x_1} = 0, \quad \frac{\partial f}{\partial x_2} = g\, x_1, \quad \frac{\partial f}{\partial x_3} = kg \ .$$

It follows that

$$\frac{\partial^2 f}{\partial x_1 \partial x_2} = 0 = g + x_1 \frac{\partial g}{\partial x_1}$$

$$\frac{\partial^2 f}{\partial x_3 \partial x_1} = k \frac{\partial g}{\partial x_1} = 0 \ .$$

These equations can be satisfied only if $g$ is identically zero, so that no integrating factor exists in this case. This shows that the case $n = 2$ when an integrating factor always exists and the case $n \geq 3$, when an integrating factor may not exist, are very different.

## 4.6. SUMMARY

The functions $Y_j(x_1,\ldots,x_n)$ and the components of S are assumed single-valued and once differentiable.

(1) The necessary and sufficient condition for $d'F \equiv \mathbf{S}.d\mathbf{r}$ to be exact is, for $n = 3$, either

$$\text{curl } \mathbf{S} = 0$$

or some scalar function $\phi$ exists such that

$$\mathbf{S} = \text{grad } \phi$$

(see section 4.1).

(2) If $\oint d'F \neq 0$ somewhere in a domain, then $d'F$ is inexact in that domain (section 4.1).

(3) The necessary and sufficient condition for an inexact $d'F$ to have an integrating factor is, for $n = 3$, $\mathbf{S}.\text{curl } \mathbf{S} = 0$ everywhere in the domain considered (section 4.2). The heat increment $d'Q$ has an integrating factor for general $n$ if all points in phase space are so-called $i$-points (see section 5.2, below).

(4) If surfaces, such that for any line in a surface $d'F = 0$, fill a domain in phase space without intersecting, then $d'F$ is exact or has an integrating factor in that domain (section 4.4).

(5) If $d'F = \sum_{j=1}^{n} Y_j(x_1,\ldots,x_n)dx_j$, then $d'F$ has an integrating factor for $n=2$, but need not have such a factor for $n \geq 3$ (sections 4.3, 4.5).

PROBLEMS

4.1. Where precisely, and why, is there an error in the following argument?

$$d'Q = \left(\frac{\partial Q}{\partial t}\right)_v dt + \left(\frac{\partial Q}{\partial v}\right)_t dv = dU + p dv = \left(\frac{\partial U}{\partial t}\right)_v dt + \left\{p + \left(\frac{\partial U}{\partial v}\right)_t\right\} dv .$$

Therefore

$$\left(\frac{\partial Q}{\partial t}\right)_v = \left(\frac{\partial U}{\partial t}\right)_v , \quad \left(\frac{\partial Q}{\partial v}\right)_t = p + \left(\frac{\partial U}{\partial v}\right)_t .$$

Therefore

$$\frac{\partial^2 Q}{\partial t \partial v} = \frac{\partial^2 U}{\partial t \partial v} = \left(\frac{\partial p}{\partial t}\right)_v + \frac{\partial^2 U}{\partial v \partial t} .$$

Therefore

$$\left(\frac{\partial p}{\partial t}\right)_v = 0 \quad \text{(always)} .$$

4.2. Prove the condition $\nabla \times \mathbf{S} = 0$ of section 4.1 to be sufficient for the exactness of $d'F = 0$.

4.3. Verify that the following equation is exact and find the general solution, $a$, $b$, and $h$ being constants,

$$ax + hy + (hx + by)y' = 0 .$$

4.4. Prove that, for any two vectors $\mathbf{a}, \mathbf{b}$, $\mathbf{a}.\mathbf{b}\times\mathbf{a} = 0$. If $\mathbf{b}$ is the differential operator $\nabla$, show by example that this identity need no longer be satisfied. (The condition $\mathbf{S}.\nabla\times\mathbf{S} = 0$ of section 4.2 is therefore a constraint.)

4.5. Show that if

$$d'Q = \sum_{j=1}^{n} X_j(x_1, \ldots, x_n)\, dx_j$$

has one integrating factor then it has an infinity of such factors.

# 5. The second law

## 5.1. SOME QUALITATIVE FORMULATIONS OF THE LAW

The second law of thermodynamics has received many formulations. One, due to Lord Kelvin, states that it is impossible to convert an amount of heat completely into work by a cyclic process without at the same time producing other changes. Consider an engine based on such a cyclic process. Such an engine, if it existed, could cool the oceans and convert all the heat thus extracted into mechanical work. This does not violate energy conservation, but it would nonetheless provide a large amount of energy. Because this energy would be almost unlimited, such an engine is called a perpetual motion engine 'of the second kind'. The second law of thermodynamics declares that such engines are impossible. This form of the second law is due to Wilhelm Ostwald. There are still other statements of the second law of thermodynamics, but for the moment it suffices to observe that these statements are remarks about the *macroscopic* world. We now infer from them an important *microproperty* of thermodynamic phase space.

Suppose phase space contains a point P which possesses some open connected neighbourhood N, all points of which can be reached adiabatically from P. Let the co-ordinates of the space be the internal energy $U$ which acts as the 'thermal co-ordinate' and 'deformation' co-ordinates $v_1, v_2, \ldots$ (volumes, magnetic

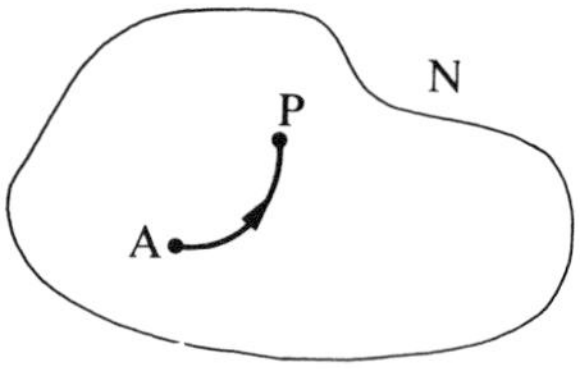


Fig.5.1. The neighbourhood N of the point P.

fields, etc.) which act as non-thermal or mechanical co-ordinates. Keeping these fixed, let us take the system from a point A in N to P where $U_A < U_P$. Since the $v_i$ are constant, no mechanical work is done. The increase in internal energy is due to heat $Q > 0$ having been supplied. Because A is in N, the system can be returned adiabatically from P to A using appropriate changes of deformation co-ordinates. The drop in internal energy is now entirely due to mechanical work $W$ which is done by the system. Conservation of energy during the cycle APA demands that $W = Q$, so that heat has been completely converted into work contrary to Kelvin's principle. It follows that points such as P cannot exist. Hence it becomes desirable to define an i-point; this is a point such that there exists in *every* neighbourhood of it some point which is adiabatically inaccessible from it. The above argument has then shown that

| |
|---|
| All points in a domain $\gamma$ of thermodynamic phase space are i-points |

(as judged by the quasi-static process).

The phrase in brackets holds if the above processes can be thought of only if they are quasi-static. In fact, however, the two linkages of A and B can occur via non-equilibrium states so long as $Q$ and $W$ are well defined for the process. The phrase in brackets can therefore be omitted. This makes the statement a much stronger one. It now states that even if non-static adiabatic processes are used, there will still remain at least one point in N which is adiabatically inaccessible from P. This stronger statement is *Carathéodory's formulation* of the second law.

A few points may here be noted

(i) One actually knows at least some of the points which are adiabatically inaccessible from P. These are points like A which have the same deformation co-ordinates as P but have a lower internal energy. We shall not require to make use of this additional information.

(ii) The adiabatic linkage of any two points of a set $\gamma$ is not violated by what has been found. In fact, it has been seen that one can go from A to $\dot{P}$ by supplying heat. One can obtain the heating effect by internal agitation or mechanical deformations of the system equally well, provided only the final state has the same values of the deformation co-ordinates as the initial state. In this way, then, P is adiabatically accessible from A, thus establishing adiabatic linkage. The argument based on Fig.5.1 shows that the reverse process is **not** possible. Incidentally, one has here the first indication that the *direction* in which processes can go will play a part in our discussion.

From the second law it follows (see section 5.2) that $d'Q = 0$ has an integrating factor for systems which do not contain adiabatic partitions or enclosures, i.e. for systems which are in equilibrium specified by a *single* temperature. If the number of thermodynamic variables $n = 2$, this is clearly a fact which can be deduced from section 4.3. However, if $n \geq 3$ it adds to our knowledge. It further follows (see section 5.3) that there exist functions of state called entropy $S$ and absolute temperature $T$ which have a definite value at any point in a set $\gamma$ of a thermodynamic phase space **E**. Each equilibrium state represented by a point of **E** has therefore associated with it a definite value of $S$ and $T$. In a cyclic process the total change in entropy is therefore zero. In some expositions of thermodynamics the existence of the entropy function and the absolute temperature are *stipulated* at an early stage. We shall here spend a little time to make their existence plausible.†

## 5.2. PLAUSIBILITY ARGUMENT FOR THE EXISTENCE OF AN ENTROPY FUNCTION

Take the thermal co-ordinate in thermodynamic phase space **E**

†The reader may wish to assume the existence of these functions, thus saving himself a little work on a first reading of the book. He can then proceed from here to sections 5.4 and 5.5.

as the internal energy $U$, and the deformation co-ordinates will be denoted by $v_i$ ($i=1,2,\ldots,n-1$), making an $n$-dimensional space. The deformation co-ordinates can be varied arbitrarily and slowly without there being any need for a heat exchange with the surroundings. It follows that any values $(v_1', v_2')$, say, can be realized from any other values $(v_{10}, v_{20})$, say, by quasi-static adiabatic methods. One merely has to remain within the appropriate domain $\gamma$. We now introduce the following into **E** (Fig.5.2)

(i) An artificial origin P $(v_{10}, v_{20}, U_0)$.

(ii) A line $L$ of given deformation $(v_1', v_2')$ which lies at least partially in the set $\gamma$ which contains P. Let $B$ be the set of points adiabatically accessible from P. By quasi-static adiabatic changes in $v_1$ and $v_2$ one can reach the line L from P at some point to be denoted by Q. Then Q lies in $B$ and L, and hence Q lies in the set of points common to B and L (the so-called 'intersection' $B \cap L$).

(iii) The sequence of equilibrium states required to pass from P to Q. Such a sequence can clearly be described in either direction so that the inverse process $Q \to P$ is also quasi-static adiabatic. This process can therefore be given arrows pointing in *both* directions.

With appropriate continuity assumptions $B \cap L$ will include all points on *some* segment of L. The point Q has to be an end point of this segment. This can be seen as follows. If Q is not an end-point of the segment, it can be linked adiabatically to neighbouring points of $B \cap L$ by processes of the type $Q \to P \to Q'$ and $Q \to P \to Q''$ (see Fig. 5.2). That such a process is adiabatic follows from (ii) and (iii) above. However, Q can be linked adiabatically to additional neighbouring points by arbitrary variations of the deformation co-ordinates, but then Q would have a neighbourhood in which *all* points would be adiabatically accessible from Q, contrary to Carathéodory's principle. The point Q has, therefore, to be an end-point of the segment on L.

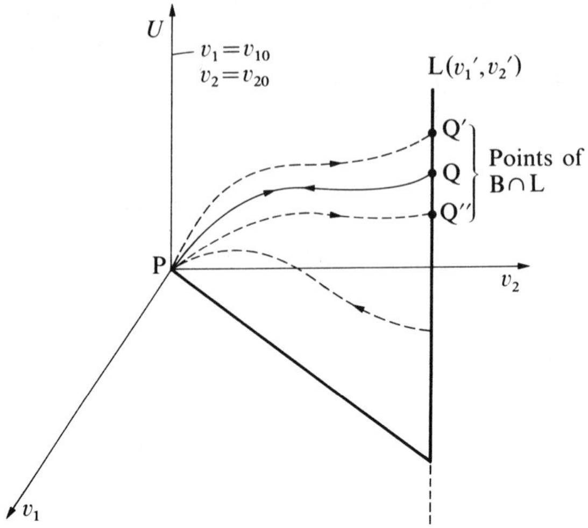


Fig.5.2. A diagram helping to visualize the construction of level surfaces in phase space.

There is an ambiguity here if the section of the line L adiabatically accessible from P has two end-points, Q(P) and Q'(P) say. Could both be reached from P by quasi-static adiabatic processes? Or could the point which is accessible by quasi-static adiabatic processes from P be Q(P) (the lower say) for some initial states P of the given system and Q'(P) (the upper, say) for other initial states P of the same system? All these possibilities are ruled out here by assuming that adequate continuity conditions are satisfied.

Now repeat this argument for displaced lines L', L", etc., parallel to L. The point Q then traces out a surface on which P itself must lie. Now repeat the argument for different initial states P, and the thermodynamic phase space is found to be decomposed into a family of non-intersecting surfaces. These can be labelled by a continuously varying parameter $\phi$ (which can be called the *empirical entropy*) such that distinct values of $\phi$ refer to distinct surfaces, and conversely.

Two explanations are now in order. First, one may ask:

Since we are in a set β, any two points of which are linked adiabatically, how is this linkage effected between P and the points of L which lie below Q(P)? The answer is to note that these points are not adiabatically accessible from P. The only way of linking them with P which remains is that P must be adiabatically accessible from *them*. So one sees that the point Q(P) divides the line L into two parts: the (possible non-static) adiabatic processes proceed from P to L for points above Q(P), and they proceed from L to P for points below Q(P). These processes are shown by dashed curves in Fig.5.2. Q(P) will be assumed to lie at the lower end of the points in B∩L.

Secondly one may ask: How can we be sure that the level surfaces which have been constructed do not intersect within a domain β? The answer can be given by supposing that two surfaces do intersect at some point $X$ as shown in Fig.5.3.

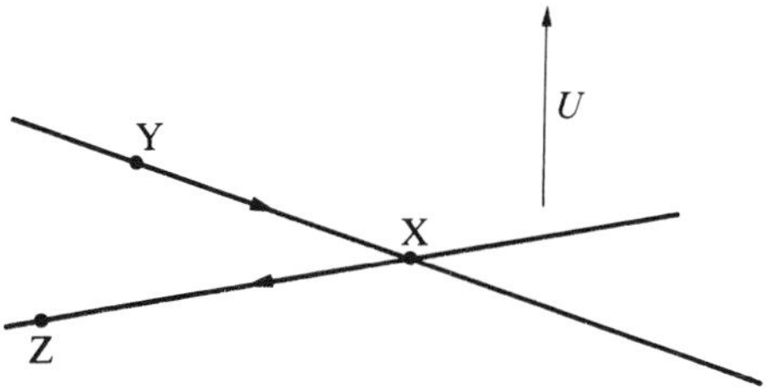


Fig.5.3. Can two of the generated surfaces intersect?

One then considers a pair of points, Y and Z say, which lie in the surfaces and correspond to the same values of the deformation co-ordinates. Thus Y and Z differ only in the value of the corresponding internal energies. The process Y → X → Z is quasi-static adiabatic with mechanical work, $W$ say, given out by the system. The process Z → Y requires thermal energy, $Q$ say, but no work. Hence in the cyclic process Z → Y → X → Z heat is completely converted into work, which is against Kelvin's principle. Hence there can be no intersection of the surfaces.

Two important consequences follow:

(*a*) The method of construction of the level surfaces implies that each quasi-static adiabatic process lies in one such surface. We now use the fact that the decomposition of phase space into level surfaces is mathematically equivalent to saying that there exists a function $f(t,v_1,v_2,...)$ of the thermodynamic variables such that for quasi-static processes $d'Q = fd\phi$, where $d\phi$ is an exact differential (see section 4.4). The function $\phi$ is of course not defined uniquely; one can write $dQ = f^*d\phi^*$ where $f^* = fg$ and $d\phi^* = d\phi/g$ and $g$ is some function of the variables. Any function $\phi$ which can occur in the equation $d'Q = fd\phi$ is called an empirical entropy. In particular, for quasi-static adiabatic processes $d\phi = 0$ since such processes are confined to surfaces of constant $\phi$.

(*b*) One can now draw an additional conclusion from the fact that the points Q always lie at an *end-point* of the appropriate set $B \cap L$. This means that the $\phi$-value of the point $Q \in B \cap L$ has either the smallest or the largest value of the $\phi$-values represented in the set $B \cap L$. From this there follows the rule that the empirical entropy $\phi$ changes monotonically in adiabatic processes. For example, if $\phi$ has been associated with the surfaces in the manner shown in Fig.5.4(*a*), it is clear that the empirical entropy tends always to increase in general (i.e. usually non-static) adiabatic processes. The possibility of Fig.5.4(*b*) leads to a rather unconventional definition of an empirical entropy function and will not be considered (though it can also be incorporated into thermodynamics without arriving at inconsistent results). Thus for an increment of a general adiabatic process $\delta'Q = 0$ (the adiabatic condition) and $\delta\phi \geq 0$.

## *5.3. THE EXISTENCE OF AN ABSOLUTE TEMPERATURE AND AN ABSOLUTE ENTROPY

The following sections 5.3.1 to 5.3.6 can be omitted at first reading. They are inserted here because we wish to be careful in our arguments. However, the reader should not lose sight of the key point, which is as follows: the possible integrating factors $\lambda$ in $d'Q = \lambda d\phi$ can be

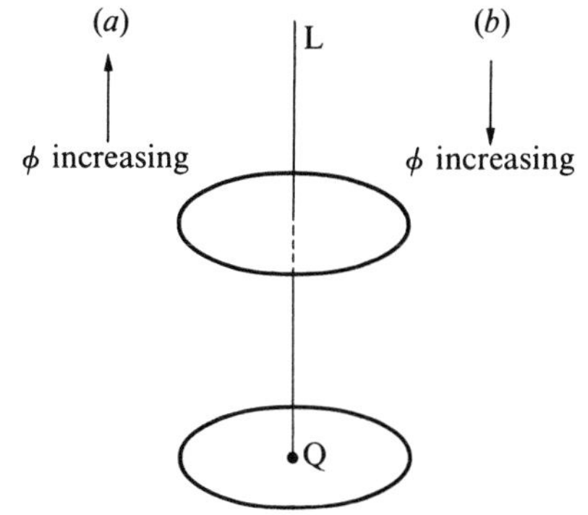


Fig.5.4. The labelling of the level surfaces.

arranged to depend on the empirical temperature as the only variable, thus yielding $d'Q = T(t)dS$. The essential steps to achieve this are given in equations (5.7), (5.8), (5.14) and (5.15).

*5.3.1. The possibility of substituting empirical temperature for one of the mechanical variables*

According to Chapter 2 the $n$ independent thermodynamic variables consist of the internal energy $U$ and $n$-1 other variables $(v_2,\ldots,v_n)$. We now wish to show that the empirical temperature $t(U,v_2,\ldots,v_n)$ must in fact depend on at least one of the $v$'s. Certainly $t$ must depend on at least one of the $n$ variables indicated because the equation $t(U,v_2,\ldots,v_n) =$ constant, for the specified range of variation of the independent variables, would express a relation between these variables and they would then not all be independent.

Suppose next that the empirical temperature is a function only of the internal energy. In that case arbitrary variations in the remaining variables leave the empirical temperature and therefore the internal energy unaltered. These arbitrary variations are not restricted if the system is adiabatically enclosed. In that case we must have $d'Q = 0$, $dU = 0$ in the equation $d'Q = dU + d'W$, for all changes of the $v$'s. It follows that $d'W = 0$ for all such changes. This implies that there is a relation among the supposedly independent remaining variables, which is against the hypothesis. It follows that any empirical temperature must depend on at least one of the remaining variables which will be taken to be $v_2$. In expressions involving $v_2$, the empirical temperature can be introduced by virtue of this relation.

*5.3.2. The possibility of substituting the empirical entropy for the internal energy as thermal variable*

Let us now write $d'Q = \lambda d\phi$, where $\lambda$ is an integrating factor and $d\phi$ an exact differential. Then

$$d'Q = dU + \sum_{2}^{n} p_t dv_t = \lambda \frac{\partial\phi}{\partial U} dU + \lambda \sum_{2}^{n} \frac{\partial\phi}{\partial v_j} dv_j \; .$$

Hence

$$\lambda \frac{\partial\phi}{\partial U} = 1, \qquad \lambda \frac{\partial\phi}{\partial v_j} = p_j \; . \tag{5.1}$$

Equation (5.1) shows that $\partial\phi/\partial U \neq 0$. The equation $\phi = \phi\ (U, v_2, \ldots, v_n)$ can therefore be solved for $U$, and $\phi$ can be used instead of $U$ as one of the independent thermodynamic variables.

*5.3.3. Factorization of the integrating factor so that the temperature dependence is factored out*

Consider now the thermal equilibrium of two systems having respectively $m$ and $n$ independent thermodynamic variables. The variables corresponding to the two systems will be distinguished by superscripts, and a set of these variables is given by

$$t^{(1)}, \phi^{(1)}, v_3^{(1)}, \ldots, v_m^{(1)}; \quad t^{(2)}, \phi^{(2)}, v_3^{(2)}, \ldots, v_n^{(2)} \; . \tag{5.2}$$

Thermal equilibrium imposes the relation $t^{(1)} = t^{(2)}$. The common empirical temperature will be denoted by $t$. Now apply the result that an integrating factor exists for each system separately and also for the joint system. Using $d'Q = d'Q^{(1)} + d'Q^{(2)}$ for the increment of thermal energy supplied to the combined system by quasi-static processes, one finds

$$\lambda d\phi = \lambda^{(1)}\ d\phi^{(1)} + \lambda^{(2)}\ d\phi^{(2)} \; , \tag{5.3}$$

omitting superscripts for the combined system. Now $d\phi$ can be expressed in terms of the $m + n - 1$ independent variables:

$$d\phi = \frac{\partial\phi}{\partial t}\,dt + \frac{\partial\phi}{\partial\phi^{(1)}}\,d\phi^{(1)} + \frac{\partial\phi}{\partial\phi^{(2)}}\,d\phi^{(2)} + \sum_{3}^{m}\frac{\partial\phi}{\partial v_r^{(1)}}\,dv_r^{(1)} + \sum_{3}^{n}\frac{\partial\phi}{\partial v_s^{(2)}}\,dv_s^{(2)} \,. \tag{5.4}$$

From (5.3) and (5.4)

$$\frac{\partial\phi}{\partial\phi^{(j)}} = \frac{\lambda^{(j)}}{\lambda}\ (j = 1,2), \quad \frac{\partial\phi}{\partial t} = \frac{\partial\phi}{\partial v_r^{(1)}} = \frac{\partial\phi}{\partial v_s^{(2)}} = 0 \,. \tag{5.5}$$

$$(r = 3,4,\ldots,m;\ s = 3,4,\ldots,n) \,.$$

It follows that

$$\frac{\partial^2\phi}{\partial\phi^{(j)}\partial t} = 0, \quad \text{that is } \frac{\partial}{\partial t}\left(\frac{\lambda^{(j)}}{\lambda}\right) = 0 \quad (j = 1,2) \,. \tag{5.6}$$

This implies that

$$\frac{1}{\lambda^{(1)}}\frac{\partial\lambda^{(1)}}{\partial t} = \frac{1}{\lambda^{(2)}}\frac{\partial\lambda^{(2)}}{\partial t} = \frac{1}{\lambda}\frac{\partial\lambda}{\partial t} \,. \tag{5.7}$$

Since $\lambda^{(1)}$ depends on $t$ and on variables which apply to the first system only and $\lambda^{(2)}$ depends on $t$ and variables which apply to the second system only, it follows that each expression (5.7) is a function of $t$ only, $g(t)$ say, and $g(t)$ is independent of the specific properties distinguishing system 1 from system 2. It is instead a universal function of the chosen empirical temperature scale. Now if a positive function $\lambda$ is an integrating factor, then a corresponding negative function $-\lambda$ is also an integrating factor. Both choices lead to the same function $g(t)$. No special assumption will be made concerning the sign of $\lambda$.

Integration shows that the integrating factor has the form

$$\lambda^{(j)}(t,\phi^{(j)},v_3^{(j)},v_4^{(j)},\ldots)$$

$$= \Phi^{(j)}(\phi^{(j)},v_3^{(j)},\ldots)\exp\left\{\int_{t_0}^{t} g(t)\,dt\right\} \quad (j = 1,2) \,. \tag{5.8}$$

Here $t_0$ is a standard empirical temperature, which is assumed to be the same for all physical systems. The value of the integrating factor $\lambda^{(j)}$ at this temperature occurs as a constant of integration, and has been denoted by

$$\lambda^{(j)}(t_0,\phi^{(j)},v_3^{(j)},v_4^{(j)},\ldots) \equiv \Phi^{(j)}(\phi^{(j)},v_3^{(j)},v_4^{(j)},\ldots) \,. \tag{5.9}$$

*5.3.4. The possibility of using a single entropy function for a system consisting of two subsystems*

The integrating factor *for the combined systems* may also be written in the form (5.8), as can be seen by supposing the combined systems to be placed in thermal contact with some other system and regarding these two systems now as the part-systems which occur in the preceding paragraphs. It might be thought, however, that this argument cannot be taken over exactly because the part-systems which occur in the preceding paragraphs have as independent thermodynamic variables: one empirical temperature, *one* $\phi$-function, and remaining variables $v_3,v_4,\ldots$ . The combined system, regarded as a part-system in the new argument, however, contains among its independent variables one empirical temperature, *two* $\phi$-functions, and remaining variables. The properties of the combined system would thus depend on two $\phi$-functions and other variables. It is easily shown, however, that the function $\Phi$ involves the functions $\phi^{(1)}$ and $\phi^{(2)}$ only in the combination $\phi(\phi^{(1)},\phi^{(2)})$, so that in this respect it is similar to the part-systems considered in the preceding paragraphs. By way of proof, note that by (5.3) and (5.8) and if $\phi$ is some function of $\phi^{(1)},\phi^{(2)}$,

$$\Phi\, d\phi = \Phi^{(1)}d\phi^{(1)} + \Phi^{(2)}d\phi^{(2)} \,. \tag{5.10}$$

Hence

$$\Phi \frac{\partial\phi}{\partial\phi^{(j)}} = \Phi^{(j)} \qquad (j = 1,2)\,. \tag{5.11}$$

Differentiating again,

$$\frac{\partial\Phi}{\partial\phi^{(2)}}\frac{\partial\phi}{\partial\phi^{(1)}} + \Phi\frac{\partial^2\phi}{\partial\phi^{(1)}\partial\phi^{(2)}} = 0 = \frac{\partial\Phi}{\partial\phi^{(1)}}\frac{\partial\phi}{\partial\phi^{(2)}} + \Phi\frac{\partial^2\phi}{\partial\phi^{(1)}\partial\phi^{(2)}}$$

that is, using the notation for Jacobians,

$$\frac{\partial\Phi}{\partial\phi^{(1)}}\frac{\partial\phi}{\partial\phi^{(2)}} = \frac{\partial\phi}{\partial\phi^{(1)}}\frac{\partial\Phi}{\partial\phi^{(2)}} \quad \text{or} \quad \frac{\partial(\Phi,\phi)}{\partial(\phi^{(1)},\phi^{(2)})} = 0 \tag{5.12}$$

This is just the condition for $\Phi$ to depend on $\phi^{(1)}$ and $\phi^{(2)}$ only through $\phi(\phi^{(1)},\phi^{(2)})$, as shown in section 5.7.

*5.3.5. The existence of absolute temperature and absolute entropy*

If now not only $t_0$ but also a constant $C$ is given for a whole class of systems

$$d'Q = \lambda \, d\phi = T \, dS \tag{5.13}$$

where

$$T(t) \equiv C \exp\left\{\int_{t_0}^{t} g(t) \, dt\right\} \tag{5.14}$$

$$S(v_2, v_3, \ldots) \equiv C^{-1} \int_{\alpha}^{\phi} \Phi(\phi, v_2, v_3, \ldots) \, d\phi \tag{5.15}$$

and $\alpha$ refers to a standard value of $\phi$ for the particular system under discussion. The value $\phi = \alpha$ has the property $S(\alpha) = 0$. $T$ is called the *absolute temperature*, and depends only on $t_0$ and the empirical temperature $t$, while $S$ is called the *entropy* which depends on the variables $v_2, v_3, \ldots$, the $\phi$-function, and the value $\phi = \alpha$. The $\phi$-function has been called the *empirical* entropy. We shall assume that, by adjustment of the sign of the empirical temperature if necessary, $g(t)$ has been arranged to be positive. Then, since $g(t)$ is real, $T$ and $dT/dt$ both have the sign of $C$. It follows that the absolute temperature is a strictly increasing or a strictly decreasing function of the empirical temperature, which is therefore also true for each of the $n$ thermodynamic variables when the remaining $n$-1 variables are kept fixed.

One way of fixing the value of $C$ is to choose two empirical temperatures, $t_0$ and $t_1$ say, and make a convenient convention regarding the difference $T(t_1) - T(t_0)$ . For instance, if this difference is made equal to 100 'degrees', $C$ is a solution of the equation

$$\ln \frac{100}{C} + 1 = \int_{t_0}^{t_1} g(t) \, dt \, .$$

This result can be derived from the observation that

$$\int_{t_0}^{t_1} g(t) dt = \ln \frac{T(t_1)}{C} = \ln \frac{T(t_0)+100}{C}$$

and the fact that $T(t_0) = C$.

*5.3.6. Some properties of the entropy*

We return now to (5.3), using (5.13), to find for quasi-static processes

$$T\,\mathrm{d}S = T\,\mathrm{d}S^{(1)} + T\,\mathrm{d}S^{(2)} = T\,\mathrm{d}(S^{(1)} + S^{(2)}) \ . \tag{5.16}$$

Since $T$ has by (5.14) a fixed sign, the value $T = 0$ lies on the frontier $F(\gamma)$, and therefore does not belong to $\gamma$. $T$ and $S$ have been defined only for sets of points $\gamma$, so that (5.16) implies

$$\mathrm{d}S = \mathrm{d}S^{(1)} + \mathrm{d}S^{(2)}$$

that is

$$S = C^{-1}\int_{\alpha}^{\phi} \Phi\,\mathrm{d}\phi = C^{-1}\int_{\alpha}^{\phi} (\Phi^{(1)}\mathrm{d}\phi^{(1)} + \Phi^{(2)}\mathrm{d}\phi^{(2)}) \tag{5.17}$$
$$= S^{(1)} + S^{(2)} + A \ .$$

The additive constant $A$ depends only on the standard entropy states selected for the given substances:

$$AC = \int_{\alpha}^{\alpha^{(1)}} \Phi^{(1)}\mathrm{d}\phi^{(1)} + \int_{\alpha}^{\alpha^{(2)}} \Phi^{(2)}\mathrm{d}\phi^{(2)} \ . \tag{5.18}$$

$A$ can therefore be made zero by a suitable convention. The entropy of a system consisting of two part-systems in thermal contact is therefore obtained (except, possibly, for an additive constant) by adding the entropies of the individual systems. If the part-systems are each in equilibrium, but their temperatures are different, then the total system must contain an adiabatic partition and has therefore no unique temperature function. An entropy function has so far not been defined for this system, but it can be defined to be the sum of the individual entropies. The expression for an increment of thermal energy added to the total system by quasi-static processes is still $\mathrm{d}'Q = T^{(1)}\mathrm{d}S^{(1)} + T^{(2)}\mathrm{d}S^{(2)}$, but if $\mathrm{d}S^{(1)} \neq 0$, $\mathrm{d}S^{(2)} \neq 0$, one cannot put $\mathrm{d}'Q = T\,\mathrm{d}S$ . It is convenient in such a case to define an increment of the entropy of the total system by

$$dS = dS^{(1)} + dS^{(2)} \tag{5.19}$$

bearing in mind that the equation $d'Q = T\,dS$ does not hold in this case. However, $dS$ defined in (5.19) gives $d'Q$ for quasi-static processes if $T$ is interpreted as a weighted average of $T^{(1)}$ and $T^{(2)}$:

$$d'Q = \left( T^{(1)} \frac{dS^{(1)}}{dS} + T^{(2)} \frac{dS^{(2)}}{dS} \right) dS \ . \tag{5.20}$$

This assumption of entropy increments for systems containing adiabatic partitions can so far be checked only against the limiting cases $T^{(1)} = T^{(2)}$ or $dS^{(1)} = 0$ or $dS^{(2)} = 0$. In all these cases (5.19) is satisfactory, so that this assumption will be adopted. Its reasonableness is connected with the *extensive property* of the entropy.

## 5.4. MATHEMATICAL FORMULATIONS OF THE SECOND LAW

The second law implies two mathematical statements:

(i) For quasi-static processes there exist functions $T$ and $S$ (or $\lambda$ and $\phi$) such that in any domain $\gamma$ of phase space $d'Q = T dS = \lambda d\phi$. Here $T$ depends only on the empirical temperatures.

(ii) In a general adiabatic process the entropy of a system satisfies $\delta\phi \geq 0$, $\delta S \geq 0$ where equality holds only if the process is quasi-static. If the inequality holds the process is said to be non-static or dissipative.

One can give a thermodynamic expression for the integrating factor $\lambda$, by noting that

$$d'Q = dU + d'W = \left(\frac{\partial U}{\partial \phi}\right) d\phi + \sum_{j=1}^{n-1} \left(\frac{\partial U}{\partial v_j}\right) dv_j + \sum_j p_j \, dv_j$$

$$= \left(\frac{\partial U}{\partial \phi}\right) d\phi + \sum_j \left(\frac{\partial U}{\partial v_j} + p_j\right) dv_j \ . \tag{5.21}$$

For an adiabatic process $d'Q = 0$, $d\phi \geq 0$; if it is also quasi-static then $d\phi = 0$. Assuming the deformation co-ordinates to have $dv_j \neq 0$, one has

$$p_j = -\left(\frac{\partial U}{\partial v_j}\right)_{\phi, v_1, \ldots, v_{j-1}, v_{j+1}, \ldots, v_{n-1}} \qquad (j = 1,2,\ldots) .$$

One can now identify $\lambda$ from

$$d'Q = \lambda d\phi = \left(\frac{\partial U}{\partial \phi}\right)_{v_1, \ldots, v_{n-1}} d\phi$$

in terms of thermodynamic variables as

$$\lambda = \left(\frac{\partial U}{\partial \phi}\right)_{v_1, v_2, \ldots} . \tag{5.22}$$

Once absolute temperature $T$ and absolute entropy $S$ are defined, one can choose $\lambda$ as $T$ and $\phi$ as $S$, so that

$$d'Q \le T\, dS \quad \text{and} \quad \left(\frac{\partial S}{\partial U}\right)_{v_1, v_2, \ldots} = \frac{1}{T} \tag{5.23}$$

Also the condition $\delta S \ge 0$ means that in linking two points A,B of the phase space **E**

$$S_B - S_A \ge \int_A^B \frac{d'Q}{T} \tag{5.24}$$

where the equality signs hold only if the process is quasi-static. In particular if A and B coincide, as in a cyclic process,

$$0 \ge \oint \frac{d'Q}{T} . \tag{5.25}$$

This is called the *Clausius inequality*.

The definitions of the heat capacities in section 3.1 can now be rewritten as

$$C_v = T\left(\frac{\partial S}{\partial T}\right)_v , \quad C_p = T\left(\frac{\partial S}{\partial T}\right)_p$$

and these expressions are often used. They yield a method of estimating entropy differences via equations of the type

$$S_2 - S_1 = \int_{T_1}^{T_2} \frac{C_p}{T}\, dT$$

provided $C_p$ is known as a function of temperature and the pressure is fixed.

The various processes considered in this section are illustrated in Fig.5.5. PQ represents a quasi-static adiabatic process, and it lies wholly in a level surface $S$ = constant. PR and R'P represent non-static adiabatic processes

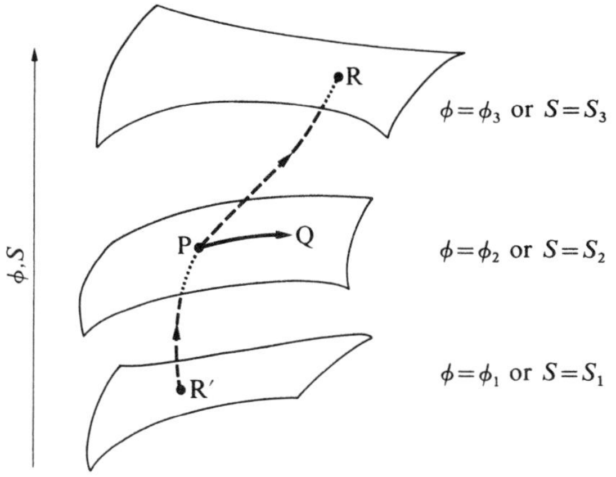


Fig.5.5. Various adiabatic processes.

and so lie, at least in part, outside all sets of points $\gamma$. The state R is typical of states adiabatically accessible from P. R' is typical of states adiabatically inaccessible from P. In fact, the adiabatically inaccessible points are all those points of a set $\gamma$ which lie below the surface $S = S_2$, thus ensuring that any neighbourhood of P, however small, contains points which are adiabatically inaccessible from P. This makes P an i-point, as required.

If black holes exist, statement (ii) above must be amended. One proposed amendment is: 'The sum of the black-hole entropy and the ordinary entropy in the black-hole exterior never decreases'. However, in the limit when no black hole is present, a reference to adiabatic processes is required. Strictly speaking, the above proposal must therefore be re-worded.

## 5.5 THE INTUITIVE MEANING OF ENTROPY

In section 5.4 the meaning of (i) is that a function of state $S$ exists for equilibrium states of thermodynamic systems. The meaning of (ii) is that *this function increases or stays constant for an adiabatically isolated system, and therefore for a completely isolated system.* A low initial entropy has thus the advantage over a larger initial entropy that many more states can be reached adiabatically from it. Precisely how does this remark receive a meaning if entropy is defined only for equilibrium states?

To answer this question consider two systems at temperatures $T_h$ and $T_c (T_h > T_c)$ completely isolated from each other but conceived as a compound system. Each system and hence the compound system is in equilibrium and hence has an entropy. If the part-systems can communicate an amount of heat $Q$ to each other for a short period only, new equilibrium states establish themselves. The entropy sum is now greater than it was, since the cooler system has gained entropy $Q/T_c$ while the warmer system has lost entropy $Q/T_h$. This assumes that the period of interaction was short enough for the temperatures of the part systems to remain effectively unchanged during the interaction, so that

$$\mathrm{d}S = \int \frac{\mathrm{d}'Q}{T} = \frac{1}{T}\int \mathrm{d}'Q \equiv \frac{Q}{T} \quad .$$

By (5.20) the entropy of the compound system is the sum of the entropy of the parts. Hence the increase in the entropy of the compound system is $Q\{(1/T_c)-(1/T_h)\} > 0$. This is obtained by comparing equilibrium states before and after the temporary removal of a constraint, and so gives a precise meaning to the above italicized remark.

Other formulations of the second law can now also be understood. For example, if an isolated system contains two subsystems A and B at the same temperature $T_A = T_B$, then it is not possible for system A to gain energy from system B so that $T_A > T_B$ after the process. The reason is clearly that this process would lead to a reduction in the entropy of the system.

The entropy ceases to increase when the system has

reached its equilibrium condition. It is in this sense that the following popular remark, which is often made, is to be understood: 'the entropy of an adiabatically isolated system attains its maximal value in equilibrium.'

In order to fix ideas we shall work out the entropy change which takes place when two heat reservoirs of initial temperatures $T_h$, $T_c$ ($T_h > T_c$) come to equilibrium with each other. Let the heat capacities be constant at $C_1$ and $C_2$ respectively, and let the final temperature be $T_f$. Then the first law says that the heat lost by the hot reservoir equals the heat gained by the cold reservoir, and this furnishes an equation for $T_f$:

$$C_1(T_h - T_f) = C_2(T_f - T_c) \; .$$

This yields

$$T_f = aT_h + bT_c \left( a \equiv \frac{C_1}{C_1+C_2} \; , \; b \equiv \frac{C_2}{C_1+C_2} \; , \; a + b = 1 \right) \; .$$

We now consider the entropy. Since the initial and final states are equilibrium states, and are therefore represented by points in a set $\gamma$, any quasi-static linkage of the two states can be used to calculate the entropy change; entropy has, after all, a definite value at any point in phase space. We shall therefore use a straightforward application of the equation $d'Q = TdS$ for quasi-static processes. The entropy lost by the hot reservoir is

$$\int dS = \int \frac{d'Q}{T} = C_1 \int_{T_f}^{T_h} \frac{dT}{T} = C_1 \ln \frac{T_h}{T_f} \; .$$

The entropy gained by the cold reservoir is $C_2 \ln(T_f/T_c)$ . Hence the gain in entropy of the whole system is

$$C_2 \ln \frac{T_f}{T_c} + C_1 \ln \frac{T_f}{T_h} = \ln \frac{T_f^{C_1+C_2}}{T_h^{C_1} T_c^{C_2}}$$

$$= (C_1+C_2) \ln \frac{T_f}{T_h^a T_c^b} = (C_1+C_2) \ln \left( \frac{aT_h + bT_c}{T_h^a T_c^b} \right) \; .$$

Since the system as a whole is isolated its entropy can go up but cannot go down. It follows that thermodynamics suggest that for all positive numbers $a$, $b$ such that $a+b = 1$, and for all positive $T_h, T_c$, one must have

$$aT_h + bT_c \geq T_h^a \, T_c^b \quad .$$

If this inequality were untrue thermodynamics would be inconsistent. For $a = b = \frac{1}{2}$ one has here the theorem that the arithmetic mean of two positive numbers cannot be less than their geometric mean.

In science generally there are very few quantities which are distinguished by this *one-way behaviour*. For example, classical mechanics is time-reversible. This means, for example, that a sequence of photographs of an ideal swinging pendulum cannot be ordered uniquely in accordance with the time at which the photographs were taken. Any sequence is possible, because it would be in agreement with the laws of mechanics. The same applies to a perfectly elastic ball bouncing on a rigid plane. Dissipation of mechanical energy is needed to give one time irreversibility. This is introduced for example by the friction in the support of the pendulum or by the heating of the bouncing ball which could occur if it is not ideally elastic. Thus, while the notion of time does not enter into the thermodynamics of quasi-static processes, it provides an 'arrow for the time axis' when dissipative processes are considered.

The expansion of the universe (which is widely believed to be taking place) supplies a cosmological arrow of time. Life processes provide a third, biological, arrow; black holes, if they exist, would provide a fourth. It is a semi-philosophical scientific problem to understand why all these arrows point in the same direction. This question is still without a generally accepted answer.

One can calculate the dissipation of energy in our example of the hot and cold reservoir, and this will now be explained. The two bodies can reach equal temperatures without any overall change in entropy of the system provided quasi-static processes only are employed (Fig.5.6*b*). This can be

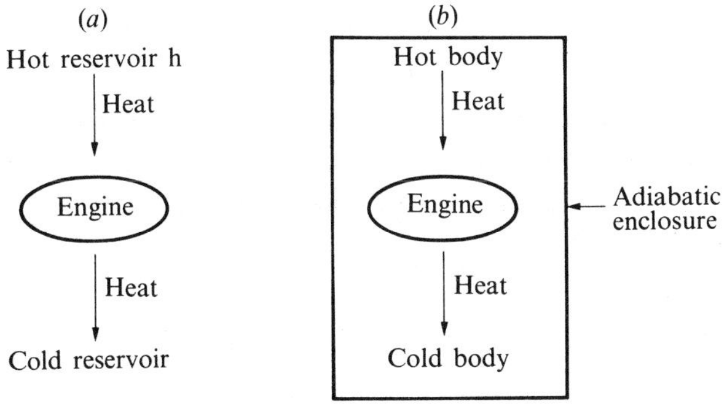


Fig.5.6. (*a*) Normal engine whose reservoirs are infinite so that their temperatures remain constant. The overall effect is that the system of reservoirs gains entropy. (*b*) A quasi-static adiabatic method of equalizing temperatures using a quasi-statically acting engine. The entropy of the enclosed system remains constant during the process.

done by using a quasi-statically working engine which, acting cyclically, withdraws heat from the hot body and rejects heat to the cold body in each cycle and uses the difference in energy to perform work. This could be the Carnot engine discussed in problem 5.5, and it ceases to function when a final temperature $T_f^*$ (say) has established itself for both reservoirs. For this process $T_f^*$ is given by

$$C_1 \ln(T_h/T_f^*) = C_2 \ln(T_f^*/T_c) \ .$$

Hence

$$T_f^* = T_h^a \, T_c^b$$

and the total work done by the engine is by energy conservation

$$W^* = (C_1+C_2)(aT_h+bT_c-T_f^*) \ .$$

In a general process the work done is less, since the non-static part of the process generates heat leading to a final temperature in excess of $T_f^*$:

$$W = (C_1+C_2)(aT_h+bT_c-T_f)$$

where $T_f$ is the appropriate final temperature. (In the extreme case of heat leakage discussed above $W = 0$.) Thus the circumstance that the process is not quasi-static leads to a dissipation of a pre-existing temperature difference which might have been used to perform more work. This energy dissipation is

$$W_{\text{diss}} \equiv W^* - W = (C_1+C_2)(T_f - T_h^a\, T_c^b) \;.$$

It is non-negative since $T_f \geq T_f^*$ quite generally.

## *5.6. AN INTEGRATING FACTOR FOR $d'Q$ THROUGH EQUIVALENCE RELATIONS

Let us consider the relation ($R$ say) of mutual quasi-static adiabatic attainability of two equilibrium states $x,y$. It has already been noted in section 5.2 that

$$(\forall x,y)\ (x\ R\ y \Rightarrow y\ R\ x) \quad \text{symmetry.} \tag{5.26}$$

Thus if $y$ can be reached from $x$ by a quasi-static adiabatic process, the reverse is also the case: $x$ can be reached from $y$ by the same type of process and this holds for all $x$ and $y$ in **E**. This means that the relation $R$ is symmetrical.

It has also been implied that the relation $R$ is transitive, so that if one can go from $x$ to $y$ and from $y$ to $z$ then one can also go from $x$ to $z$ by a quasi-static adiabatic process, and this holds for all $x,y,z \in$ **E**. Symbolically:

$$(\forall x,y,z)\ (x\ R\ y \text{ and } y\ R\ Z \Rightarrow x\ R\ z) \quad \text{transitivity.} \tag{5.27}$$

In fact any (quasi-static or non-static) adiabatic process has this property.

Lastly, each state can be reached from itself by a quasi-static adiabatic process:

$$(\forall x)\ (x\ R\ x) \quad \text{reflexivity.} \tag{5.28}$$

This statement really means that 'no process' shall also be counted as a special case of a 'quasi-static adiabatic process'. This property also holds for general (quasi-static or non-static) adiabatic processes.

Thus it is only (5.26) which is the *additional* property which distinguishes quasi-static adiabatic processes from general adiabatic processes.

A relation $R$ subject to (5.26) to (5.28) is called an *equivalence relation*. Let us use it to define subsets $T_{\mathrm{a}}$ in $\mathbf{E}$ as follows:

$$a,\ a_i \in T_{\mathrm{a}} \quad \text{if} \quad a\ R\ a_i;\ b,\ b_j \in T_{\mathrm{b}} \quad \text{if} \quad b\ R\ b_j; \quad \text{etc.} \qquad (j=1,2,..)$$

Then each state is in at least one subset $T_{\mathrm{a}}$, for one could introduce a subset for each element by itself, if necessary. It follows that the union of all subsets is the original set of all states in $\mathbf{E}$. However, there is a further characteristic of this subdivision into subsets. This is that any two distinct subsets have no common elements. For if $c \in T_{\mathrm{a}}$ and $c \in T_{\mathrm{b}}$ ($T_{\mathrm{a}} \neq T_{\mathrm{b}}$), then for $a_i \in T_{\mathrm{a}}$, $b_j \in T_{\mathrm{b}}$

$$a_i\ R\ c \text{ and } c\ R\ b_j \quad (\text{all } i,j) \tag{5.29}$$

by (5.26), so that

$$a_i\ R\ b_j \quad (\text{all } i,j) \tag{5.30}$$

by (5.27). It follows that all elements of $T_{\mathrm{a}}$ and $T_{\mathrm{b}}$ are in the same subset and $T_{\mathrm{a}} = T_{\mathrm{b}}$, contrary to hypothesis. Thus one sees that any equivalence relation partitions a set into disjoint subsets, i.e. they have no common elements. This is expressed formally as $T_{\mathrm{a}} \cap T_{\mathrm{b}} = 0$ for $T_{\mathrm{a}} \neq T_{\mathrm{b}}$. The subsets $T_{\mathrm{a}}, T_{\mathrm{b}}$ ... are called *equivalence classes*.

Let $a_i \to b_j$ mean that one can go from state $a$ to state $b$ by an adiabatic process. If $T_{\mathrm{a}} \neq T_{\mathrm{b}}$ such a process *cannot be quasi-static* since this would imply $T_{\mathrm{a}} = T_{\mathrm{b}}$ by an argument like that which involves (5.29) and (5.30). Furthermore, if $\mathrm{a}_1 \to \mathrm{b}_j$ then $\mathrm{b}_k \to \mathrm{a}_2$ is impossible. For it would lead to

$$b_j\ R\ b_k, \quad b_k \to a_2, \quad \text{and} \quad a_2\ R\ a_1 .$$

By transitivity, this implies $b_j \to c_1$ in addition to the hypothesized relation $a_1 \to \mathrm{b}_j$. It follows that $\mathrm{a}_1$ and $\mathrm{b}_j$ are mutually quasi-statically adiabatically accessible to each other, so that we can write $\mathrm{a}_1\ R\ \mathrm{b}_j$ and

and hence $T_a = T_b$ as before. This contradicts our assumption. Hence if $a_1 \to b_j$ and if $T_a \neq T_b$, then

$$a_i \to b_k \quad \text{for all} \quad a_i \in T_a \quad \text{and} \quad b_k \in T_b \ .$$

One may therefore associate an adiabatic linkage, in one direction only, with any two subsets $T_a$, $T_b$. Suppose therefore that the labelling is such that

$$T_a \to T_b \to T_c \to \dots \tag{5.31}$$

If one makes ample continuity assumptions, then one can imagine each equivalence class $T_a$ to represent a level surface or hyperplane in **E** such as was introduced in section 5.2. The fact that $T_a \cap T_b = 0$ for $T_a \nparallel T_b$ corresponds to the fact that these surfaces do not intersect. Hence we arrive again at the notion that a domain in **E** is subdivided into an infinity of level surfaces. Thus one arrives at an integrating factor for $d'Q$ as before. Alternatively one can associate values $\phi_a < \phi_b < \phi_c < \dots$ with the equivalence classes (or surfaces) (5.26) and obtain a direct interpretation for the empirical entropy $\phi$. The real numbers $\phi$ are then ordered in such a way that adiabatic processes are possible, starting from states in an equivalence class $T_a$, provided the terminal states are in equivalence classes whose $\phi$-values are equal to those for states in $T_a$, or larger. Thus, although the precise numerical values of $\phi$ are not important, their ordering is crucial, and one sees that the *empirical entropy increases in adiabatic processes*. Any alternative ordering which leads to *decreasing* empirical entropy values can be changed to increasing values by multiplying all the entropy values by -1. This can be done since only the ordering of the $\phi$-values is important.

## *5.7. DISCUSSION OF EQUATION (5.12)

This equation may be proved as follows. If

$$\Phi = F(\phi)$$

then

$$\frac{\partial \Phi}{\partial \phi^{(i)}} = F'(\phi) \frac{\partial \phi}{\partial \phi^{(i)}} \ , \qquad (i = 1,2) \ .$$

Eliminating $F'(\phi)$,

$$\frac{\partial\Phi}{\partial\phi^{(1)}} = \frac{\partial\Phi/\partial\phi^{(2)}}{\partial\phi/\partial\phi^{(2)}} \frac{\partial\phi}{\partial\phi^{(1)}}$$

which is the required relation. The converse is needed here, and is also valid. For if $\Phi$ and $\phi$ are known functions of the $\phi^{(i)}$, one can eliminate $\phi^{(2)}$ and obtain a relation of the form $\Phi = f(\phi,\phi^{(1)})$. Then

$$\frac{\partial\Phi}{\partial\phi^{(1)}} = \frac{\partial f}{\partial\phi}\frac{\partial\phi}{\partial\phi^{(1)}} + \frac{\partial f}{\partial\phi^{(1)}}$$

$$\frac{\partial\Phi}{\partial\phi^{(2)}} = \frac{\partial f}{\partial\phi}\frac{\partial\phi}{\partial\phi^{(2)}} .$$

Hence

$$\frac{\partial(\Phi,\phi)}{\partial(\phi^{(1)},\phi^{(2)})} = \left(\frac{\partial f}{\partial\phi}\frac{\partial\phi}{\partial\phi^{(1)}} + \frac{\partial f}{\partial\phi^{(1)}}\right)\frac{\partial\phi}{\partial\phi^{(2)}} - \frac{\partial f}{\partial\phi}\frac{\partial\phi}{\partial\phi^{(2)}}\cdot\frac{\partial\phi}{\partial\phi^{(1)}}$$

and this vanishes identically, so that

$$\frac{\partial f}{\partial\phi^{(1)}}\frac{\partial\phi}{\partial\phi^{(2)}} = 0 .$$

If $\partial f/\partial\phi^{(1)} = 0$ then $f$ is independent of $\phi^{(1)}$ and the proof is complete. Also $\partial\phi/\partial\phi^{(2)}$ does not normally vanish, since $\phi$ normally involves $\phi^{(1)}$ and $\phi^{(2)}$.

Both proofs fail if $\partial\phi/\partial\phi^{(2)}$ vanishes. If it does, we note for the forward theorem that, since $\phi$ is a function only of $\phi^{(1)}$, so is $\Phi$ and

$$\frac{\partial\Phi}{\partial\phi^{(1)}}\frac{\partial\phi}{\partial\phi^{(2)}} - \frac{\partial\Phi}{\partial\phi^{(2)}}\frac{\partial\phi}{\partial\phi^{(1)}} = 0 \tag{5.32}$$

holds again. For the converse theorem note that if $\partial\phi/\partial\phi^{(2)} = 0$, then $\partial\phi/\partial\phi^{(1)} \neq 0$ since $\phi$ is not a constant. Hence $\partial\Phi/\partial\phi^{(2)} = 0$, by (5.32) and both $\Phi$ and $\phi$ depend only on $\phi^{(1)}$, and one can by elimination express $\Phi$ in terms of $\phi$.

PROBLEMS

5.1. By considering $\boxed{dU = TdS - pdv,\ dF \equiv d(U-TS),\ dH \equiv d(U+pv)}$

$dG \equiv d(U + pv - TS)$ establish

$$\left(\frac{\partial T}{\partial v}\right)_S = -\left(\frac{\partial p}{\partial S}\right)_v , \quad \left(\frac{\partial T}{\partial p}\right)_S = \left(\frac{\partial v}{\partial S}\right)_p$$

$$\left(\frac{\partial T}{\partial v}\right)_p = -\left(\frac{\partial p}{\partial S}\right)_T , \quad \left(\frac{\partial T}{\partial p}\right)_v = \left(\frac{\partial v}{\partial S}\right)_T .$$

($F$ is called the Helmholtz free energy, $G$ is called the Gibbs free energy, and $H$ is the enthalpy. The above are called 'Maxwell's relations', after James Clerk Maxwell.)

5.2. Show that (i) $C_p - C_v = Tv\alpha_p^2/K_T$, (ii) $K_T - K_a = Tv\alpha_p^2/C_p$, and (iii) $C_p = (\partial H/\partial T)_p$.

5.3. Two simple fluids which are completely insulated from each other are treated as a single system. The heat capacities at constant volume are $C_1$ and $C_2$. Explain the equation $d'Q = C_1 dT_1 + C_2 dT_2 + p_1 dv_1 + p_2 dv_2$ for the combined system, and discuss the existence of an integrating factor. Discuss this condition in the additional cases: (i) pressure-coupled components ($p_1 = p_2$); (ii) thermally coupled components ($T_1 = T_2$).

5.4. (*a*) Consider $d'Q = \sum_{j=1}^{3} Y_j \, dx_j$ given that $d'Q = 0$ has no integrating factor. Show, from a book on differential equations, that by changing to three new variables $z_1$, $z_2$, $z_3$ one can write

$$d'Q = dz_1 + z_2 dz_3$$

(*b*) Given any two points $A = (z_{1A}, z_{2A}, z_{3A})$, $B = (z_{1B}, z_{2B}, z_{3B})$ in the new phase space, show that they can be connected by a curve along which $d'Q = 0$.

(This shows that, if $d'Q = 0$ has no integrating factor, then thermodynamic space has no i-points: any two points can be connected not only adiabatically but even quasi-statically as well, and hence in both directions.)

5.5. In a general quasi-static cycle, a working fluid receives heat at various temperatures and gives up heat at various temperatures. The efficiency $\eta$ of the cycle is defined as the mechanical work done divided by the sum of all the positive increments of heat gained. Prove that

$$\boxed{\eta \le \frac{T_1 - T_2}{T_1} \quad (\equiv \eta_C)}$$

where $T_1$ is the highest temperature at which heat is gained, and $T_2$ is the lowest temperature at which heat is given up.

(The Clausius inequality can be useful; $\eta_C$ is the *Carnot efficiency*.)

5.6. A fluid as in problem 5.5 works between isothermals at temperatures $T_1$ and $T_2$ separated by an adiabatic expansion and an adiabatic compression. Show that the efficiency $\eta_C$ is attained in this case.

(This is the *Carnot cycle*.)

5.7. A fluid has zero coefficient $\alpha_p$ of volume expansion for a certain range on an isothermal surface. Show that the states of this range of values cannot be reached by adiabatic expansion or compression of the fluid if $C_v > 0$, $K_T > 0$ in this region of phase space.

5.8. Establish the two relations

$$\left(\frac{\partial^2 p}{\partial T^2}\right)_v = \left\{\frac{\partial}{\partial T}\left(\frac{C_v \Gamma}{v}\right)\right\}_v = \frac{1}{T}\left(\frac{\partial C_v}{\partial v}\right)_T$$

and deduce the compatibility relation

$$\left(\frac{\partial C_v}{\partial v}\right)_T = \frac{T}{v}\left\{\Gamma\left(\frac{\partial C_v}{\partial T}\right)_v + C_v\left(\frac{\partial \Gamma}{\partial T}\right)_v\right\} .$$

(Note the implications:

($C_v$ is a function of $T$ only) and ($\Gamma$ is a function of $v$ only) $\Rightarrow$ ($C_v$ is a constant) $\Rightarrow$ ($\Gamma$ is a function of $v$ only).)

5.9. Establish the compatibility relation

$$\frac{1}{C_v}\left(\frac{\partial C_v}{\partial T}\right)_S = \frac{1}{T} + T\Gamma\left(\frac{\partial}{\partial T}\frac{1}{T\Gamma}\right)_v \quad .$$

(Note the implication:

($\Gamma$ is a function of $v$ only) $\Rightarrow$ ($C_v$ is a function of $S$ only).)

5.10. Show that for *any* simple fluid

$$pv = AT$$

where $A$ is in general a variable given by

$$A = -\left\{\frac{\partial S}{\partial \ln p}\right\}_H \quad .$$

5.11. Decide the following finer points of thermodynamics, giving reasons:

(*a*) Is an isolated system an adiabatic system, and conversely?

(*b*) How good a statement is 'The entropy of an isolated system cannot decrease with time'?

(*c*) Two equilibrium states of a system are linked by a non-static process. The change in entropy is worked out by integrating $dS = d'Q/T$ for an appropriate path. Is this procedure satisfactory in view of the inequality $\delta S \geq \delta' Q/T$?

# 6. Absolute and empirical temperature scales

## 6.1. A GENERAL CONNECTION BETWEEN THE TWO SCALES

We are now able to give a correct formulation of the argument in problem 4.1 by using the fact that one may choose the absolute temperature instead of $t$ and also by using the fact that for quasi-static processes we have by (5.21) and (5.23)

$$d'Q = dU + p dv = T dS .$$

Hence

$$\left(\frac{\partial Q}{\partial T}\right)_v = T\left(\frac{\partial S}{\partial T}\right)_v , \quad \left(\frac{\partial Q}{\partial v}\right)_T = T\left(\frac{\partial S}{\partial v}\right)_T$$

so that

$$\left(\frac{\partial S}{\partial T}\right)_v = \frac{1}{T}\left(\frac{\partial U}{\partial T}\right)_v , \quad \left(\frac{\partial S}{\partial v}\right)_T = \frac{1}{T}\left\{p + \left(\frac{\partial U}{\partial v}\right)_T\right\} . \tag{6.1}$$

Hence

$$\frac{\partial^2 S}{\partial T \partial v} = \frac{\partial}{\partial v}\left\{\frac{1}{T}\left(\frac{\partial U}{\partial T}\right)_v\right\}$$

$$= \frac{\partial}{\partial T}\left[\frac{1}{T}\left\{p + \left(\frac{\partial U}{\partial v}\right)_T\right\}\right]$$

so that, since $T$ and $v$ are the independent variables,

$$\frac{1}{T}\frac{\partial^2 U}{\partial T \partial v} = -\frac{1}{T^2}\left\{p + \left(\frac{\partial U}{\partial v}\right)_T\right\} + \frac{1}{T}\left(\frac{\partial p}{\partial T}\right)_v + \frac{1}{T}\frac{\partial^2 U}{\partial v \partial T} .$$

There follows a relation which depends on the use of the absolute temperature scale:

$$\boxed{T\left(\frac{\partial p}{\partial T}\right)_v = p + \left(\frac{\partial U}{\partial v}\right)_T .} \tag{6.2}$$

Relation (6.2) is very useful. For example, it leads one

at once to the result that for an ideal quantum gas (which is defined by $pv = gU$) the internal energy satisfies the functional relation

$$U = v^{-g} f(Tv^g)$$

where $f$ is some undetermined function. This is shown in problem 6.6 . In particular, if $g = \frac{1}{3}$ and $f(z) = az^4$, where $a$ is a constant,

$$U = avT^4 \ .$$

This is a celebrated relation for black-body radiation at temperature $T$. Why chose $g = \frac{1}{3}$ and $f(z) = az^4$? Thermodynamics does not tell us, and, as the philosophy of our subjects suggests, one has to go to statistical mechanics for a more specific model. This is discussed in Chapters 11 and 12.

An application of (6.2) to show that different empirical temperature scales yield the same absolute temperature scale will be discussed next. Clearly

$$\frac{(\partial p/\partial T)_v}{p+(\partial U/\partial v)_t} = \frac{1}{T} \ .$$

Multiplying both sides by $(\mathrm{d}T/\mathrm{d}t)\ \mathrm{d}t$ and integrating from $t_0$ to $t_i$, corresponding to $T_0$ and $T_i$,

$$\int_{T_0}^{T_i} \frac{\mathrm{d}T}{T} = \int_{t_0}^{t_i} \frac{(\partial p/\partial t)v}{p+(\partial U/\partial v)_t}\ \mathrm{d}t \equiv K_i \qquad (i = 1,2,\ldots) \tag{6.3}$$

where $t_0$ is a standard temperature and $t_i$ are arbitrary temperatures. Note that $K_i$ involves the empirical, but not the absolute, temperature scale. Hence

$$\ln \frac{T_i}{T_0} = K_i \ . \tag{6.4}$$

By international agreement $T_0 = 273.16$ K refers to the triple point of water (see p. 86). Now

$$T_2 = e^{K_2} T_0 = e^{K_2}\left(\frac{T_1 - T_0}{e^{K_1} - 1}\right) = (T_1 - T_0)\,\frac{e^{K_2}}{e^{K_1} - 1}\;. \qquad (6.5)$$

If $t_1 - t_0 = 100$ °C we define the corresponding value of $T_1 - T_0$ to be 100 K (100 degrees Kelvin or 100 degrees absolute). Then for any other empirical temperature $t_2$ equation (6.5) furnishes the absolute temperature.

Suppose that on a second empirical temperature scale $t_i$ corresponds to $t_i^*$ and $K_i$ is replaced by $K_i^*$. Then

$$K_i^* = \int_{t_0^*}^{t_i^*} \frac{(\partial p/\partial t^*)_v \mathrm{d}t^*}{p + (\partial U/\partial v)_{t^*}} = \int_{t_0}^{t_i} \frac{(\partial p/\partial t)_v \mathrm{d}t}{p + (\partial U/\partial v)_t} = K_i \qquad (i = 1, 2, \ldots)\;.$$

It follows that the absolute temperature corresponding to $t_2^*$ is, by equation (6.5),

$$T_2^* = (T_1^* - T_0^*)\,\frac{e^{K_2}}{e^{K_1} - 1} = T_2\,\frac{T_1^* - T_0^*}{T_1 - T_0}\;.$$

Hence if the basic intervals $t_1 - t_0 = 100° = t_1^* - t_0^*$ correspond to 100 K in both cases, then $T_2 = T_2^*$ and the choice of the empirical temperature scale does not affect the absolute temperature scale.

For dilute gases it is often true that *Joule's law*

$$\left(\frac{\partial U}{\partial v}\right)_t = 0 \qquad (6.6)$$

holds, and also for the Celsius scale $t_c$ these systems satisfy

$$\frac{p - p_0}{p_0} = \lambda t_c \quad (v \text{ constant}) \qquad (6.7)$$

where $\lambda = 1/273{\cdot}15$ and where $p_0$ is the gas pressure at 0°C. It follows that if such a system is chosen as the thermometric substance, then

$$K_i = \int_{t_{c0}}^{t_{ci}} \frac{p_0\lambda}{p_0(1+\lambda t_c)}\, dt_c = \ln \frac{1+\lambda t_{ci}}{1+\lambda t_{c0}} .$$

Since $T_1 - T_0 = t_{c1} - t_{c0} = 100$, equation (6.5) yields

$$T_2 = (T_1 - T_0)\,\frac{1+\lambda t_{c2}}{(1+\lambda t_{c1}) - (1+\lambda t_{co})}$$

that is

$$T_2 = \frac{T_1 - T_0}{t_{c1} - t_{c0}}\left(\frac{1}{\lambda} + t_{c2}\right) = \frac{1}{\lambda} + t_{c2} = 273{\cdot}15 + t_{c2} . \qquad (6.8)$$

Thus all empirical temperature scales lead to $T = 0$ at $t_c = -1/\lambda = -273{\cdot}15$ °C. This is the so-called *absolute zero of temperature.*

It is a consequence of (6.2) and (6.6) that

$$\frac{(\partial p/\partial t)_v}{p} = \frac{(\partial p/\partial T)_v}{p} = \frac{1}{T} .$$

It follows by integration that there exists a function $g$ of volume such that

$$g(v)p = T \qquad \text{(Joule's law for fluids)}$$

for dilute gases.

## 6.2. APPLICATION TO THE VAN DER WAALS GAS AND THE IDEAL CLASSICAL GAS

In this section we shall discuss an ideal gas and the more realistic van der Waals gas.

The *van der Waals gas* differs from an ideal gas in two respects. Firstly, it takes account of the finite size of the molecules so that the volume available for the motion of molecules is less than $v$, the *observed* volume. The unavailable volume can be estimated by considering a collision between two spherical molecules A and B. If A is regarded as fixed, the centre of B must lie, at collision, on a sphere of radius $2r$, where $r$ is the radius of a molecule. The unavailable volume is therefore of the order of

$$\frac{4\pi}{3}(2r)^3 = 8v_{\mathrm{m}}$$

where $v_{\mathrm{m}}$ is the volume of a molecule. The unavailable volume *per molecule* is therefore of the order of $4v_{\mathrm{m}}$. The effective available volume to be used is therefore $v-b$, where $b \sim 4v_{\mathrm{m}}N$ and $N$ is the number of molecules in the gas. Secondly, the force of attraction between the molecules is taken into account. This effect cancels out in the bulk of the gas since each molecule is attracted equally in all directions. The pressure in the bulk will be denoted by $p_{\mathrm{b}}$. At the wall of the gas the effect is, however, noticeable, and causes an attraction of each molecule back into the interior. Let this lower pressure at the walls be denoted by $p$ since it is the pressure normally measured. The value $p_{\mathrm{b}}-p$ can be estimated as proportional to the number of molecules colliding with the wall per second multiplied by the number of molecules with which each such surface molecule interacts. Each of these numbers is proportional to the density of the gas, so that for a fixed mass of gas

$$p_{\mathrm{b}} - p \propto \left(\frac{N}{v}\right)^2 .$$

The pressure $p_{\mathrm{b}}$ to be used in the equation is therefore

$$p_{\mathrm{b}} = p + \frac{a}{v^2} .$$

This leads to the van der Waals equation of state

$$\boxed{\left(p + \frac{a}{v^2}\right)(v-b) = ct ,} \tag{6.9}$$

where $a \propto N^2$, $b \propto N$, and $c \propto N$ are constants for given $N$. In an ideal gas one neglects $a$ and $b$.

Using some empirical temperature scale $t$, or the absolute temperature scale $T$, *dilute* gases usually obey Joule's law

$$\left(\frac{\partial U}{\partial v}\right)_t = 0 \quad \text{or equivalently} \quad \left(\frac{\partial U}{\partial v}\right)_T = 0 \ . \tag{6.10}$$

This states that such a system retains a constant internal energy in isothermal expansion. Broadly speaking, constant temperature means that the average kinetic energy of a molecule remains constant. Expansion under these conditions could lead to a rise in the internal energy as the molecules become more widely separated. Because of the mutual attraction between the molecules, the potential energy increases in this process, and the internal energy rises. An example is provided by the van der Waals gas, as will now be shown. By equations (6.2) and (6.9)

$$\left(\frac{\partial U}{\partial v}\right)_t = T\left(\frac{\partial p}{\partial T}\right)_v - p$$

$$= \frac{cT}{v-b}\frac{\mathrm{d}t}{\mathrm{d}T} - \left(\frac{ct}{v-b} - \frac{a}{v^2}\right) = \frac{ct}{v-b}\left(\frac{T}{t}\frac{\mathrm{d}t}{\mathrm{d}T} - 1\right) + \frac{a}{v^2} \ . \tag{6.11}$$

It follows that even if the absolute temperature scale is adopted, so that $T = t$,

$$\left(\frac{\partial U}{\partial v}\right)_T = \frac{a}{v^2} \ . \tag{6.12}$$

The constant $a$ which takes account of interparticle forces vanishes for an ideal classical gas, since these forces are treated as negligible in this limit.

The above equations illustrate an important point concerning the definition of an *ideal classical gas*. One can define such a system in two equivalent ways as a system which satisfies either

$$\boxed{\left(\frac{\partial U}{\partial v}\right)_t = 0 \quad \text{and} \quad pv = ct} \tag{6.13}$$

or

$$\boxed{pv = cT \ .} \tag{6.14}$$

Putting $a = b = 0$ in (6.11) one sees that $pv = ct$ still leads to a violation of Joule's law if the absolute temperature scale is *not* adopted. Hence for this system (6.10) has then to be stipulated separately. This is the definition (6.13). However, Joule's law is implied if the definition (6.14), is used.

A further consequence of these results is that, integrating (6.11),

$$U = c\left(T\frac{dt}{dT} - t\right)\ln(v-b) - \frac{a}{v} + f(t) . \qquad (6.15)$$

Hence, by (3.14)

$$C_v = \left(\frac{\partial U}{\partial t}\right)_v = cT\frac{d^2t}{dT^2}\frac{dT}{dt}\ln(v-b) + \frac{df}{dt} \qquad (6.16)$$

and

$$\left(\frac{\partial C_v}{\partial v}\right)_t = \frac{c}{v-b}T\frac{dT}{dt}\frac{d^2t}{dT^2} . \qquad (6.17)$$

This shows that the van der Waals gas *and* the ideal classical gas have one important property in common. If the absolute temperature scale is adopted $dt/dT = 1$, $d^2t/dT^2 = 0$ and therefore

$$\left(\frac{\partial C_v}{\partial v}\right)_T = 0 . \qquad (6.18)$$

This is a useful result since it shows that $C_v$ can be taken to be a function of temperature only.

What is this function? It is thermodynamically unspecified. All one can say is that at infinite dilution the van der Waals $C_v$ will go over into the $C_v$ for the corresponding ideal gas. Contrary to what is sometimes stated, this does not mean that the van der Waals $C_v$ is also independent of temperature.

## 6.3. VARIOUS FORMS OF THE THIRD LAW OF THERMODYNAMICS

### *6.3.1. Unattainability from the second law?*

The set of equilibrium states of a physical system such that any two can be adiabatically linked are represented by a set of points (in a thermodynamic phase space) which has been denoted by $\beta$ in section 2.3. The internal points of $\beta$ are obtained by removing the frontier points $F(\beta)$ from $\beta$ and the resulting connected sets have been denoted by $\gamma$ (section 2.3). Absolute temperature and entropy are then defined for these open sets $\gamma$. Whether or not the points $F(\gamma)$ on the frontier of $\gamma$ lie in a set $\beta$, i.e. whether or not they are adiabatically linked with the points of $\gamma$, is a matter which has so far been left open, and a third law of thermodynamics which deals with it is therefore essential. This matter has to be settled differently for different types of states, and in this section the set of points, $\zeta$ say, at the absolute zero of temperature will be studied. In fact we shall show first of all that *for any set* $\beta$ *the set* $\beta\cap\zeta$ *is so poorly populated that it is impossible to draw a continuous curve in it.* This follows already from the second law of thermodynamics.

As a preliminary observation, we note that absolute temperatures can be associated with points in $F(\gamma)$. For if $T(g_n)$ be the absolute temperature of a state $g_n$ in $\gamma$, then one can always define the temperature of a point f of $\zeta$, when it is on the boundary of the set $\gamma$ under consideration, by $T(f) = \lim_{n\to\infty} T(g_n)$ (provided the limit exists and is independent of the path used to approach f). In this way, $T$ and $S$ can be defined for points on $F(\gamma)$, and we are in order to talk about the set of points $\zeta$ for which $T = 0$ even though these points lie on $F(\gamma)$.

To come to the actual argument leading to the italicized statement suppose that $\beta\cap\zeta$ contains a continuous curve. This is then made the isothermal $T_2 = 0$ in the Carnot cycle of problem (5.6) whose efficiency is

$$\eta_c = \frac{T_1 - T_2}{T_1} = 1 \ .$$

This, however, implies complete conversion of heat into mechanical work and a *perpetuum mobile* of the second kind. As this is contrary to the observation in section 5.1, continuous curves cannot exist in $\beta\cap\zeta$.

It will be appreciated that while $F(\gamma)\cap\zeta$ can contain curves at $T = 0$, points on $F(\gamma)$ are not necessarily in the corresponding set $\beta$, i.e. they are not necessarily linked adiabatically with the points in $\gamma$. The statement 'sets $\beta\cap\zeta$ do not contain continuous curves' asserts this absence of an adiabatic linkage.

It can be argued that an isothermal at the absolute zero does not make sense on general physical grounds. How can one vary the volume of a system at $T = 0$ and keep it at that temperature? Questions of this type are justified, but if one puts oneself in the position of consolidating one's knowledge of thermodynamics into laws and into explicit assumptions, one must imagine that one is approaching the physical situation at $T = 0$ with an unprejudiced mind, ready to treat a process at $T = 0$ like any process at $T > 0$. With this attitude the maximum information concerning conditions at $T = 0$ can be deduced, and this was the procedure adopted above.

Experiment suggests that states at the absolute zero of temperature are not among the equilibrium states of physical systems which are adiabatically linked with other states: $\beta\cap\zeta = 0$. This is the natural generalization to which the above italicized observation points, but strictly speaking it is a new statement. In fact, a stronger statement can be made: States of physical systems at absolute zero are not attainable. This is the principle of the unattainability of the absolute zero due to W. Nernst in 1912. It is rather general: there is no mention of equilibrium states, nor of adiabatic linkage.

It is a usual implicitly made assumption that other boundary points of sets $\gamma$ are capable of being adiabatically linked with points of $\gamma$. For these points, therefore, $\beta\cap F(\gamma) = F(\gamma)$.

*6.3.2. Nernst's theorem from unattainability?*

One can now ask: Given the unattainability of the absolute zero, what properties of the entropy may be deduced from it? Fig. 6.1 shows typical possible behaviours of the entropy curves $S(x,T)$ where $x$ is one of the external parameters such as volume, magnetic field, etc., which are variable within certain ranges $x_1 > x > x_2$, $x_3 > x > x_4$, ... etc. Thus we have that the values $x_1,\ldots,x_4,\ldots$ of $x$ yield points lying

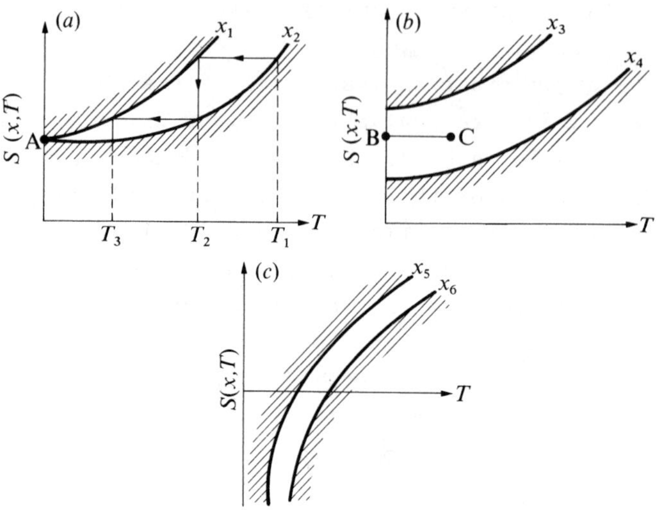


Fig.6.1. Possible entropy-temperature curves for different configurations of a system as specified by the external parameter $x$. Such curves are compatible with all laws of thermodynamics, excepting the third law. Shaded regions lie outside the appropriate sets of points $\gamma$.

on the boundary of $\gamma$. For example, phase transitions may occur at these values of $x$ so that a different thermodynamic phase space is applicable when $x$ lies outside the specified ranges. Normally the heat capacities are non-negative (see equation (7.21), and the curves are then non-decreasing. For example, entropy functions which diverge to $+\infty$ in the limit $T \to 0$ are then not permitted. Negative heat capacities present different problems (section 13.6) and are not considered here.

In the case of Fig.6.1($a$) only the state A has to be treated as unattainable. If the system is represented by Fig.6.1($b$) a horizontal line CB can be drawn. It represents a quasi-static adiabatic process to the absolute zero. It is consistent with the unattainability of the absolute zero only if a discontinuity in thermodynamic properties occurs infinitely close to the absolute zero so as to prevent its attainment. Fig. 6.1($c$), like Fig.6.1($b$) does not occur in practice. Hence one may say that the unattainability of the absolute zero implies entropy diagrams of the type shown in Figs. 6.1(*a*) or 6.2 provided it is assumed that (*a*) entropies do not diverge as $T \to 0$ and (*b*) thermodynamic properties do not approach a discontinuity infinitely close to the absolute zero. Entropy curves as shown in Fig.6.2 are still compatible with these assumptions. The parts $\gamma_1$, $\gamma_2$, $\gamma_3$ of the set $\gamma$ will be called *branches* of $\gamma$. It is now assumed that (*c*) one is

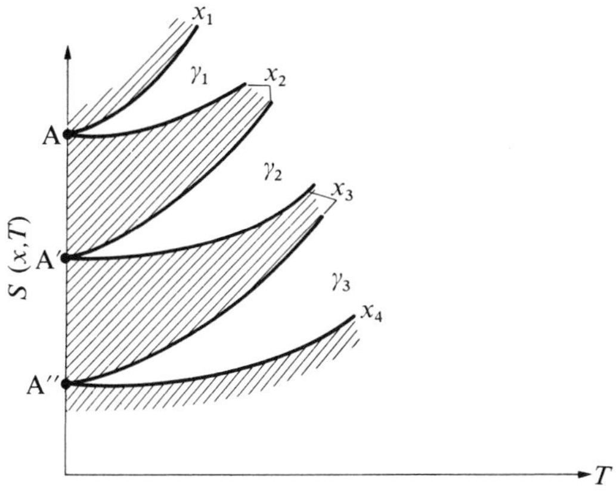


Fig.6.2. Possible entropy curves compatible with the unattainability of absolute zero.

concerned with only one such branch. With these assumptions one finds that the entropies $S(x,T)$, $S(x',T)$ of two configurations of a physical system, specified by $x$ and $x'$, have the property

$$\lim_{T\to 0} \{S(x,T) - S(x',T)\} = 0 \ . \tag{6.19}$$

This is Nernst's heat theorem (1906).Contrary to what has sometimes been said, it is not equivalent to the unattainability principle since the additional assumptions (*a*), (*b*), and (*c*) are needed in the above argument. There is considerable evidence for the unattainability principle and for Nernst's theorem. We have no space to discuss these matters here and the reader is referred to a book on physical chemistry.

*6.3.3. Unattainability from Nernst's theorem?*

Can the unattainability of the absolute zero be deduced from Nernst's theorem (6.19)? One may be tempted to suppose that this is so by considering the process shown in Fig.6.1(*a*) for two configurations for which $S \propto T$ will be assumed:

$$S(x_1,T) = a_1 T, \qquad S(x_2,T) = a_2 T, \qquad (0 < a_2 < a_1)$$

where $S(x,0) = 0$ has been chosen for convenience.
Given a heat reservoir at temperature $T_1$, the state $(x_2,T_1)$ is cooled to $(x_1,T_2)$ by quasi-static adiabatic process, so that

$$a_2 T_1 = a_1 T_2 ,$$

that is

$$T_2 = \frac{a_2}{a_1} T_1 .$$

A heat reservoir at the lower temperature $T_2$ is now prepared and the process is repeated by chosing the path $(x_2,T_2) \rightarrow (x_1,T_3)$, where

$$T_3 = \left(\frac{a_2}{a_1}\right) T_2 = \left(\frac{a_2}{a_1}\right)^2 T_1 .$$

Thus one can reach, after $n$ steps, a temperature which decreases as $n$ increases:

$$T_{n+1} = \left(\frac{a_2}{a_1}\right)^n T_1 .$$

However, the absolute zero cannot be attained in a finite number of steps.

This argument can be generalized to other forms of the functions $S(x,T)$, but it still does not prove the complete unattainability of the absolute zero since the existence of *other* processes of lowering temperatures has not been investigated in this argument. Even if it had been investigated, the possible existence of so far undiscovered processes of lowering temperatures must mean that a rigorous deduction of the unattainability principle from Nernst's theorem is not possible.

PROBLEMS

6.1. An *ideal classical gas* of $N$ molecules is defined to be a fluid which satisfies (cf. equation (6.13))

(i) $\left(\frac{\partial U}{\partial v}\right)_t = 0$ (Joule's law)

(ii) $pv = Nkt$ (the gas equation)

in its equilibrium states. Here $k$ is Boltzmann's constant equal to $1{\cdot}3807 \times 10^{-23}$ J K$^{-1}$ or $1{\cdot}3807 \times 10^{-16}$ erg K$^{-1}$. Show that, for any arbitrary quasi-static change from a state A to a state B,

$$S_A^B \equiv \int_A^B \frac{d'Q}{t} = \int_A^B \frac{C_v}{t}\,dt + Nk \ln \frac{v_B}{v_A} .$$

Why does $d'Q$ have an integrating factor in this case, although there has been no appeal to the second law, and what is the form of $S_A^B$ when $C_v$ is a constant?

6.2. Show that $l_p = -T(\partial v/\partial T)_p$ and hence show that for a general fluid

$$\int_A^B \frac{d'Q}{T} = \int_A^B \frac{C_p}{T}\,dT - \int_A^B v\alpha_p\,dp .$$

6.3. (*a*) Show that, for an ideal classical gas, $C_p - C_v = Nk$. (*b*) For the same system as in problem 6.1 show that the entropy variation satisfies

$$S(p,T) = S(p_0,T) - Nk \ln (p/p_0)$$

$$S(v,T) = S(v_0,T) + Nk \ln (v/v_0) .$$

6.4. What is the entropy change in a system consisting of two reservoirs at temperatures $T_c$ and $T_h$ when a small amount of heat $Q$ leaks from one to the other? On what general principles would you expect the result to indicate an increase or decrease in the entropy?

6.5. (i) Two dilute non-reacting gases A and B occupy volumes $v_A$ and $v_B$ and consist of $N_A$ and $N_B$ identical molecules respectively. Their entropies have the forms

$$S_j = kN_j\{c + \ln(v_j/N_j)\} \qquad (j = A, B)$$

where $c$ is a constant. They are at the same temperature and initially separated. Both gases may later occupy the total volume $v = v_A + v_B$. If $p \equiv v_A/v$, $q \equiv N_A/N$ where $N = N_A + N_B$, show that the increase in the entropy of the total system is

$$S_1 = -kN\{q \ln p + (1-q)\ln(1-p)\}$$

or

$$S_2 = kN\left\{q \ln\left(\frac{q}{p}\right) + (1-q)\ln\left(\frac{1-q}{1-p}\right)\right\}$$

depending on whether the molecules of A are distinguishable from those of B or not.

(ii) Show that $S_1 \geq kN \ln 2$, $S_2 \geq 0$ and $S_1 - S_2 \geq kN \ln 2$, the minimum values being reached for $p=q=\frac{1}{2}$ in the case of $S_1$, for $p=q$ in the case of $S_2$, and for $p=q=\frac{1}{2}$ in the case of $S_1 - S_2$.

(iii) Does the entropy change drop *abruptly* from $S_1$ to $S_2$ as the molecules of A are *gradually* made identical with those of B? [This is the *Gibbs paradox*.]

6.6. (*a*) Establish the result $p = T\left(\frac{\partial p}{\partial T}\right)_v - \left(\frac{\partial U}{\partial v}\right)_T$.

(*b*) Show that the internal energy $U$ of an *ideal quantum gas* defined by $pv = gU$ ($g$ is a constant) satisfies

$$U = T\left(\frac{\partial U}{\partial T}\right)_v - \frac{v}{g}\left(\frac{\partial U}{\partial v}\right)_T$$

and hence show that a function $f$ exists such that

$$Uv^g = f(Tv^g)$$

(*c*) Show that a function $h(z)$, $z = Tv^g$, exists such that the entropy of this system satisfies

$$S = h(z), \quad \frac{dh}{dz} = z^{-1}\frac{df}{dz} .$$

(*d*) Show that the following quantities are constant in quasi-static adiabatic changes of this system

$$pv^{1+g}, \quad Tv^g, \quad \frac{T^{1+g}}{p^g} .$$

6.7. The third law can by p. 71 be stated as: '$S(x,T)$ is finite, continuous, and unique as $T \to 0$.' Hence:

(*a*) Show that the quantities $C_v$, $C_p$, $l_v$, $l_p$, $m_v$, $m_p$ all tend to zero as $T \to 0$.

(*b*) Show that an *ideal classical gas* cannot exist near $T = 0$.

(*c*) Show that an *ideal quantum gas* is compatible with (*a*).

6.8. If $T_0$ is the absolute zero of temperature in °C and if $\alpha \equiv e^2/\hbar c$ is the fine-structure constant, obtain the equation for $\alpha$,

$$T_0 = -\left(\frac{2}{\alpha} - 1\right) .$$

(This 'problem' is one of the few practical jokes which got past the referees. For a 'solution' see Beck, Bethe, and Riezler in *Naturwissenschaften* of 9th January 1931. The equation yields $\alpha = 1/137$ in good agreement with the usual value for $\alpha$.)

6.9. Adopt the absolute temperature scale for a van der Waals gas (6.9). Hence show that

(i) $U = \int C_v \mathrm{d}T - \frac{a}{v} + B_1$

(ii) $S = \int \frac{C_v}{T} \mathrm{d}T + c \ln (v-b) + B_2$

(iii) $F = \int C_v \mathrm{d}T - T \int \frac{C_v}{T} \mathrm{d}T - cT \ln (v-b) - B_2 T + B_1 - \frac{a}{v}$

(iv) The work required to produce quasi-static isothermal expansion from $v_1$ to $v_2$ is

$$\frac{a}{v_2} - \frac{a}{v_1} + cT \ln \frac{v_2 - b}{v_1 - b} \, .$$

# 7. Variable particle numbers

## 7.1. EXTENSIVE AND INTENSIVE VARIABLES

We shall now make good the promise, given in section 3.4, that a formal definition of extensive variables can be given. To achieve this a 'simple' system will be treated for which *one* entropy $S$, *one* internal energy $U$, $r$ external parameters or 'deformation co-ordinates' $v_1, v_2, \ldots, v_r$ and $\chi$ additional variables $n_1, n_2, \ldots, n_\chi$ are required. Of these $r+\chi+2$ variables only $r+\chi+1$ are independent, the other variable (the entropy in our case) being regarded as a function of them. For such a system, the formal definition of an extensive variable can be given by considering what happens to a thermodynamic function when it is multiplied by some positive number $a$. We shall give the definition in terms of the entropy, although the volume, the internal energy, etc., could also be used. Since the heat capacity $C$ is extensive, as already noted in section 3.4, so is the entropy $C\mathrm{d}T/T$. The entropy is extensive if for all positive $a$:

$$S_a \equiv S(aU, av_1, \ldots, av_r, an_1, \ldots, an_\chi) = aS(U, v_1, \ldots, v_r, n_1, \ldots, n_\chi)$$

$$\equiv aS_1 \; . \qquad (7.1)$$

The integers $r$ and $\chi$ are fixed for a given domain. The new variables $n_j$ can be thought of as describing the internal state of the system. It is by no means necessary to do so, but it helps if one thinks of the $n_j$, for definiteness, as giving the number of molecules of component $j$ in the system. They could also describe processes of pressure equalization proceeding within the system.

The fact that there is *a single* energy $U$ implies that adiabatic partitions are not part of the system, so that there is also a single temperature function $T$. If a simple system is doubled (i.e. $a = 2$) by doubling $U$, the $v_i$, and the $n_j$, then (7.1) asserts that the entropy is also doubled, $S_2 = 2S_1$. Functions with this property are called *extensive* functions.

In fact, if one has a function $f$ and independent variables $X_i$ such that $f = f(X_1, X_2, \ldots)$ can be solved for $X_i$,

$$X_i = g_i(X_1, \ldots, X_{i-1}, f, X_{i+1} \ldots) ,$$

then clearly a condition of the type (7.1) means that also $af = f(aX_1, aX_2, \ldots)$ can be solved for

$$aX_i = g_i(aX_1, \ldots, af, \ldots) .$$

It then follows that

$$g_i(aX_1, \ldots, aX_{i-1}, af, aX_i, \ldots) = aX_i = ag_i(X_1, \ldots, X_{i-1}, f, X_{i+1}, \ldots)$$

Thus if $f$ is an extensive variable, so are all the $X_i$'s. In particular, (7.1) shows now that the following variables are all extensive in character:

$$S,\ U,\ v_i,\ n_j \quad (i = 1, 2, \ldots, r;\ j = 1, 2, \ldots, \chi) .$$

In Chapters 1 to 6, systems have been considered which can possibly exchange work or heat (or both or neither) with their surroundings, but which could not exchange matter. Such systems are said to be *closed* with regard to matter and *open* with regard to work and heat transfer. It will be supposed that thermodynamics also holds for systems which are open with regard to the transfer of matter, hence the occurrence of the $n_j$, with their interpretation as the number of molecules of type $j$, as variables in (7.1). This is a distinct extension of the theory and is in agreement with what is found experimentally

In this way, the thermodynamic phase space **E** introduced in section 2.3 is extended into a larger phase space **F** for any one physical system. Of such systems it will be assumed that they are free of adiabatic partitions so that *one* $S$ and *one* $U$ function suffice, that in the states of interest the system is either in equilibrium or homogeneous or both, that such states are represented by points in **F**, and that (7.1) applies. The domains $\gamma$ of **E** are now replaced by (generally larger) domains $\delta$ (say) of **F**, and some description of non-equilibrium processes and of systems which gain or lose matter becomes possible within this framework.

The domains $\delta$ of interest in **F** shall share with the domains $\gamma$ of **E** the following two simple properties :

(*a*) Any two points in $\delta$ can be linked by processes which can be represented by curves in $\delta$, as well as by other processes.

(*b*) All points of $\delta$ are internal points (the domain is open; it contains no boundary points).

Not all non-equilibrium systems or open systems can be simply represented in a phase space **F** ; one need only think of turbulent flow, when the representation of the motion of all the molecules would become excessively complicated. We shall therefore restrict attention to the systems, called 'simple', systems, introduced in (7.1).

A basic assumption involved in the argument so far is that a thermodynamic system free of adiabatic partitions which is in its states of interest either in equilibrium or homogeneous (or both) is a simple system. This makes (7.1) applicable and a description of the main thermodynamic variables as either extensive or intensive becomes possible. This has been called by the present author the *Fourth Law of Thermodynamics*. As one has occasionally to work with general powers of extensive variables, variables are conceivable (e.g. $v^{1/4}$) which are neither extensive nor intensive.

Note that (7.1) implies that, for any increment represented in a domain $\delta$,

$$T\mathrm{d}S = T\left(\frac{\partial S}{\partial U}\right)_{v_i,n_j}\mathrm{d}U + \sum_i T\left(\frac{\partial S}{\partial v_i}\right)_{v_i',n_j,U}\mathrm{d}v_i + \sum_j T\left(\frac{\partial S}{\partial n_j}\right)_{v_i,n_j',U}\mathrm{d}n_j \tag{7.2}$$

where $(\ )_{v_i'}$ means that all $v$'s are kept constant except for $v_i$. If the $n$'s are constants, one returns to the phase space

E and hence the equation

$$T\mathrm{d}S = \mathrm{d}U + \sum_i p_i \, \mathrm{d}v_i ,$$

whence

$$\left(\frac{\partial S}{\partial U}\right)_{v_i,n_j} = \frac{1}{T}, \quad \left(\frac{\partial S}{\partial v_i}\right)_{v_i',n_j,U} = \frac{p_i}{T},$$

$$\left(\frac{\partial S}{\partial n_j}\right)_{v_i,n_j',U} \equiv -\frac{\mu_j}{T}. \tag{7.3}$$

The last equation is a definition of a quantity $\mu_j$ called the *chemical potential* of component $j$, which is analogous to temperature and pressure in the following respects:

(*a*) The doubling of the system does not affect any of these quantities. They are therefore called *intensive* to distinguish them from the extensive quantities.

(*b*) They have uniform values in equilibrium.

This last point is established in section 7.3. The sign in the definition of $\mu_j$ follows convention. If the $n_j$'s are more general variables, $x_j$ say, the $\mu_j$ will also become more general quantities, $y_j$ say.

## 7.2. THE MEANING OF CHEMICAL POTENTIAL

The physical meaning of chemical potential may be discussed as follows. A semi-permeable membrane is a theoretical device which can, however, often be approximated in practice. It is a dividing wall which allows one particular component of a system to pass freely in both directions, while stopping the flow of all other components. An example is provided by hot palladium which is semi-permeable to hydrogen. Thus a multi-component gaseous system with one semi-permeable membrane in its wall for each component can yield a measurement for the pressure of each component in the system, as shown in Fig. 7.1. This pressure may be called the *partial or escaping pressure* of the component. In equilibrium, the pressure $p$ is

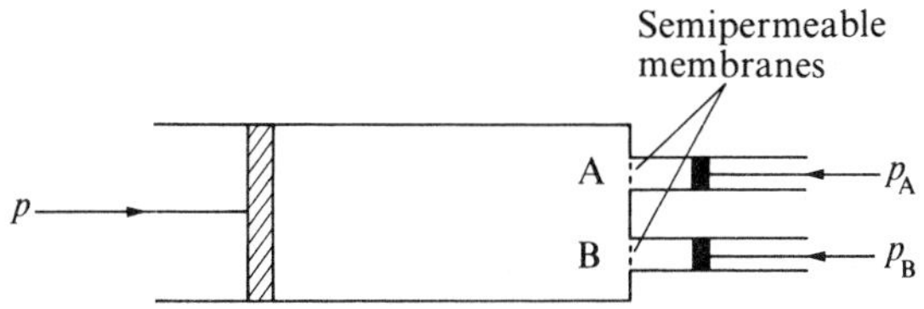


Fig.7.1. The escaping pressure.

equal to $p_A + p_B$ only in special cases. It clearly is *not* the case when the system contains a third component C. Even if the system contains only two components the equality holds in general only if A and B are non-reacting, dilute (i.e. ideal classical) gases. In that case $p_A$ and $p_B$ are the pressures which the gases would exert were they to occupy the whole volume of the system on their own. This interpretation is, however, clearly not correct in a real or non-ideal gas. In this case the molecular interactions depend on the mean distance between the molecules and so change with the volume, thus affecting the pressure exerted by the gas. It is in the ideal gas limit only that one can expect *Dalton's law of partial pressures* to hold, which states, in effect, $p = p_A + p_B$.

For a given pure component C at a given temperature in a (possibly multi-component) system A, we know that (*a*) the addition of C to A increases its chemical potential and (*b*) the addition of C to A increases its escaping pressure. It follows that the above 'apparatus' can be calibrated to read the chemical potential of component C. This is certainly so for liquids and gases, and even for those solids whose vapour pressure is readily measured. It is analogous to measuring temperature by noting the length of a column of mercury.

There are two further applications of the above idea. One can regard the device of Fig.7.1 as a quasi-static separator of a multi-component system. One then has to supply one semi-permeable membrane for each component and to collect the components as the main piston is brought slowly to the right-hand end of the chamber.

The second application is to subject a system D to a specific chemical potential for a component C by connecting it via a semi-permeable membrane to a reservoir of this component. By choosing a reservoir with the correct escaping pressure one can impose the corresponding chemical potential on the system D. This is analogous to the use of a heat reservoir to impose a temperature and to the way a system can be subjected to an external pressure.

In these ways the chemical potential can be thought of as analogous to temperature and pressure.

## 7.3. THE GIBBS FREE ENERGY AND THE GIBBS-DUHEM RELATION

Next differentiate (7.1) with respect to $a$ to find, using $\mathrm{d}(aU)/\mathrm{d}a = U$, $\mathrm{d}(av_i)/\mathrm{d}a = v_i$, etc.,

$$\left\{\frac{\partial S_a}{\partial(aU)}\right\}_{v_i,n_j} U + \sum_i \left\{\frac{\partial S_a}{\partial(av_i)}\right\}_{v_i',n_j,U} v_i + \sum_j \left\{\frac{\partial S_a}{\partial(an_j)}\right\}_{v_i,n_j',U} n_j = S_1 .$$

As the ratio of two extensive quantities we have

$$\left\{\frac{\partial S_a}{\partial(aU)}\right\}_{v_i,n_j} = \left(\frac{\partial S_1}{\partial U}\right)_{v_i,n_j} = \frac{1}{T}$$

$$\left\{\frac{\partial S_a}{\partial(av_i)}\right\}_{v_i',n_j,U} = \frac{p_i}{T} .$$

Hence, writing again $S$ for $S_1$, one has the important result

$$\boxed{U - TS + \sum_i p_i v_i = \sum_r \mu_r n_r \equiv G .} \tag{7.4}$$

Each term in this sum is extensive and the symbol $G$ denotes the Gibbs free energy.

One now has from (7.2) and (7.4) respectively

$$\mathrm{d}U - T\mathrm{d}S + \sum_i p_i\, \mathrm{d}v_i - \sum_j \mu_j\, \mathrm{d}n_j = 0 \quad \text{(the Gibbs equation)} \tag{7.5}$$

$$\mathrm{d}U - T\mathrm{d}S - S\mathrm{d}T + \sum_i (p_i\, \mathrm{d}v_i + v_i\, \mathrm{d}p_i) - \sum_j (\mu_j\, \mathrm{d}n_j + n_j\, \mathrm{d}\mu_j) = 0$$

from which it follows that the intensive variables, $T$, $p_i$, $\mu_j$ are not all independent. They are connected by the so-called *Gibbs-Duhem equation*

$$\boxed{S\mathrm{d}T - \sum_i v_i\,\mathrm{d}p_i + \sum_j n_j\,\mathrm{d}\mu_j = 0\ .} \tag{7.6}$$

One sees that a simple system with $r$ deformation coordinates and $\chi$ components has $r + \chi + 1$ intensive variables $T$, $p_i$, $\mu_j$ of which $r + \chi$ are independent. Other intensive variables can be constructed, for example, as ratios of extensive variables. These ratios determine the intensive properties of the system but leave its extent unspecified, since doubling the system leaves these $r + \chi$ independent ratios unchanged.

It is worth memorizing the simplest forms of the important relations (7.4) and (7.6), which are

$$\boxed{\begin{aligned} U - TS + pv &= \mu n = G \\ S\mathrm{d}T - v\mathrm{d}p + n\mathrm{d}\mu &= 0 \end{aligned}}$$

It is to be emphasized that the relations obtained add extra terms involving $\mu$ and $n$ to thermodynamic relations already derived. We have seen earlier (for example in solving problem 5.1) that

$$\mathrm{d}G = \mathrm{d}(U - TS + pv) = \mathrm{d}U - T\mathrm{d}S - S\mathrm{d}T + p\mathrm{d}v + v\mathrm{d}p$$

reduces by virtue of

$$\mathrm{d}'Q = T\mathrm{d}S = \mathrm{d}U + p\mathrm{d}v$$

to

$$\mathrm{d}G = -S\mathrm{d}T + v\mathrm{d}p\ .$$

However, by (7.4) and the Gibbs-Duhem relation, we have in the generalized theory that

$$dG = d(\mu n) = \mu dn + n d\mu = \mu dn + (v dp - S dT) .$$

The additional term on the right-hand side means that

$$dU = d(G + TS - pv) = \mu dn + T dS - p dv .$$

We find again an extra term on the right-hand side (see problem 7.1).

## 7.4. EQUILIBRIUM CONDITIONS AND THE PHASE RULE

For an isolated system of two of the three phases solid, liquid, and vapour, one has that, in an incremental virtual (i.e. imagined) change of material from one phase to another,

$$\delta U_1 + \delta U_2 = \delta n_1 + \delta n_2 = \delta v_1 + \delta v_2 = 0 .$$

These equations express the constancy in the total number of molecules, the total energy, and the total volume. An equilibrium condition is that the entropy of the state has its maximum possible value, so that it cannot be increased by the small virtual changes. Hence, using (7.5),

$$0 = \delta S_1 + \delta S_2 = \frac{1}{T_1}(\delta U_1 + p_1 \delta v_1 - \mu_1 \delta n_1) + \frac{1}{T_2}(\delta U_2 + p_2 \delta v_2 - \mu_2 \delta n_2)$$

$$= \left(\frac{1}{T_1} - \frac{1}{T_2}\right) \delta U_1 + \left(\frac{p_1}{T_1} - \frac{p_2}{T_2}\right) \delta v_1 - \left(\frac{\mu_1}{T_1} - \frac{\mu_2}{T_2}\right) \delta n_1 .$$

Since $\delta U_1$, $\delta v_1$, and $\delta n_1$ are arbitrary small changes, this condition implies that

$$\left.\begin{aligned} T_1 &= T_2 \quad \text{(thermal equilibrium)} \\ p_1 &= p_2 \quad \text{(mechanical equilibrium)} \\ \mu_1 &= \mu_2 \quad \text{(chemical equilibrium)} . \end{aligned}\right\} \tag{7.7}$$

This shows the importance of the chemical potential.

Suppose there are $\pi$ phases and each phase contains the same $\chi$ non-reacting constituents. Suppose also that the total mass of each constituent present is determined. For a system of arbitrary size one has the following intensive variables available: $\pi$ temperatures, $\pi$ pressures, and $\chi$ chemical potentials for each phase. This corresponds to $r=1$ and the total number of intensive variables is

$$\pi + \pi + \pi\chi = \pi(\chi + 2).$$

Because one Gibbs-Duhem equation exists among the $\chi+2$ intensive variables of each phase, this number is reduced to $\pi(\chi+1)$.

Let $\mu_{ij}$ denote the chemical potential of component $j$ in phase $i$. Then the temperatures, pressures, and masses in the various phases adjust themselves in equilibrium to satisfy $(\pi-1)(\chi+2)$ relations:

$$T_1 = \ldots = T_\pi \qquad \pi-1 \text{ relations}$$

$$p_1 = \ldots = p_\pi \qquad \pi-1 \text{ relations}$$

$$\mu_{11} = \ldots = \mu_{\pi 1} \qquad \pi-1 \text{ relations}$$

$$\vdots$$

$$\mu_{1\chi} = \ldots = \mu_{\pi\chi} \qquad \pi-1 \text{ relations}\ .$$

The total number of independent variables, the number of 'degrees of freedom' of the system, is therefore

$$\boxed{f = \pi\chi + \pi - (\pi\chi + 2\pi - \chi - 2) = \chi - \pi + 2}$$

(The Gibbs phase rule). (7.8)

If there are $R$ independent relations among the $\chi$ chemical potentials, as there would be if $R$ independent chemical reactions take place, $\chi$ in (7.8) has to be replaced by

$\chi' = \chi - R$. There are no degrees of freedom for a single component system if $\pi = 3$. Thus vapour, liquid, and solid of a single substance can be in equilibrium only at one point in thermodynamic phase space. This is called the triple point $T_0$. However, two phases of one substance can be in equilibrium along a line in phase space obtained by varying temperature or pressure. The typical phase diagram (Fig.7.2) of a simple *metal* results, $C$ being the critical point. For $T > T_c$ the

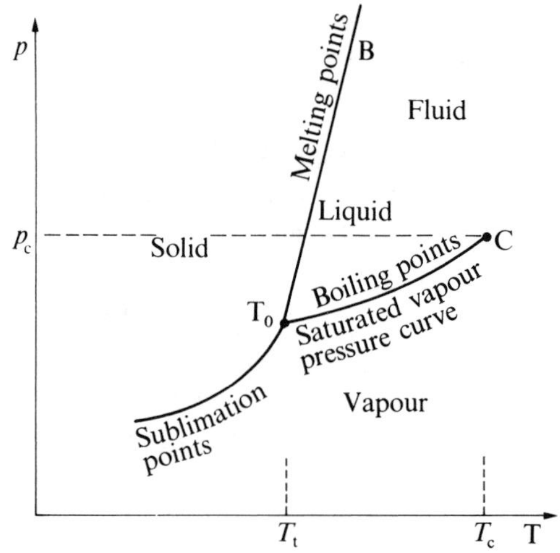


Fig.7.2. A $p$-$T$ phase diagram for fixed volume. The lines shown are called *isochores*, i.e. the volume is constant.

vapour cannot be liquified however much one increases the pressure. The liquid phase cannot exist below the triple point represented by the line $T_0' \; T_0''$ in Fig.7.3.

For any fluid, one has, for temperature differences which are not too great,

$$p(v,t) - p(v,t_0) = \left\{\left(\frac{\partial p}{\partial t}\right)_v\right\}_{t=t_0} (t-t_0) = \left(\frac{\alpha_p}{K_t}\right)_{t=t_0} (t-t_0) .$$

Since $K_t$, $\alpha_p > 0$ for normal substances, the solid-liquid isochore $T_0B$ (which extends beyond B, Fig.7.2) has been shown with a positive slope. However, since ice contracts on melting at normal pressures, water is anomalous and the curve $T_0B$ for

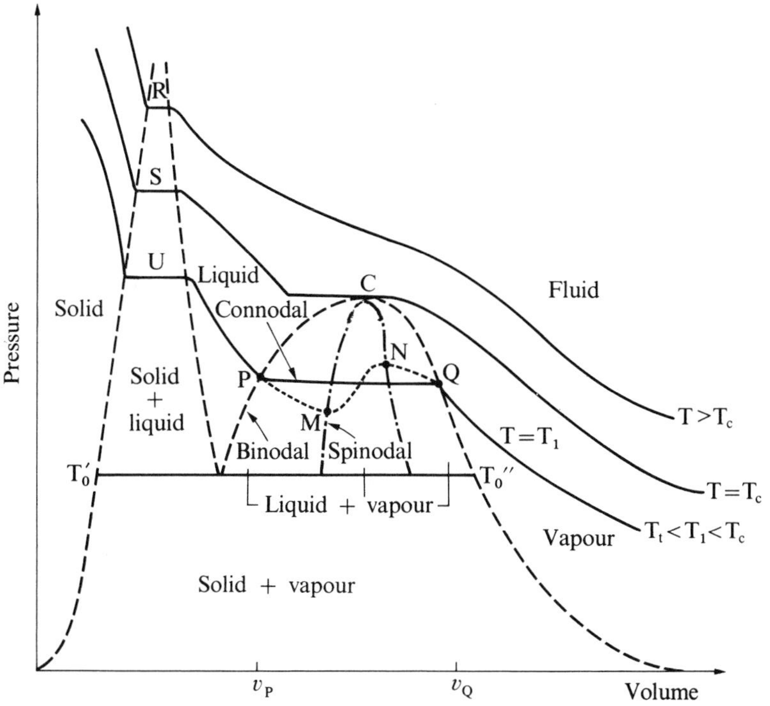


Fig.7.3. A $p$-$v$ phase diagram. The states of liquid and vapour equilibrium are inside the binodal and above the triple temperature isotherm. The distinction between liquid and vapour states is not precise for $T > T_c$.

water has a negative slope in certain parts.

Fig.7.3 shows a number of additional phenomena for a normal substance, since the $(p,v)$-diagram is used there. Isothermal compression can solidify a liquid and this transition is represented by the horizontal steps R, S, U. The critical point C is shown to lie on the critical isotherm. As the vapour is compressed at temperature $T_1$, liquid begins to form at Q when $v=v_Q$. It coexists with the vapour until the volume $v_P$ is reached at P, when the gas phase disappears and further isothermal compression requires a considerable increase in pressure.

For $T > T_c$ vapour can become liquid without discontinuity and there is only a formal distinction between liquid $(v<v_c)$ and gas $(v>v_c)$. The fluid solidifies on compression. No critical points for solid-liquid equilibrium appear to exist.

Thermodynamic theories usually do not give the horizontal lines PQ but only the dotted lines PMNQ. The part MN is necessarily unstable and the curve MCN, defining the unstable region, is called the *spinodal curve*. The region within the *binodal curve* PCQ also includes the so-called metastable states. An example is provided by a supercooled liquid which exists as a liquid below its freezing point but solidifies at once when slightly disturbed. The horizontal line PQ is called a *connodal curve*. From the theoretical curve PMNQ the horizontal line PQ may be constructed by ensuring that the area above the minimum at M to the connodal is equal to the area below the maximum at N. From this construction one predicts that P and Q should represent the volumes of liquid and vapour in equilibrium at the temperature and pressure appropriate to the curve. This is called the *equal-area rule* and was first proposed by J.C. Maxwell.

To understand this construction, imagine that an equation of state is given and that it yields as a typical isotherm a curve $v(p)$ as shown in Fig.7.4. The wave in this curve can be deduced from the van der Waals equation and is a

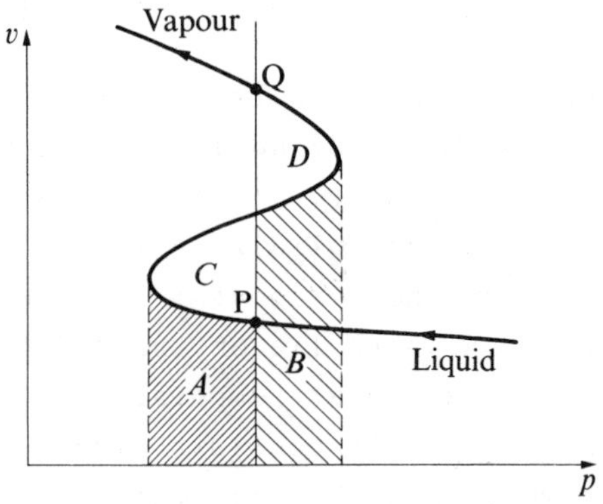


Fig.7.4. The Maxwell construction.

more general characteristic of equations of state. It raises the question of what state on the lower (liquid) limb can be in equilibrium with a state on the upper (vapour) limb, for one would then expect that the actual change of state from liquid to vapour would take place along the line thus constructed. By condition (7.7) the pressures in these two states are the same, so that the line is vertical. Let us denote the states, whose position on the isothermal is to be determined, by P and Q. Then, if $n$ is the number of molecules involved, the Gibbs free energy at P is $\mu_l n$ and at Q it is $\mu_v n$. By condition (7.7) we have $\mu_l = \mu_v$ and therefore $G_{\mathrm{P}} = G_{\mathrm{Q}}$. Next observe

$$\mathrm{d}G = -S\mathrm{d}T + v\mathrm{d}p + \sum_{i=v,l} \mu_i \mathrm{d}n_i ,$$

where the last term vanishes because

$$\mu_l \mathrm{d}n_l + \mu_v \mathrm{d}n_v = \mu_l \mathrm{d}(n_l + n_v)$$

is zero by particle conservation. Integrating from P to Q along the isotherm one finds

$$G_{\mathrm{Q}} - G_{\mathrm{P}} = \boxed{0 = \int_{\mathrm{P}}^{\mathrm{Q}} v\mathrm{d}p .}$$

Introducing the areas $A, B, C$ and $D$ as shown in Fig.7.4 and counting them in the usual way as negative when the variable $p$ decreases, one finds

$$(-A) + (A+B+C) - (B+D) = 0 .$$

It follows that $C = D$, and this is the equal-area rule.

One of the earliest model systems which exhibits a binodal and a spinodal curve and which describes the liquid-vapour transition is the van der Waals gas. This is defined by the equation of state (6.9) already discussed. It is subject to the criticism that it leads to an isothermal compressibility which is negative for small volumes:

$$K_T = -\frac{1}{v}\left(\frac{\partial v}{\partial p}\right)_T = \frac{1}{v}\left\{\frac{ct}{(v-b)^2} - \frac{a}{v^3}\right\}^{-1} .$$

In fact for any ordinary fluid $K_T$ must be positive since an increase of pressure causes a diminution of the volume (see section 7.6). The volume of a gas molecule is also not given very accurately. The equation is improved, however, if the constants $a$ and $b$ are eliminated so as to yield a 'reduced' equation of state (cf. problem 7.5*c*.)

*EULER'S FORMULA (see problem 7.7 )
It is interesting to observe that if $f$ is the number of vertices, $\chi$ the number of edges, and $\pi$ the number of faces of any convex polyhedron, then these quantities are related also through (7.8) (Euler's formula). Similarly, the partitioning of a spherical surface into $\pi$ countries which border along $\chi$ edges, yielding $f$ corners (where at least three countries meet) is also subject to (7.8). As an application of this result we shall prove that the partitioning of a spherical surface by a regular hexagonal net is not possible. If there are $n$ hexagons, this would require $\pi = n$ and

$$\chi = \text{number of edges} = \frac{6n}{2} = 3n$$

$$f = \text{number of corners} = \frac{6n}{3} = 2n$$

whence $f = \chi - \pi$, which contradicts (7.8).

## 7.5. CHANGES INTERNAL TO A SYSTEM

The second law in a form which allows for non-static (or irreversible) processes linking equilibrium states in a set $\beta$ is

$$T\delta S \geq \delta U + \sum_i p_i\, \delta v_i \quad \text{or} \quad \delta U - T\delta S + \sum_i p_i\, \delta v_i \leq 0 \qquad (7.9)$$

(closed simple equilibrium system).

For processes linking states represented by points in sets $\delta$, (7.5) holds, i.e.

$$\delta U - T\delta S + \sum_i p_i\, \delta v_i - \sum_j y_j\, \delta x_j = 0 \qquad (7.10)$$

(closed simple system)

where the more general notation $y\,\mathrm{d}x$ instead of $\mu\,\mathrm{d}n$ emphasizes that $-y_j\mathrm{d}x_j$ is just like the work term $p_i\mathrm{d}v_i$, but it is internal to the system. Examples are as follows.

(*a*) A membrane confines one gas of the system to part of the container. Upon withdrawing it, this gas can spread throughout the system. This corresponds to an internal equalization of the chemical potential of this component.

(*b*) An analogous internal pressure equalization when a rigid partition is withdrawn.

(*c*) An analogous temperature equalization when an adiabatic partition is withdrawn.

Relations (7.9) and (7.10) can be combined into

$$\boxed{\delta U - T\delta S + \sum_i p_i \delta v_i = \sum y_j \delta x_j \leq 0} \tag{7.11}$$

(closed simple system)

for a closed equilibrium system. However, (7.11) holds even for a closed simple system, i.e. when the end-points of the increments may not be equilibrium states. To see this, consider a closed simple system A for which, therefore,

$$\delta' Q = \delta U + \sum_i p_i\, \delta v_i \tag{7.12}$$

is valid. $\delta' Q$ is the heat transferred to A from its surroundings B in an incremental process linking states in a set $\delta$ of system A. The entropy gain of A is $\delta S$, so that by the second law for the isolated total system A + B

$$\delta S - \delta' Q/T_{\mathrm{B}} \geq 0 \tag{7.13}$$

where $T_{\mathrm{B}}$ is the temperature of B. By (7.12) and (7.13), and if $T_{\mathrm{A}} \leq T_{\mathrm{B}}$ is the temperature of A,

$$\delta'Q - T_{\mathrm{B}}\delta S = \delta U - T_{\mathrm{A}}\delta S + \sum_i p_i \delta v_i - (T_{\mathrm{B}}-T_{\mathrm{A}})\ \delta S \le 0$$

This holds even if $T_{\mathrm{A}} \sim T_{\mathrm{B}}$ so that (7.9) holds again, this time just for a closed simple system:

$$\delta U - T\delta S + \sum p_i \delta v_i = (T_{\mathrm{B}}-T_{\mathrm{A}})\ \delta S \le 0.$$

Writing the centre term as a general internal change $\Sigma_j\, y_j\, \delta x_j$, one finds (7.11).

For systems at negative absolute temperatures, see section 10.4, entropy is still maximal in equilibrium. However, relations (7.9) and (7.11) become

$$T\delta S \le \delta U + \sum_i p_i \delta v_i \qquad (T < 0)$$

$$-\delta U + T\delta S - \sum_i p_i \delta v_i = \sum_j y_j\ \delta x_j \le 0 \qquad (T < 0). \quad (7.11')$$

As an application of (7.11), consider a system of two phases each of which contains all $\chi$ (non-reacting) chemical components. Let the term for the internal changes be

$$\sum_j y_j\ \mathrm{d}x_j \to -p_1\ \delta v_1 - p_2\ \delta v_2 + \sum_{j=1}^{\chi} \left(\mu_{1j}\delta n_{1j} + \mu_{2j}\delta n_{2j}\right) \le 0\ .$$

If the composition of each phase and the volume of the whole system is kept fixed, one has in addition

$$\delta v_1 + \delta v_2 = 0, \qquad \delta n_{1j} = \delta n_{2j} = 0 \qquad (j = 1,\ldots,\chi)\ .$$

The infinitesimal process must therefore satisfy (for positive or negative temperatures)

$$(p_2 - p_1)\ \delta v_1 \le 0\ . \qquad (7.14a)$$

However, if the volumes of the phases are kept fixed and only the amount of the $i$th component in each phase can change, then

$$\delta v_1 = \delta v_2 = 0, \qquad \delta n_{1j} = \delta n_{2j} = 0\ (j \ne i), \qquad \delta n_{1i} + \delta n_{2i} = 0\ .$$

One now finds, for positive or negative temperatures,

$$(\mu_{1i} - \mu_{2i})\ \delta n_{1i} \le 0\ . \qquad (7.14b)$$

One learns from (7.14*a*) that, in an equilibrating internal process, the phase whose pressure is greater expands at the expense of the phase whose pressure is smaller. Similarly, (7.14*b*) implies that the component $i$ tends to move to regions of lower chemical potential. If the thermodynamically permitted changes take place, then these processes of equalization terminate when the equilibrium conditions (7.7) are satisfied. This leads, normally, to the smallest or largest attainable value of some thermodynamic function or other (see problem 7.2 ), depending on the nature of the process.

The word 'potential' in the term 'chemical potential' reflects the tendency for systems to adopt states of the lowest admissible chemical potential, there being an analogy with potential energy.

## *7.6. INTERNAL LOCAL STABILITY OF A PHASE KEPT AT GIVEN $p$ AND $T$

Suppose that a phase of a system is in equilibrium in a state 1 where it is characterized by a pressure $p_1$ and a temperature $T_1 > 0$. Its thermodynamic potential is then the Gibbs free energy $G(T_1,p_1)$ of (7.4) and it has its least value compatible with the external constraints (problem 7.2 ). If one imagines a small displacement from this state, for which external temperature and external pressure are kept constant, a non-equilibrium state 2 is reached. It will be supposed that this state can be represented in a domain $\delta$ of a thermodynamic phase space $\mathbf{F}$. It is then clear that $G$ has not decreased:

$$U_2 - U_1 - T_1(S_2 - S_1) + p_1(v_2 - v_1) \ge 0\ .$$

In order to convert state 2 to an equilibrium state, the temperature and pressure have to be adjusted to $T_2$ and $p_2$ (say). Now take this equilibrium state as an initial state, and make a small displacement to a non-equilibrium state, represented in a domain $\delta$, in which $U$, $S$, and $v$ have the values of state 1. The change in $G$ is again positive:

$$U_1 - U_2 - T_2(S_1 - S_2) + p_2(v_1 - v_2) \ge 0\ .$$

The addition of these two inequalities leads to

$$(T_2 - T_1)(S_2 - S_1) - (p_2 - p_1)(v_2 - v_1) \geq 0 . \qquad (7.15)$$

This condition incorporates the equilibrium condition $\delta G = 0$ which states that $G$ is extremum and the stability condition which states that this extremum is a minimum. Condition (7.15) is a necessary and sufficient condition for the stability of a phase. From it one can deduce the following conditions:

Dividing (7.15) by $(v_2 - v_1)^2$, passing to the limit when states 1 and 2 lie infinitely close together, and considering constant entropy

$$-\left(\frac{\partial p}{\partial v}\right)_S = \frac{1}{vK_s} \geq 0 . \qquad (7.16a)$$

Similarly, dividing by $(S_2 - S_1)^2$ and considering constant $v$,

$$\left(\frac{\partial T}{\partial S}\right)_v = \frac{T}{C_v} \geq 0 \qquad (7.16b)$$

(see equation (7.21), below). However, one can also obtain alternative (but not equivalent) forms as follows. Dividing by $(v_2 - v_1)^2$ at constant $T$ and by $(S_2 - S_1)^2$ at constant $p$ one also finds

$$-\left(\frac{\partial p}{\partial v}\right)_T = \frac{1}{vK_T} \geq 0 \qquad (7.16c)$$

$$\left(\frac{\partial T}{\partial S}\right)_p = \frac{T}{C_p} \geq 0 . \qquad (7.16d)$$

The following is the physical meaning of conditions (7.16). If $T > 0$ and $C_v$ is negative throughout the system, then a small temperature fluctuation would be magnified: a low-temperature region would gain heat and entropy from its surroundings and hence it would further drop in temperature. This instability is ruled out by (7.16b), and the other conditions have a similar interpretation.

The graphical interpretation of the condition $(\partial T/\partial S)_v = (\partial^2 U/\partial S^2)_v \geqslant 0$ at a given state is clear from Fig.7.5 in which $(\partial U/\partial S)_v$ has to be positive since it represents $T$ for any given state. However, the second derivative can change sign.

Since $T > 0$ has been assumed, the above results imply $C_v$, $K_s \geq 0$,

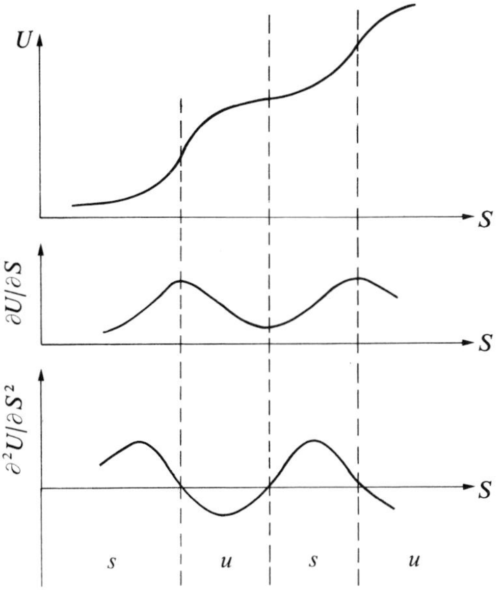


Fig.7.5. The condition $(\partial^2 U/\partial S^2)_v \geq 0$ of stability (schematic). Regions of local stability are denoted by s; other regions are denoted by u. Vertical lines represent states at which the U-S curve has a maximum or minimum slope.

from which $C_p$, $K_T \geq 0$ can be deduced using problem 5.2. The case $T < 0$ is discussed in section 10.4, p. 165.

On departing from the given state 2 by a larger displacement, one may of course find a lower minimum of $G$ even though (7.16) is satisfied *near* state 1. The conditions obtained are in that sense *local* stability conditions rather than *global* ones.

Additional insight can be derived from the following argument. Suppose a phase i (for 'initial') is in equilibrium at temperature $T_i$, pressure $p_i$ and chemical potential $\mu_i$ so that equation (7.4) yields

$$U_i - T_i S_i + p_i v_i - \mu_i n_i = 0 .$$

Suppose two coexisting phases 1 and 2 at $T_i$, $p_i$, and $\mu_i$ represent a competing equilibrium state. One can then define

$$X_j \equiv U_j - T_i S_j + p_i v_j - \mu_i n_j \quad (j = 1,2) \; (T_j = \frac{\partial U_j}{\partial S_j}, \text{ etc.})$$

with

$$S_1 + S_2 = S_i, \quad v_1 + v_2 = v_i, \quad n_1 + n_2 = n_i .$$

The change in internal energy upon the break-up of the initial phase is

$$X_1 + X_2 = U_1 + U_2 - TS_i + p_i v_i - \mu_i n_i = U_1 + U_2 - U_i .$$

If $X_1 + X_2 > 0$ one can regard the virtual change from phase i to phases 1 and 2 to be energetically unfavourable, and phase i is locally stable. If $X_1 + X_2 < 0$, phase i is unstable, and if $X_1 + X_2 = 0$ the equilibrium is neutral. We shall again include equality in our condition for stability $X_1 + X_2 \geq 0$.

A Taylor expansion of $U_j = U(S_j, v_j, n_j)$ about $U_i = U(S_i, v_i, n_i)$ is as follows:

$$U_j = U_i + \left(\frac{\partial U}{\partial S}\right)_{v,n} (S_j - n_i) + \left(\frac{\partial U}{\partial v}\right)_{S,n} (v_j - v_i) + \left(\frac{\partial U}{\partial n}\right)_{S,v} (n_j - n_i) + A_j$$

$$= (U_i - T_i S_i + p_i v_i - \mu_i n_i) + T_i S_j - p_i v_j + \mu_i n_j + A_j .$$

Here $A_j$ represents the second and higher terms in the expansion. One sees that $X_j = A_j$ and the condition for stability of phase $i$ is $A_1 + A_2 \geq 0$. This condition must hold for small, but arbitrary, virtual changes from $S_i$ to $S_1$, $S_2$, from $v_i$ to $v_1$, $v_2$, and from $n_i$ to $n_1$, $n_2$. Hence only the second-order terms need to be considered in $A_1$ and $A_2$. Writing down only typical terms,

$$A_j = \left(\frac{\partial^2 U}{\partial S^2}\right)_i (S_j - S_i)^2 + 2\left(\frac{\partial^2 U}{\partial S \partial v}\right)_i (S_j - S_i)(v_j - v_i) + \dots .$$

If $A_1$ is non-negative for arbitrary virtual displacements, $S_1 - S_i$, $v_1 - v_i$, $n_1 - n_i$, then so is $A_2$ since the displacements are arbitrary and the derivatives are the same for $j = 1$ and $j = 2$, and it is sufficient therefore to consider only $A_1$ and $X_1$. The condition $A_1 \geq 0$ is

$$\sum_{j=1}^{3} \frac{\partial^2 U}{\partial x_j^2} (\delta x_j)^2 + 2 \sum_{\substack{j<k \\ j,k=1}}^{3} \frac{\partial^2 U}{\partial x_j x_k} \delta x_j \, \delta x_k \geq 0$$

where $x_1 = S$, $x_2 = v$, $x_3 = n$. Indeed, this change of notation shows that for $\chi$ components the condition can be generalized to

$$\sum_{j,k=1}^{\chi+2} \frac{\partial^2 U}{\partial x_j \partial x_k} \delta x_j \, \delta x_k \geq 0.$$

Writing superscript T to denote a transpose,

$$\delta \mathbf{x} \equiv \begin{pmatrix} \delta x_1 \\ \vdots \\ \delta x_{\chi+2} \end{pmatrix}, \quad \delta \mathbf{x}^T = (\delta x_1, \ldots, \delta x_{\chi+2}) ,$$

$$H = \begin{pmatrix} \partial^2 U/\partial x_1^2 & & \partial^2 U/\partial x_1 \partial x_{\chi+2} \\ \vdots & & \\ \partial^2 U/\partial x_{\chi+2} \, \partial x_1 & \ldots & \end{pmatrix} ,$$

we have a vector, its transpose, and a 'Hessian' matrix of $U$. The condition is now

$$\sum_{j,k=1}^{\chi+2} \delta x_j h_{jk} \delta x_k \equiv \delta \mathbf{x}^T H \, \delta \mathbf{x} \geq 0 .$$

The left-hand side is a quadratic form and the condition is that it be 'positive semi-definite' (i.e. $\geq 0$). The relevant theorem is that a real quadratic form is positive semi-definite only if $H$ is singular and the principal minors are non-negative. Thus

$$h_{11} \geq 0 , \quad \begin{vmatrix} h_{11} & h_{12} \\ h_{21} & h_{22} \end{vmatrix} \geq 0 , \ldots, \quad \begin{vmatrix} h_{11} & & \\ & & \\ & & h_{\chi+2\ \chi+2} \end{vmatrix} \geq 0 ,$$

so that for $\chi$ components $\chi + 2$ conditions have to be satisfied. These are

$$\left(\frac{\partial^2 U}{\partial S^2}\right)_{v,n_1,\ldots} = \left(\frac{\partial T}{\partial S}\right)_{v,n_1,\ldots} = \frac{T}{C_v} \geq 0 ,$$

$$\begin{vmatrix} (\partial T/\partial S)_{v,n_1,\ldots} & (\partial T/\partial v)_{S,n_1\ldots} \\ -(\partial p/\partial S)_{v,n_1,\ldots} & -(\partial p/\partial v)_{S,n_1\ldots} \end{vmatrix} = -\frac{\partial(T,p)}{\partial(S,v)} = \frac{\partial(T,p)}{\partial(v,S)} = \frac{T}{vC_vK_T} \geq 0$$

where the result of problem (7.9) has been used. Also

$$\begin{vmatrix} (\partial T/\partial S)_{v,n_1,\ldots} & (\partial T/\partial v)_{S,n_1,\ldots} & (\partial T/\partial n_1)_{v,S,n_2,\ldots} \\ -(\partial p/\partial S)_{v,n_1,\ldots} & -(\partial p/\partial v)_{S,n_1,\ldots} & -(\partial p/\partial n_1)_{v,S,n_2,\ldots} \\ (\partial \mu/\partial S)_{v,n_1,\ldots} & (\partial \mu/\partial v)_{S,n_1,\ldots} & (\partial \mu/\partial n_1)_{v,S,n_2,\ldots} \end{vmatrix} \geq 0$$

etc.; the last condition to be written down is $|H| = 0$.

We now show that the last condition is always satisfied. To see this, put

$$T = y_1 \; , \quad -p = y_2 \; , \quad \mu_1 = y_3 \; , \; \ldots, \; \mu_\chi = y_{\chi+2} \; .$$

Then the Gibbs-Duhem equation (7.6) is

$$\sum_{j=1}^{\chi+2} x_j \, dy_j = 0 \; .$$

The Hessian determinant is

$$|H| = \begin{vmatrix} \partial y_1/\partial x_1 & \ldots & \partial y_1/\partial x_{\chi+2} \\ \vdots & & \vdots \\ \partial y_{\chi+2}/\partial x_1 & \ldots & \partial y_{\chi+2}/\partial x_{\chi+2} \end{vmatrix} = \frac{1}{x_1 x_2 \ldots x_{\chi+2}} \begin{vmatrix} x_1 \partial y_1/\partial x_1 \ldots x_1 \partial y_1/\partial x_{\chi+2} \\ \vdots \\ x_{\chi+2} \partial y_{\chi+2}/\partial x_1 \quad \ldots \end{vmatrix} .$$

Adding the last $\chi+1$ rows to the first row does not alter $|H|$ and produces zeros along the first row by the Gibbs-Duhem relation, and $|H|$ is seen to vanish. Thus $H$ is a singular matrix. If it has at least one non-singular submatrix of $r$ rows and columns, but none of $r + 1$ rows and columns, a matrix is said to have rank $r$. We see here that the above Hessian matrix normally has rank $\chi + 1$.

Since there are only $\chi + 1$ relations, one sees that for $\chi = 1$ we have recovered the two independent conditions of relations (7.16).

*7.7. STABILITY OF EQUILIBRIUM OF TWO PARTS OF AN ISOLATED SYSTEM

Suppose two macroscopic systems 1,2, free of external forces, are in thermal and mechanical equilibrium and isolated from the rest of the world. It is assumed that each system has one fixed constituent so that $\chi = 0$ and one volume-type co-ordinate so that $r = 1$. There are then $2(1+1) = 4$ independent extensive variables. We take their entropies $S_1, S_2$ and volumes $v_1, v_2$ to be the four variables; they are reduced in number to two independent ones (say to $v_1$ and $S_1$) by fixing the total volume and energy, $v = v_1 + v_2$ and $U = U_1 + U_2$, so that

$$\left.\begin{aligned} v_1 + v_2(v_1, S_1) &= v \\ U_1(v_1, S_1) + U_2\{v_2(v_1, S_1),\ S_2(v_1, S_1)\} &= U. \end{aligned}\right\} \qquad (7.17)$$

We first find the equilibrium condition between the fluids by finding the condition for an *extremum* of $S = S_1 + S_2(v_1, S_1)$. If $v_1$ is considered to be the variable first, this is

$$\left(\frac{\partial S}{\partial v_1}\right)_{S_1} = \left(\frac{\partial S_2}{\partial v_1}\right)_{S_1} = 0 .$$

Now by (7.17)

$$\left.\begin{aligned} 1 + \left(\frac{\partial v_2}{\partial v_1}\right)_{S_1} &= 0 \\ \left(\frac{\partial U_1}{\partial v_1}\right)_{S_1} + \left(\frac{\partial U_2}{\partial v_2}\right)_{S_2}\left(\frac{\partial v_2}{\partial v_1}\right)_{S_1} + \left(\frac{\partial U_2}{\partial S_2}\right)_{v_2}\left(\frac{\partial S_2}{\partial v_1}\right)_{S_1} &= 0 . \end{aligned}\right\} \qquad (7.18a)$$

Using both equations (7.18*a*) it follows that

$$0 = \left(\frac{\partial S_2}{\partial v_1}\right)_{S_1} = -\left(\frac{\partial S_2}{\partial U_2}\right)_{v_2}\left\{\left(\frac{\partial U_1}{\partial v_1}\right)_{S_1} - \left(\frac{\partial U_2}{\partial v_2}\right)_{S_2}\right\}$$

$$= -\frac{1}{T_2}(p_1 - p_2) . \qquad (7.19a)$$

Also, the second derivative must be negative for a *maximum*. Thus, by (7.18*a*),

$$\left(\frac{\partial^2 U_1}{\partial v_1^2}\right)_{S_1} - \left(\frac{\partial^2 U_2}{\partial v_2^2}\right)_{S_2} \left(\frac{\partial v_2}{\partial v_1}\right)_{S_1} - \frac{\partial^2 U_2}{\partial S_2 \partial v_2} \left(\frac{\partial S_2}{\partial v_1}\right)_{S_1} +$$

$$\left\{\left(\frac{\partial^2 U_2}{\partial S_2^2}\right)_{v_2} \left(\frac{\partial S_2}{\partial v_1}\right)_{S_1} + \frac{\partial^2 U_2}{\partial S_2 \partial v_2} \left(\frac{\partial v_2}{\partial v_1}\right)_{S_1}\right\} \left(\frac{\partial S_2}{\partial v_1}\right)_{S_1} + \left(\frac{\partial U_2}{\partial S_2}\right)_{v_2} \left(\frac{\partial^2 S_2}{\partial v_1^2}\right)_{S_1} = 0 .$$

Using (7.18*a*) and (7.19*a*) this yields

$$\left(\frac{\partial^2 U_1}{\partial v_1^2}\right)_{S_1} + \left(\frac{\partial^2 U_2}{\partial v_2^2}\right)_{S_2} + \left(\frac{\partial U_2}{\partial S_2}\right)_{v_2} \left(\frac{\partial^2 S_2}{\partial v_1^2}\right)_{S_1} = 0$$

that is

$$\left(\frac{\partial^2 S}{\partial v_1^2}\right)_{S_1} = \left(\frac{\partial^2 S_2}{\partial v_1^2}\right)_{S_1} = -\frac{1}{T_2}\left\{\left(\frac{\partial^2 U_1}{\partial v_1^2}\right)_{S_1} + \left(\frac{\partial^2 U_2}{\partial v_2^2}\right)_{S_2}\right\}$$

$$= -\frac{1}{T_2}\left\{\left(\frac{\partial p_1}{\partial v_1}\right)_{S_1} + \left(\frac{\partial p_2}{\partial v_2}\right)_{S_2}\right\}$$

$$= -\frac{1}{v_1 T_2 K_{S_1}} - \frac{1}{v_2 T_2 K_{S_2}} < 0 \tag{7.20a}$$

where the compressibilities have been introduced.

Consider now that $S_1$ is the variable. Then (7.17) yields

$$\left.\begin{aligned} &\left(\frac{\partial v_1}{\partial S_1}\right)_{v_2} + \left(\frac{\partial v_2}{\partial S_1}\right)_{v_1} = 0 \\ &\left(\frac{\partial U_1}{\partial S_1}\right)_{v_1} + \left(\frac{\partial U_2}{\partial S_2}\right)_{v_2} \left(\frac{\partial S_2}{\partial S_1}\right)_{v_1} + \left(\frac{\partial U_2}{\partial v_2}\right)_{S_2} \left(\frac{\partial v_2}{\partial S_1}\right)_{v_1} = 0 \end{aligned}\right\} \tag{7.18b}$$

so that each term in the first equation vanishes. The condition for an *extremum* of $S$ is from $S = S_1 + S_2$

$$\left(\frac{\partial S}{\partial S_1}\right)_{v_1} = 1 + \left(\frac{\partial S_2}{\partial S_1}\right)_{v_1} = 0$$

so that

$$- 1 = - \frac{(\partial U_1/\partial S_1)_{v_1}}{(\partial U_2/\partial S_2)_{v_2}} = - \frac{T_1}{T_2} \quad . \tag{7.19$b$}$$

The second derivative is negative for a *maximum*, so that from (7.18*b*)

$$\left(\frac{\partial^2 U_1}{\partial S_1^2}\right)_{v_1} + \left\{ \left(\frac{\partial^2 U_2}{\partial S_2^2}\right)_{v_2} \left(\frac{\partial S_2}{\partial S_1}\right)_{v_1} + \frac{\partial^2 U_2}{\partial S_2 \partial v_2} \left(\frac{\partial v_2}{\partial S_1}\right)_{v_1} \right\} \left(\frac{\partial S_2}{\partial S_1}\right)_{v_1} + \left(\frac{\partial U_2}{\partial S_2}\right)_{v_2} \left(\frac{\partial^2 S_2}{\partial S_1^2}\right)_{v_1} = 0 \; .$$

Using (7.18*b*) and (7.19*b*)

$$\left(\frac{\partial^2 U_1}{\partial S_1^2}\right)_{v_1} + \left(\frac{\partial^2 U_2}{\partial S_2^2}\right)_{v_2} + \left(\frac{\partial U_2}{\partial S_2}\right)_{v_2} \left(\frac{\partial^2 S_2}{\partial S_1^2}\right)_{v_1} = 0$$

that is

$$\left(\frac{\partial^2 S}{\partial S_1^2}\right)_{v_1} = \left(\frac{\partial^2 S_2}{\partial S_1^2}\right)_{v_1} = - \frac{1}{T_2} \left\{ \left(\frac{\partial^2 U_1}{\partial S_1^2}\right)_{v_1} + \left(\frac{\partial^2 U_2}{\partial S_2^2}\right)_{v_2} \right\}$$

$$= - \frac{1}{T_2} \left\{ \left(\frac{\partial T_1}{\partial S_1}\right)_{v_1} + \left(\frac{\partial T_2}{\partial S_2}\right)_{v_2} \right\} = - \frac{1}{T_2} \left(\frac{T_1}{C_{v_1}} + \frac{T_2}{C_{v_2}}\right) < 0 \tag{7.20$b$}$$

It is quite clear, therefore, that the equilibrium (i.e. *extremum*) conditions are, by (7.19),

$$p_1 = p_2, \quad T_1 = T_2 \quad \text{(cf. (7.7))}.$$

The conditions for *stability*, i.e. that the extremum be a maximum, are, by (7.20),

$$\frac{1}{v_1 K_{S_1}} + \frac{1}{v_2 K_{S_2}} > 0, \quad \frac{1}{C_{v_1}} + \frac{1}{C_{v_2}} > 0 \tag{7.21}$$

The first relation refers to *mechanical stability*, the second to *thermal stability*. The third equilibrium condition (7.7), $\mu_1 = \mu_2$, leads in a similar way to a condition for *diffusional stability* and is beyond the scope of this book. At critical points, the compressibilities and the heat capacities can diverge (section 7.8) and the stability conditions are then violated. In connection with black holes we shall apply the thermal stability condition for a negative heat capacity in section 13.6.

The conditions derived above express the important requirements that the specific heat is positive and that increased volume leads to smaller pressures (mechanical stability). For two phases and several components the discussion is more involved, but the conditions (7.21) still feature in these cases.

The presence of long-range forces causes important amendments to thermodynamics, some of which are not fully investigated as yet. However, one knows as one example of such amendments that gravitational forces can give rise to negative heat capacities (sections 15.8 and 18.2). In this case (7.21) still holds for each macroscopic (but small) volume, but the gravitational interaction energy has to be added to the energies in any two elements. The energy $U$, the entropy $S$, therefore cease to be extensive qualities, and the body ceases to be homogeneous. An obvious example is a column of liquid which is denser near the bottom. This opens up the possibility that (7.21) for macroscopic but small volume elements is compatible with a negative heat capacity for the body as a whole. The fourth law of thermodynamics (section 7.1) fails in these cases, and even entropy maximization for equilibrium is dubious, see p. 369.

## *7.8. THE VAN DER WAALS (FIRST-ORDER) TRANSITION

For a van der Waals gas, reduced variables can be defined as follows:

$$\pi = \frac{p}{p_c}, \qquad \Phi \equiv \phi - 1 = \frac{v}{v_c} - 1, \qquad T \equiv \tau - 1 \equiv \frac{t}{t_c} - 1 \tag{7.22}$$

where $p_c$, $v_c$, and $t_c$ are pressure, volume, and temperature at the critical point. The reduced equation of state is (problem 7.5)

$$\pi = \frac{8\tau}{3\phi - 1} - \frac{3}{\phi^2} \ . \tag{7.23}$$

Thus all van der Waals systems can be described by universal curves (7.23) in reduced variables. If two of the reduced variables for two such systems are the same, the third must also be the same by (7.23). The two systems are then said to be in corresponding states and this result is the *law of corresponding states*.

One finds, from the equal-area rule, that, near the critical point but below the critical temperature ($T < 0$), the volumes $v_1$, $v_2$ at the end of the isothermal constructed by that rule are such that

$$\Phi_2 = (-T)^{\frac{1}{2}}, \qquad \Phi_1 = -(-T)^{\frac{1}{2}} \tag{7.24}$$

(problem 7.8 ). These relations show how the $\Phi_1$, $\Phi_2$ approach zero with T as the critical point is approached. This corresponds to the coincidence at C of the points P and Q in Fig.7.3.

The reduced pressure in this region is

$$\pi = \frac{8(T+1)}{3\Phi+2} - \frac{3}{(\Phi+1)^2}$$

$$= 4(1+T)\left(1+\frac{3\Phi}{2}\right)^{-1} - 3\ (1+\Phi)^{-2}$$

$$= 4(1+T)\left(1-\frac{3\Phi}{2}+\frac{9\Phi^2}{4}-\frac{27}{8}\Phi^3\ \ldots\right) -$$

$$3\ (1-2\Phi + 3\Phi^2 - 4\Phi^3 + \ldots)$$

$$= 1 + 4T - 6\Phi T + \{+\ 9\Phi^2 T\} - \frac{3}{2}\Phi^3 + \ldots \qquad (7.25)$$

where the term in braces can be neglected compared with the $\Phi T$-term.

In problem (7.5*g*) we have one method of showing that the compressibility of the van der Waals gas

$$K_T \equiv -\frac{1}{v}\left(\frac{\partial v}{\partial p}\right)_T = -\frac{1}{v}\left(\frac{\partial^2 G}{\partial p^2}\right)_T = \frac{1}{3c}\,\frac{4}{T-T_c} \quad \text{at} \quad v = v_c \qquad (7.26)$$

diverges as the critical point is approached. An analogous result can be obtained from (7.25) as follows:

$$K_T = -\frac{1}{p_c\phi}\left(\frac{\partial\phi}{\partial\pi}\right)_T = \frac{1}{6p_c\phi T} = \frac{4b}{3c}\,\frac{v_c}{v}\,\frac{1}{T-T_c} \qquad (7.27)$$

whence

$$(K_T)_{\Phi=0} = \frac{9b^2}{2a}\,\frac{1}{T}\ . \qquad (7.28)$$

The horizontal isotherm constructed by the equal-area rule implies coexistence of gas and liquid, the length of the horizontal portion being determined by the values $\Phi_1$, $\Phi_2$ of (7.24). The possibility of the coexistence of the phases disappears as T and $\Phi$ tend to zero. This occurs as the temperature is raised. Multiplying (7.23) by $-1/3\phi$ and introducing $\nu \equiv 1/\phi$ as the variable of interest, it is easily seen that $\phi$ satisfies the following cubic equation:

$$\pi - \alpha\nu + 3\nu^2 - \nu^3 = 0 \qquad \alpha \equiv (8\tau + \pi)/3\ . \qquad (7.23')$$

We shall return to this result in section 7.11 to explain why the phase transition exhibited by the van der Waals system below its critical temperature is said to be of *first order*, while stability questions are considered briefly in problem 7.10.

## *7.9. THE FERROMAGNETIC (SECOND-ORDER) TRANSITION

One can use the same approach as in section 7.8 to describe the loss of magnetization $M$ in zero magnetic field $B$† as the temperature of a ferromagnetic material is raised. The incremental work done on such a material is $BdM$ (instead of $-pdv$ as in the case of a fluid). Among various possible schemes one can chose the replacement $p \to B, \quad v \to -M$ to formulate the thermodynamics of magnetic materials. It is found that below some critical temperature the isotherms plotted on a $B$-$M$ diagram have a finite horizontal portion which can be thought of as corresponding to a region in which the system can have 'up' or 'down' magnetization. The system is then not homogeneous : physical regions of vapour and liquid are here replaced by regions of non-zero magnetization which, even in the absence of an applied field, have $M > 0$ or $M < 0$. Since translational symmetry is lost as the temperature is lowered, this illustrates symmetry breaking as an accompaniment of phase transitions. As the temperature is raised, then just as $\Phi \to 0$ in the van der Waals case, so $M_{B=0} \to 0$ in the case of a ferromagnetic material. One can speak of the volume $v$ (or $\Phi$) and of $M$ as *order parameters* of a *phase transition* which occurs at a characteristic temperature; these parameters are non-zero below that temperature. $\Phi$ applies to a fluid, and $M$ applies to a magnetic body. In ferromagnetism, in fact, the disappearance of the magnetization in zero field above a certain temperature is a very intuitively clear phenomenon. As the temperature is raised, the entropy of a system is raised and this encourages the disappearance of 'order' (in some imprecise sense) and the order parameters vanish. The compressibility (7.26) has as its analogue the susceptibility

$$\chi_T \equiv \frac{1}{v}\left(\frac{\partial M}{\partial B}\right)_T \tag{7.29}$$

and it diverges as $(T-T_c)^{-1}$, as does (7.26). It gives the 'response' of the order parameter $\Phi$ or $M$ to changes in the 'corresponding field' $p$

† As the symbol $H$ has been used for the enthalpy, we use $B$ to denote the magnetic field.

or $B$. These analogies lead one to the so-called *Weiss theory of ferromagnetism*.

This is expressed in a simple way by the Weiss equation for a ferromagnetic transition (as deduced from statistical mechanics as equation (10.22) below). It states that if $a_1'$ and $a_3$ are slowly varying functions of temperature the following result holds near $T = T_c$

$$B = (T-T_c)a_1' \, M + a_3 M^3 \, . \tag{7.30}$$

In the limit of zero applied field, $B \to 0$, one has $M = 0$ as the only solution if $T > T_c$. For $T < T_c$ a second (stable) solution appears:

$$M = \pm \sqrt{\{(T_c-T)a_1'/a_3\}} \, . \tag{7.31}$$

This yields spontaneous magnetization for $T < T_c$, and hence a non-zero order parameter ($M = 0$ being an unstable solution for $T < T_c$).

## *7.10. CHEMICAL REACTIONS AND PHASE TRANSITIONS

We now add a third illustration of these ideas. The gas-liquid transition (section 7.8) and the ferromagnetic transition (section 7.9) were our first two examples.

Consider the abstract chemical reactions

$$A + X \underset{k_1'}{\overset{k_1}{\rightleftharpoons}} B, \qquad P + 2X \underset{k_2'}{\overset{k_2}{\rightleftharpoons}} 3X + Q \, . \tag{7.32}$$

The rate constants, or *reaction constants*, have also been shown. The concentrations of these species will be denoted by corresponding lower-case letters, and it will be assumed that $a$, $b$, $p$, and $q$ are arranged to be constant. Then the rate of change of the concentration of X is for dilute systems (as explained more fully in connection with equation (8.14), below)

$$\dot{x} = (-\, k_1 a x + k_1' b) + (k_2 p x^2 - k_2' q x^3) \tag{7.33}$$

where the first and second expressions arise respectively from the first and second reaction (7.32). If the units are chosen such that

$$k_2' q = 1, \quad k_2 p = 3$$

we are back essentially at van der Waals equation (7.23'), except that the analogy holds only if the chemical reaction is considered in the steady state. The control parameters $\pi$ and $\alpha$ are here $k_1'b$ and $k_1a$ and van der Waals-type curves can be traced if the concentrations of species A and B can be varied. In the range of the cubic where all three roots are admissible one should therefore find the reaction rate equivalent to a first-order phase transition. The requirement that a cubic equation be obtained for a transition is related to the fact that the second reaction (7.32) is *autocatalytic* in the sense that the presence of component X leads to more X being formed.

## *7.11. GENERALIZATIONS AND CRITICAL EXPONENTS

In the abstract formulation this kind of macroscopic description of a phase transition depends on finding a function (e.g. a free energy) which can be expanded in an order parameter $\Phi$

$$\Lambda = a_0\Phi + \frac{1}{2}a_1\Phi^2 + \frac{1}{3}a_2\Phi^3 + \frac{1}{4}a_3\Phi^4 + \ldots \tag{7.34}$$

about the value $\Lambda = 0$ at $\Phi = 0$ which corresponds to the point at which a phase transition occurs. The coefficients $a_i$ may be functions of each other; for example, $a_1, a_2$ and $a_3$ may all depend on $a_0$. Now $\Lambda$ is a function which assumes an extremum in equilibrium so that the order parameter in equilibrium is a solution of

$$a_0 + a_1\Phi + a_2\Phi^2 + a_3\Phi^3 = 0 \tag{7.35}$$

where the terms indicated by dots in (7.34) have now been omitted as an approximation valid in the neighbourhood of the critical point. Alternatively (7.35) may correspond to a steady state in a chemical reaction. The coefficients are functions of other system variables, notably of the temperature. This equation has the three real roots needed for a first-order phase transition if, as shown in books on algebra,

$$a_2^2 > 3a_1 a_3$$

and

$$\left|\frac{27\, a_0a_3^2 + 2\, a_2^3 - 9\, a_1a_2a_3}{3(a_2^2 - 3\, a_1a_3)^{3/2}}\right| \le 1\ .$$

For three positive roots the signs in the cubic must alternate. These conditions are satisfied for a van der Waals system below its critical points, as seen from equation (7.23'), and the phase transition is said to be of *first order*. Even if there are only two physically admissible roots a phase transition, said to be of *second order*, is possible. An interpretation of the symbols of relation (7.35) is given in Table 7.1.

TABLE 7.1

*Special cases of relation (7.35)*

| Equation | $\Phi$ | $a_0$ | $a_1$ | $a_2$ | $a_3$ |
|---|---|---|---|---|---|
| Van der Waals (7.23') | $\nu$ | $\pi$ | $-\alpha$ | 3 | -1 |
| Weiss (7.30) | $M$ | $B$ | $-(T-T_c)$ | 0 | $-a_3$ |
| Chemical reactions (7.33) | $x$ | $k_1'b$ | $-k_1a$ | $k_2p$ | $-k_2'q$ |

One can ask, using the ferromagnetic case as a guide: How does the magnetization go to zero as the temperature approaches the critical temperature from below? How does it go to zero at the critical temperature as the applied field goes to zero? How does the susceptibility diverge? Assuming power laws with the so-called *critical exponents* $\beta$, $\delta$, $\gamma$, we therefore write

$$M \propto (T_c-T)^{\beta} \qquad (B=0;\ T \to T_c-),$$

$$M \propto |B|^{1/\delta} \qquad (T = T_c)$$

$$(\partial M/\partial B)_T \propto \begin{cases} (T-T_c)^{-\gamma} & (T \to T_c+) \\ (T_c-T)^{-\gamma'} & (T \to T_c-) \end{cases}$$

More generally, using Table 7.1,

$$\left\{\lim_{a_1\to 0} \Phi\right\}_{a_0=0} \propto a_1^{\beta}$$

$$\left\{\lim_{a_0\to 0} \Phi\right\}_{a_1=0} \propto a_0^{1/\delta}$$

$$\left\{\lim_{a_1\to 0+} \left(\frac{\partial\Phi}{\partial a_0}\right)_{a_1}\right\}_{a_0=0} \propto a_1^{-\gamma}$$

$$\left\{\lim_{a_1\to 0-} \left(\frac{\partial\Phi}{\partial a_0}\right)_{a_1}\right\}_{a_0=0} \propto a_1^{-\gamma'}$$

We now find the values of $\beta$, $\delta$, $\gamma$, and $\gamma'$ for our model (7.35). Using the fact that $\Phi$ is considered small but non-zero, (7.35) for the calculation of $\beta$ is

$$a_1 + a_2\Phi + a_3\Phi^2 + \ldots = 0$$

so that

$$\beta = 1 \ (a_2 \neq 0) \ , \quad \beta = \tfrac{1}{2} \ (a_2 = 0, \ a_3 \neq 0).$$

The condition for the calculation of $\delta$ is

$$a_0 + a_2\Phi^2 + a_3\Phi^3 + \ldots = 0$$

so that

$$\delta = 2 \ (a_2 \neq 0), \quad \delta = 3 \ (a_2 = 0, \ a_3 \neq 0) \ .$$

For the calculation of $\gamma$ we have

$$1 + a_1 \frac{\partial\Phi}{\partial a_0} + \ldots = 0$$

so that

$$\gamma = \gamma' = 1 \ .$$

Other critical exponents can be defined and investigated. They can also be estimated from experiments on magnetic systems or on liquid-gas transitions. They are, for three-dimensional systems, found to be relatively independent of the detailed nature of the systems, even though these can be varied very widely. This phenomenon is known as *universality*. One finds $\beta \sim 0{\cdot}3\text{-}0{\cdot}4$, $\delta \sim 4\text{-}5$, $\gamma \sim \gamma' \sim 1{\cdot}0\text{-}1{\cdot}3$. However, if a system behaves like a d-dimensional system with d = 3, one finds that the exponents depend on the value of d.

This and other examples suggest the interest of generalised theories of phase transitions, though the details are beyond the present scope. In spite of their usefulness, they suffer from two shortcomings: (a) the assumption that a thermodynamic function (7.30) exists and is analytic in its arguments at and near the critical points fails in important cases, and indeed experiments on some substances contradict the equal-area rule. (b) The fluctuations are important near the critical point and need to be taken into account. All these questions are subjects of active research at the present time.

PROBLEMS

7.1. The Helmholtz free energy, the entropy, and the Gibbs free energy are respectively

$$F \equiv U - TS, \quad H \equiv U + pv, \quad G \equiv U - TS + pv \; .$$

Show that for incremental processes represented in a set $\delta$

$$dF = - SdT - pdv + \mu dn$$

$$dH = TdS + vdp + \mu dn$$

$$dG = - SdT + vdp + \mu dn = \mu dn + nd\mu \; .$$

7.2. A closed simple system undergoes an incremental process linking two states of a set $\delta$. Verify the entries in Table 7.2. Verify that the last four inequalities in Table 7.2 are reversed for negative temperatures.

7.3. (*a*) A fluid is in equilibrium at pressure $p_1$ and volume $v_1$. It is forced at constant pressure $p_1$ through a porous plug against a piston loaded at constant pressure $p_2$ on the other side of the plug. When the fluid has been transferred completely, and is again in equilibrium, it occupies a volume $v_2$. Let the

TABLE 7.2

| Condition of system or type of process | Equation valid for the process | Inequality (7.11) valid for the process | Equilibrium condition |
|---|---|---|---|
| Adiabatic | $\delta' Q = \delta U + p\delta v = 0$ | $\delta S \geq 0$ | $S$ has its greatest value |
| $T$ and $v$ constant | $\delta T = \delta v = 0$ | $\delta F \leq 0$ | $F$ has its least value |
| $T$ and $p$ constant | $\delta T = \delta p = 0$ | $\delta G \leq 0$ | $G$ has its least value |
| $S$ and $v$ constant | $\delta S = \delta v = 0$ | $\delta U \leq 0$ | $U$ has its least value |
| $S$ and $p$ constant | $\delta S = \delta p = 0$ | $\delta H \leq 0$ | $H$ has its least value |

internal energies be $U_1$ and $U_2$ for the two equilibrium states. Show that the work done by the system is $p_2 v_2 - p_1 v_1$, and use the first law of thermodynamics to show that $U_1 + p_1 v_1 = U_2 + p_2 v_2$.

(This type of process is called a *throttling process*.)

(*b*) Prove that the change in temperature in such an experiment is given by

$$\left(\frac{\partial T}{\partial P}\right)_H = \left\{T\left(\frac{\partial v}{\partial T}\right)_p - v\right\}/C_p = \frac{v}{C_p}(\alpha_p T - 1).$$

(*c*) Find the necessary and sufficient condition for no temperature change to be observed in this experiment.

(Part (*c*) gives an expression for the *Joule-Thomson* (or *Joule-Kelvin*) coefficient. A change in temperature, rise or fall, during this process is characteristic of gases which are not ideal classical gases, as defined in problem 6.1. To make a non-zero Joule-Thomson coefficient the basis of a definition of real or imperfect gases would, however, be misleading, since certain ideal quantum gases exhibit a non-zero Joule-Thomson coefficient.)

7.4. Show that

$$\left(\frac{\partial N}{\partial v}\right)_{\mu,T} = \left(\frac{\partial N}{\partial v}\right)_{p,T} = \frac{N}{v} .$$

Can this be generalized in any way to $(\partial E_r/\partial E_s)_x = E_r/E_s$ where the $\{E_r\}$ represent a set of extensive variables and $x$ a set of intensive variables?

7.5. (*a*) The van der Waals equation of state is

$$\left(p + \frac{a}{v^2}\right)(v - b) = At .$$

On a $(p,v)$-diagram the extrema of this equation, which are obtained by choosing various values of $t$, lie on the spinodal curve. Find the equation of this curve. (*b*) Show that the maximum of the curve found in part (*a*) is given by

$$v_c = 3b , \quad p_c = \frac{a}{27b^2} , \quad t_c = \frac{8a}{27Ab}$$

so that $At_c/p_c v_c = 2{\cdot}667$.

(This point is called the critical point.)

(*c*) Show that the van der Waals equation of state can be written

$$\left(\pi + \frac{3}{\phi^2}\right)(3\phi - 1) = 8\tau$$

where $\phi \equiv v/v_c$, $\pi \equiv p/p_c$, and $\tau \equiv t/t_c$.

(*d*) One sometimes writes a general equation of state in the form

$$pv = E_1 + E_2 p + E_3 p^2 + \ldots$$

where $E_1$, $E_2$, $E_3$ ... are functions of $t$ and are called first, second, third ... virial coefficients. If $a$ and $b$ are small enough, show that a van der Waals gas has an approximate second virial coefficient

$$E_2 = b - \frac{a}{At} \, .$$

(Various definitions of virial coefficients are in use.)

(*e*) The Boyle temperature $t_B$ of a fluid is defined by $\{\partial(pv)/\partial p\}_{p=0} = 0$ so that $E_2 = 0$ and the fluid approximates a Boyle's law gas (for small $E_n$, when $n \geq 3$) in the neighbourhood of this temperature. Show that for a van der Waals gas

$$\frac{t_B}{t_c} = 3 \cdot 375 \, .$$

(*f*) Show that $C_v$ for a van der Waals gas is independent of volume (see also (6.18), p. 67).

(*g*) Show that at $v = v_c$

$$K_t = \frac{4b}{3A(t-t_c)} \, , \quad \alpha_p = \frac{2}{3(t-t_c)}$$

so that these quantities diverge at $t = t_c$.

(*h*) Obtain the Joule-Thomson coefficient $(\partial T/\partial p)_H$ for a van der Waals gas and hence show that the *inversion curve* $(p_i, v_i)$ which separates regions on the $p$-$v$ diagram for which $(\partial T/\partial p)_H$ has opposite signs, is given by

$$p_i = \frac{2a}{bv_i} - \frac{3a}{v_i^2} \, .$$

7.6. (*a*) If a liquid is in equilibrium with its saturated vapour, quasi-static isothermal expansion causes vaporization of more liquid and so does not alter the saturated vapour pressure $p$ (at the appropriate boiling temperature $T$). Show that the temperature dependence of the saturated vapour pressure satisfies

$$T \frac{dp}{dT} = \frac{\Lambda_T^{1\to 2}}{v_2 - v_1}$$

where the symbols on the right-hand side refer to the mass of liquid vaporized, $v_2$ being its volume in the vapour phase, $v_1$ in the liquid phase, and $\Lambda^{1\to 2}$ is the heat required for the isothermal expansion.

(*b*) By a different interpretation of this formula explain why an increase of pressure raises the boiling point and the freezing point of a normal liquid, while, in the case of water, pressure lowers the freezing point.

(*c*) $\Lambda^{1\to 2} = 80$ cal g$^{-1}$ is the latent heat of melting of ice, $v$(water) $= 1{\cdot}000$ cm$^3$ g$^{-1}$, and $v$(ice) $= 1{\cdot}091$ cm$^3$ g$^{-1}$. Estimate the numerical value of the lowering of the freezing point if the pressure is increased by 1 atm.

(Use a Maxwell equation for (*a*). The result is called the *Clausius-Clapeyron equation*. For (*b*) when solid and liquid co-exist only one independent variable is left.)

7.7. Verify the properties given in Table 7.3 for some simple structures and hence verify Euler's formula for these cases.

TABLE 7.3

| | $f$ No. of vertices | $\pi$ No. of faces | $\chi$ No. of edges | $f+\pi-\chi$ |
|---|---|---|---|---|
| Tetrahedron | 4 | 4 | 6 | 2 |
| Pyramid with a square base | 5 | 5 | 8 | 2 |
| Cube | 8 | 6 | 12 | 2 |
| Octahedron | 6 | 8 | 12 | 2 |

7.8. Apply the equal-area rule to a van der Waals system using the reduced variables of (7.22). Hence show that the rule picks out volumes lying at the end of an isothermal $t < t_c$ such that to lowest orders

$$(\Phi_2-\Phi_1)\left\{-\frac{3}{2}(\Phi_1^2+\Phi_1\Phi_2+\Phi_2^2)+\frac{33}{8}(\Phi_1+\Phi_2)(\Phi_1^2+\Phi_2^2)-6T+6T(\Phi_1+\Phi_2)\right\}$$

vanishes. Verify that this condition is satisfied for $\Phi_1 \neq \Phi_2$

by $$\Phi_1^2 = \Phi_2^2 = -4T \ .$$

7.9. Prove that

$$\frac{\partial(T,p)}{\partial(v,S)} \equiv \left(\frac{\partial T}{\partial v}\right)_S \left(\frac{\partial p}{\partial S}\right)_v - \left(\frac{\partial T}{\partial S}\right)_v \left(\frac{\partial p}{\partial v}\right)_S = \frac{T}{vC_v K_T} \ .$$

7.10. A chemical reaction for a component X yields the rate equation

$$\dot{x} = a_0 + a_1 x + a_2 x^2 + a_3 x^3$$

where the $a$'s are such that $\dot{x} = 0$ has three distinct real roots. Show that for two of the roots to be stable one requires $a_3 < 0$ If $a_3 > 0$, only one root is stable.

7.11. The attainment by a fluid of temperatures of order $10^{-3}$ K depends typically on (*a*) an adiabatic expansion below the Joule-Thomson inversion temperature where $(\partial T/\partial p)_H > 0$ (problems 7.3 and 7.5), (*b*) a Joule-Thomson expansion so that the temperature drops with pressure, and (*c*) an adiabatic demagnetization.

Using the replacement $p \to B$, $v \to -M$, show that the temperature change in process (*c*) is given by

$$\left(\frac{\partial T}{\partial B}\right)_S = -\frac{T}{C_B}\left(\frac{\partial M}{\partial T}\right)_B \ .$$

Make it plausible that this quantity is positive so that a lowering of the field induces a drop in temperature.

# 8. The mass action law and the equilibrium constants

A chemical reaction such as

$$2NiO \rightarrow 2Ni + O_2$$

states that two molecules of nickel oxide can dissociate into two atoms (or molecules!) of nickel and one molecule of oxygen, and conversely. In equilibrium the reaction takes place in the material in both directions with roughly equal rates. One can write a stoichiometric equation as

$$2Ni + O_2 - 2NiO = 0$$

(though this is rarely done), giving a negative integer to the 'reactants', and positive integers to the 'products'.

More generally a chemical reaction can be written as

$$\boxed{\sum_i \nu_i C_i = 0} \tag{8.1}$$

where the $\nu_i$ are positive or negative integers (the 'stoichiometric coefficients') and the $C_i$ denote the components. If $n_i$ molecules of component $C_i$ are present, then the equilibrium condition is

$$\sum \mu_i \, dn_i = 0$$

since for a virtual change the modification can be assumed to be carried out quasi-statically so that the equality sign holds in (7.11). The changes in $n_i$ are proportional to the number of molecules which participate in the reaction:

$$\delta n_i = a\nu_i \tag{8.2}$$

where $a$ is a constant. The equilibrium condition is therefore

$$\boxed{\sum \mu_i \nu_i = 0 .} \tag{8.3}$$

Although not needed much in this book, a number of concepts frequently used in the literature will now be introduced.

(1) The *amount of substance* is a rather vague term, and its current and precise interpretation is in terms of the number of some specified entities of that substance, such as a specified molecule, ion, or radical, or an electron, photon, etc. A convenient *unit* of *amount of substance* is the *mole* (abbreviated to mol) which refers to the number of entities which is equal to the number of carbon atoms in 12 gm (or 0·012 kg) of $^{12}C$. (This unit used to be called the gram molecule but could then be applied only to substances which existed as molecules.) The amount of substance is independent of pressure and temperature and for a *given* substance it is proportional to its mass. (This proportionality fails if different substances are compared.) For a perfect gas 1 mole contains a constant number of $6{\cdot}023 \times 10^{23}$ molecules. This is *Avogadro's constant* $L = 6{\cdot}022 \times 10^{23}\ \text{mol}^{-1}$, and has led to the *gas constant* $R = Lk = 8{\cdot}314\ \text{J K}^{-1}\text{mol}^{-1}$. Sometimes one also uses the term *molar* volume. This is intended to mean the volume divided by the amount of substance. Analogous definitions hold for other molar quantities. If one develops arguments in terms of numbers ($n$) of molecules or electrons etc., as has been done in this book, then neither the mole nor Avogadro's number is required.

(2) Let $\xi$ be a quantity which measures the advance of chemical reaction in such a way that $\nu_i d\xi$ is a measure of the *amount of component* $C_i$ participating in the reaction (8.1) as $\xi$ changes to $\xi + d\xi$. From problem 7.2 one then has at constant temperature and pressure

$$dG = \sum_i \mu_i \nu_i d\xi \le 0 \quad \text{or} \quad A d\xi \ge 0 \quad \text{where } A \equiv - \sum_i \nu_i \mu_i . \tag{8.3'}$$

The reaction is possible if it lowers $G$ or keeps it constant. Thus the $\nu$'s must be positive for the products and negative for the reactants. In order that we can assert that the *affinity* $A$ be non-negative in a possible physicochemical reaction, as suggested by (8.3'), the numbers $(-\nu_i)$ must occur in the expression for $A$. Condition (8.3) is a special case of (8.3').

(3) At constant temperature ($dT = 0$) the Gibbs free energy of a closed system ($dn = 0$) satisfies (from problem 7.1)

$$G_B - G_A = \int_A^B v dp = \int_A^B (v_g - \alpha) dp \qquad v \equiv v_g - \alpha \tag{8.4}$$

where $v_g$ is the volume which *would* be calculated from the given temperature using the ideal gas law $pv_g = nkT$. The quantity $\alpha$ allows for the fact that the system may not be an ideal gas, and we shall replace it by virtue of

$$\int_A^B \alpha dp \equiv nkT \ln \left\{ \frac{(p/f)_B}{(p/f)_A} \right\} . \tag{8.5}$$

The *fugacity* $f$, as defined in this relation, is an effective *pressure*. Since $v_g = nkT/p$,

$$G_B - G_A = nkT \left\{ \ln\left(\frac{p_B}{p_A}\right) - \ln\left(\frac{p_B f_A}{p_A f_B}\right) \right\}$$

$$= nkT \ln\left(\frac{f_B}{f_A}\right) = nkT \ln a . \tag{8.6}$$

Here

$$a \equiv f_B/f_A \tag{8.7}$$

is called the *activity* for state B if state A is the reference state at the temperature considered. For ideal gases $f = p$ and $\alpha = 0$. Fugacity and activity are useful concepts, but mainly for isothermal processes. Since gaseous systems approximate to ideal classical gases in the limit of low pressure (or large volume or great dilution; one can choose whatever condition one wishes), and the chemical potential is $\mu = G/n$, we have from (8.6)

$$\mu = \mu_0 + kT \ln\left(\frac{f_B}{f_A}\right) = \mu_0 + kT \ln a \tag{8.8}$$

and

$$\lim_{p \to 0} \left(\frac{f}{p}\right) = 1 . \tag{8.9}$$

(8.8) and (8.9) can be used as a definition of fugacity, $\mu_0$ being the chemical potential in state A which is again regarded as a standard state.

It can depend on temperature $T$ and pressure $p$. Fugacities can be calculated from the appropriate equation of state (see problem 8.1).

We return now to our main equilibrium result (8.3) for a system at temperature $T$, pressure $p$, and in a volume $v$ to find with the aid of (8.8)

$$\sum_i \nu_i \ln a_i = -\sum_i \frac{\nu_i \mu_{0i}(p,T)}{kT} \equiv \ln K(T,p) \tag{8.10}$$

where the sums are over all reacting components and $K$ is called the *equilibrium constant* of the reaction. We can rewrite (8.10) as

$$\prod_i a_i^{\nu_i} = K(p,T) \tag{8.11}$$

and this is the *mass action law*. It states that in chemical equilibrium and at constant pressure and temperature the activities of the reactants, when taken with powers given by the stoichiometric coefficients $\nu_i$, yield a quantity which is given by $p,T$, the $\nu_i$, and the standard states of the components but is otherwise independent of the activities.

For ideal gases at temperature $T$ in a volume $v$, and by Dalton's law of partial pressures,

$$\frac{p_i}{p} = \frac{p_i}{\Sigma_j p_j} = \frac{n_i}{\Sigma_j n_j} \equiv c_i \tag{8.12}$$

where $c_i$ is the relative molecular concentration of component $i$ in the system. For ideal gases in chemical equilibrium the mass action law (8.11) can therefore also be expressed by

$$\prod_i \left(p_i^{\nu_i}\right) = K_{\text{id.gas}}(p,T)$$

or

$$\prod_i \left(c_i^{\nu_i}\right) = \left(p^{-\Sigma\nu_i}\right) K_{\text{id.gas}}(p,T) \equiv K'_{\text{id.gas}}(p,T) . \tag{8.13}$$

By (8.12) and (8.13) the activity can also be regarded as an effective ideal gas relative concentration $c_i$. The *activity coefficient* is then $a_i$ divided by the actual concentration.

An example will be given in section 15.6. In the case of reactions

$$C_1 + C_2 \rightleftharpoons C_3 \quad \text{or} \quad 2C_4 \rightleftharpoons C_5 + C_6 \ , \tag{8.14}$$

for example, one finds for dilute gases the equilibrium constants to be given respectively by

$$\frac{c_3}{c_1 c_2} \quad \text{or} \quad \frac{c_5 c_6}{c_4^2} \tag{8.15}$$

One can think of the term $c_1 c_2$ as arising from the probability of a collision between species $C_1$ and $C_2$, since this is for dilute systems proportional to $c_1$ and $c_2$ and hence to their product. Similarly, if two molecules of $C_4$ are needed for the reaction, the probability of finding them is in a dilute system proportional to the square of their concentration.

The forward *rate* of the reaction $C_1 + C_2 \rightarrow C_3$ can be written as $k_r c_1 c_2$ where $k_r$ is a *rate constant*. It is found that if $E$ is an activation energy and if $A$ is approximately independent of temperature, rate constants often satisfy the *Arrhenius relation*

$$k_r = A \exp\left(-\frac{E}{kT}\right) \ . \tag{8.16}$$

The rates thus tend to increase rapidly with temperature.

One of the characteristics of thermal equilibrium, denoted by the suffix 0, is that forward and reverse reactions must proceed at equal rates so as to balance each other. That equilibrium implies this balance is in fact known as the *principle of detailed balance*. The word 'detailed' here is important; it conveys that balance must be satisfied for each individual reaction. Adopting the principle, let us apply it to the reaction (7.33). One finds

$$x_0 = \frac{k_{10}' b_0}{k_{10} a_0} = \frac{k_{20} p_0}{k_{20}' q_0} \ .$$

It follows that (7.33) can be written

$$\dot{x} = k_1' b\left(1 - \frac{\lambda}{\mu}\right) + k_2 p\left(1 - \frac{\lambda}{\nu}\right) x^2 \ ,$$

where

$$\lambda \equiv \frac{x}{x_0} \qquad \mu \equiv \frac{k_1'k_{10}a_0b}{k_{10}'k_1ab_0}, \qquad \nu \equiv \frac{k_{20}'k_2q_0p}{k_2'k_{20}qp_0}.$$

This shows that whereas equilibrium entails detailed balance the converse does not hold: for detailed balance it is sufficient to have

$$\lambda = \mu = \nu ,$$

while $\lambda = 1$ is additionally needed for equilibrium.

Statistical mechanics broadens one's understanding of thermodynamics, though it has not the same generality. In particular, the notion of reaction constants will occupy us in the statistical mechanical section 15.6.

PROBLEMS

8.1. Show that for a van der Waals gas the fugacity in a state 2 relative to a very dilute state 1 (in which the system approximates an ideal classical gas) is given approximately by

$$\ln f_2 = \ln\left(\frac{NkT}{v_2-b}\right) + \frac{b}{v_2-b} - \frac{2a}{NkTv_2} \qquad (T_1 = T_2 = T)$$

8.2. Taking the van der Waals equation of state in the form

$$p = \frac{NkT}{v-bN} - \frac{aN^2}{v^2}$$

show that if $\phi$ is a function of $T$ only,

$$U = -\frac{aN^2}{v} - N\phi + NT\phi'$$

$$TS = NkT\ln\left(\frac{v}{N} - b\right) + NkT + NT\phi'$$

$$F = NkT\ln\left(\frac{v}{N} - b\right) - NkT - \left(\frac{aN^2}{v}\right) - N\phi .$$

Hence check the forms of equations (6.16) and (6.18) and verify that $C_v$ is $NT\phi''(T)$.

# PART B: STATISTICAL MECHANICS

# 9. Two basic results in statistical mechanics

## 9.1. THE FIRST EQUATION

The purpose of this section is to elucidate two main equations on which statistical mechanics is based.

As one of the first steps in the construction of statistical mechanics one decides in what microscopic or macroscopic states the system can be supposed to be. For instance, in the case of an isolated system all states which are deemed accessible must certainly lie within the same small energy range because of energy conservation. If the system is not isolated, but in contact with a heat reservoir, a different set of states is accessible to the system, and the system is described in a different way. However, given the system and its mode of description, then to each of the accessible states a probability is assigned. The bare fact that such probabilities exist is sufficient to introduce the statistical analogue of the entropy of thermodynamics. The precise form of the probability distribution is a matter of indifference. Thus the concept of entropy, when defined with reference to probability distributions, as well as the investigation of its properties, come within the purview of probability theory.

Suppose a system has a finite number of accessible states weighted with probabilities $p_1,\ldots,p_n$. Assuming that an entropy function can be defined in terms of these probabilities what form shall it take? To answer this question, some properties of this function must be specified as basic. The unknown function will be denoted by $S(p_1,p_2,\ldots,p_n)$ in order to emphasize that it is the statistical mechanical analogue of the entropy. It is hoped that the very form and context of this statistical mechanical analogue will be sufficient to enable the reader to distinguish it from the entropy of thermodynamics, whenever such distinction appears to be desirable. However, very often we may, without further apology, identify the statistical with the thermodynamic entropy.

It is at first sight puzzling that the entropy, which has in Chapters 5 - 8 been treated as a function of thermodynamic variables, appears now as a function of probabilities. However, some such change was foreshadowed as early as Chapter 1, when it was pointed out that thermodynamic theories are incomplete.

One would expect therefore that on passing from thermodynamics to statistical mechanics, one exchanges a small number of macroscopic variables for a much larger number of microscopic variables. The probabilities for various microscopic states of the system represent some of these new variables. However, the entropy concept is here developed further in a second respect. We seek a statistical analogue for the entropy which has a meaning in non-equilibrium situations without there being any need to use explicitly the *basic trick* of thermodynamics, the essence of which is to suppose that a sufficient number of inhibitions have been imposed on the system so as to convert the non-equilibrium state to a practically indistinguishable equilibrium state of the uninhibited system.

It would have been more in accord with common usage, had we introduced a function $H(p_1,\ldots,p_n)$ such that the statistical mechanical analogue of the entropy is $-kH$. This notation goes back to Boltzmann, who made fundamental contributions to the problem of finding a statistical mechanical analogue for the entropy. However, in order to avoid the introduction of a new symbol, and a new concept, we shall deal directly with the statistical analogue of the entropy.

(i) Suppose that all we know of a given isolated system, apart from its qualitative structure, is that its energy lies in a certain small range. If $n$ quantum states are accessible to this system in this energy range, it is reasonable, in the absence of other information, to suppose that in equilibrium the $n$ quantum states are equally probable. If the system is not in equilibrium, arbitrary distributions of the $p_i$ can occur, but the approach to equilibrium with lapse of time will then imply that all $p_i$ tend towards the value $1/n$. The system being isolated, the entropy cannot decrease during this process, so that

$$S(p_1,\ldots,p_n) \leq S(1/n,\ldots,1/n) \ .$$

Since entropy is to be a *universal* function of the $p_i$, *this inequality must hold generally.* This means that entropy corresponds to a kind of statistical 'spread'; *the more uniformly the probability is spread over the states, the greater is the entropy of the system.*

(ii) The addition of a further state, which is, however, inaccessible ($p_i = 0$), can clearly not affect the entropy. Put differently, the inaccessible states should not play a part in the expression for the entropy. Hence, we shall assume that

$$S(p_1,\ldots,p_n,0) = S(p_1,\ldots,p_n) \ .$$

(iii) Suppose we have two identical systems A and A', identically described. Let us regard them as forming a new system A". If the two systems do not interact, then because entropy is an extensive variable $S(\mathrm{A}'') = S(\mathrm{A}) + S(\mathrm{A}') = 2S(\mathrm{A})$. Similarly, for two arbitrary non-interacting systems A, B we must have that the entropy of the combined system AB satisfies $S(\mathrm{AB}) = S(\mathrm{A}) + S(\mathrm{B})$. If the systems do interact, the position is more complicated, as will now be discussed.

Suppose the probabilities are $p_1,\ldots,p_n$ for system A. Suppose also that if A is in the quantum state $A_k$, a number of states $B_1, B_2, \ldots$ (with probabilities $q_{k1}, q_{k2}, \ldots$) are accessible to B. Then $S(q_{k1}, q_{k2}, \ldots)$ is the entropy of B, which is conditional on A being in state $\mathrm{A_k}$. We call this entropy the *conditional* entropy, and denote it by $S_k(\mathrm{B})$. The probability of its occurrence is $p_k$, so that the mean conditional entropy is $\Sigma_{k=1}^{k=n} p_k S_k(\mathrm{B})$. It will be denoted by $S_\mathrm{A}(\mathrm{B})$. Thus, in the case of interaction, it is reasonable to suppose that

$$S(\mathrm{AB}) = S(\mathrm{A}) + S_\mathrm{A}(\mathrm{B}) \ .$$

This is our assumption (iii) in which $S(\mathrm{A})$ has been regarded as one component in the entropy of AB. This equation is analogous to the equation of probability theory

$$p(\mathrm{AB}) = p(\mathrm{A})\ p(\mathrm{B}|\mathrm{A}) \ ,$$

where the last term is the conditional probability of an event B occurring, given that event A has occurred. Note that if there is no interaction between A and B then the $q_{kl}$ are independent of $k$, and $S_{\mathrm{A}}(\mathrm{B}) = \Sigma_k p_k S_k(\mathrm{B}) = S(\mathrm{B})\ \Sigma p_k = S(\mathrm{B})$, so that the simple entropy addition is recovered.

The form of the entropy can now be determined, and this is achieved in two steps. In order to free ourselves of the now unnecessary thermodynamic associations, we shall call a set of mutually exclusive possibilities $\mathrm{A}_1,\ldots,\mathrm{A}_n$, together with their assigned probabilities $p_1,\ldots,p_n$, a probability scheme or simply a scheme. The set of events $\mathrm{A}_k\mathrm{B}_l (k = 1,\ldots,n;\ l = 1,\ldots,m)$ is called a product scheme.

*A. Proof that* $S(1/n,\ldots,1/n) = \lambda \log n$, *where* $\lambda \geq 0$.
Let $L(n) \equiv S(1/n,\ldots,1/n)$, whence by properties (i) and (ii) $L(n)$ is a non-decreasing function of its argument:

$$L(n) = S\left(\frac{1}{n},\ldots,\frac{1}{n},0\right) \leq S\left(\frac{1}{n+1},\ldots,\frac{1}{n+1}\right) = L(n+1) \, . \qquad (9.1)$$

Consider $m$ mutually independent schemes $\mathrm{A}^{(j)} (j = 1,2,\ldots m)$, each consisting of $r$ equally likely events. Regarded as a single scheme, there are clearly $r^m$ equally likely events with entropy $L(r^m)$. Regarded as a product scheme, the entropy is by property (iii) $\Sigma_{j=1}^{m} S(A^{(j)}) = mL(r)$. Hence for all positive integers $r,m$,

$$L(r^m) = mL(r) \ . \qquad (9.2)$$

This relation has the functional form of a relation satisfied by the logarithm to any base $b$:

$$\log_b r^m = m \log_b r .$$

We now make use of this point. Let $r, s, n$ be arbitrary positive integers, with $m$ determined by

$$r^m \leq s^n \leq r^{m+1} . \tag{9.3}$$

Forming $\log x$ for each term $x$ of (9.3), one finds

$$m \log r \leq n \log s \leq (m+1) \log r$$

whence

$$\frac{m}{n} \leq \frac{\log s}{\log r} \leq \frac{m}{n} + \frac{1}{n} . \tag{9.4}$$

If, instead of $\log x$, one writes $L(x)$ for each term $x$ of (9.3), one finds an exactly similar relation, by virtue of (9.2):

$$\frac{m}{n} \leq \frac{L(s)}{L(r)} \leq \frac{m}{n} + \frac{1}{n} .$$

It follows that

$$\frac{L(s)}{L(r)} - \frac{\log s}{\log r} \leq \frac{m+1}{n} - \frac{m}{n} = \frac{1}{n} .$$

Also

$$\frac{L(s)}{L(r)} - \frac{\log s}{\log r} \geq \frac{m}{n} - \frac{m+1}{n} = -\frac{1}{n} .$$

Hence,

$$\left| \frac{L(s)}{L(r)} - \frac{\log s}{\log r} \right| \leq \frac{1}{n} . \tag{9.5}$$

The left-hand side of (9.5) is independent of $m$, while $n$ is arbitrary. Hence, upon taking $n$ arbitrarily large, one finds that

$$\frac{L(s)}{\log s} = \frac{L(r)}{\log r}$$

is a constant ($\lambda$) independent of $r$. Hence

$$L(n) = \lambda \log n \tag{9.6}$$

for all $n$. By (9.1), $\lambda \geq 0$.

*B. Proof that* $S(p_1,\ldots,p_n) = -\lambda\,\Sigma_{i=1}^{n}\,p_i \ln p_i$

We now make use of result *A* in order to derive a more general result. To do so, it is convenient to invent two interacting probability schemes A and B and to utilize assumption (iii). Scheme B is carefully designed to achieve what we want to establish. The result is, however, valid for any scheme A.

Suppose that the $n$ probabilities of scheme A have the form $p_k = g_k/g$, where the $g$'s are positive integers with $\Sigma g_k = g$. Suppose also that a dependent scheme B has $n$ groups of events such that the $k$th group has $g_k$ events. If the event $A_k$ occurs, assume that all events of the $k$th group are equally likely with probability $1/g_k$. Hence the conditional entropy $S_k(B)$ of B, given A is in state $A_k$, is $L(g_k) = \lambda \log g_k$ by result A. Hence

$$S_A(B) = \sum_{k=1}^{n} p_k S_k(B) = \lambda \sum_{k=1}^{n} p_k \log g_k \ . \tag{9.7}$$

$A_k$ occurs with probability $p_k$; the number of events $A_k B_l$ ($k$ fixed) is $g_k$, and the probability of each is $p_k/g_k = 1/g$, i.e. it is the same for all events. The scheme $A_k B_l$ ($k$ and $l$ can vary) consists therefore of $g$ equally likely events, so that using again the result *A*

$$S(AB) = \lambda \log g \ . \tag{9.8}$$

By property (iii), (9.7), and (9.8),

$$S(A) = S(AB) - S_A(B) = -\lambda \sum_{k=1}^{n} p_k(\ln g_k - \ln g)$$

so that

$$\boxed{S(A) = -\lambda \sum_{k=1}^{n} p_k \ln p_k \ ,}$$

as was to be proved.

Assuming that the entropy $S(p_1,\ldots,p_n)$ is continuous in its arguments, the result must hold not only for rational

$p_k$, but for all $p_k$. One may write $\overline{S}$, instead of $S$, to indicate that an average has been taken over a probability distribution (or over an ensemble, if ensemble language is being used). The quantity being averaged is $-k \ln p_j$. One might interpret this quantity as the entropy $S_j$ of the $j$th quantum state of the system, for the reasonable relations

$$\overline{S} = \sum_{j=1}^{n} p_j S_j, \qquad S_j \equiv -k \ln p_j \tag{9.9}$$

would then hold.

When one works out the theory of the ideal classical gas on the basis of the entropy expression $S = -\lambda \Sigma p_i \ln p_i$, one finds that for such a gas $pv = N\lambda T$. At that stage the constant $\lambda$ could conveniently be identified as Boltzmann's constant. In fact a much earlier identification of $\lambda$ is desirable in order to avoid departures from established notations. We accordingly write $k$ for $\lambda$ now.

The extensive nature of the entropy expression (9.9) may be verified directly as follows. If the quantum states of two parts of a system are labelled by suffixes $i$ and $j$ respectively, and if $p_{ij}$ is the probability that one part is in its $i$th state while the other part is in its $j$th state, then

$$S = -k \sum_{i,j} p_{ij} \ln p_{ij} \quad .$$

If the probabilities for the two parts of the system are independent so that one can write

$$p_{ij} = P_i Q_j, \qquad \Sigma\, P_i = 1 = \Sigma\, Q_j \quad ,$$

it follows that

$$S = S_1 + S_2$$

where

$$S_1 \equiv -k \Sigma P_i \ln P_i, \qquad S_2 \equiv -k \Sigma Q_j \ln Q_j \quad .$$

With an analogous notation one can write for the mean energy

$$U = \sum_{i,j} e_{ij} p_{ij} \,.$$

If again $e_{ij} = E_i + F_j$ and $p_{ij} = P_i Q_j$, as for non-interacting subsystems, one finds

$$U = U_1 + U_2$$

where

$$U_1 \equiv \Sigma E_i P_i, \quad U_2 \equiv \Sigma F_j Q_j \,.$$

One can proceed similarly for other extensive thermodynamic variables.

It is instructive to note the following result. For any continuous strictly monotonic convex function $f(x)$ it is known† that for real $a_k \geq 0$ and for real $x_k \geq 0$ such that $\sum_k x_k = 1$ one can write

$$\sum_k a_k f(x_k) \geq f(\sum_k a_k x_k) \,.$$

Interpreting $f(x)$ as $x \ln x$, it follows that

$$\sum_k p_k q_{kl} \ln q_{kl} \geq (\sum_k p_k q_{kl}) \ln(\sum_k p_k q_{kl}) = q_l \ln q_l \,.$$

We have here written $q_l$ for the total probability $\Sigma_k p_k q_{kl}$ of finding the event $B_l$ in scheme B. Upon summing over $l$ and multiplying by $-k$ the left-hand side is seen to be

$$-k \sum_k p_k \sum_l q_{kl} \ln q_{kl} = S_A(B)$$

while the right-hand side is $S(B)$. There follows the inequality

$$S(AB) = S(A) + S_A(B) \leq S(A) + S(B)$$

which is sometimes associated with Shannon's name. It holds whether the schemes A and B are independent or not. It states that the conditional entropy of scheme B, knowing the state of scheme A and averaged over all states of A, is less than the entropy of scheme B in the absence

†G.H.Hardy, J.E.Littlewood, and G.Polya (1934), *Inequalities*, Theorem 86, Cambridge University Press, Cambridge.

of such knowledge. It illustrates the general rule, valid in statistical mechanics as well as information theory, that additional information can only decrease the entropy. One of the oldest and most celebrated examples of the operation of this principle is Maxwell's demon. From a knowledge of the velocity of each approaching molecule, this demon opens or closes a sliding partition so as to separate the faster from the slower molecules. In this way he has used his information to establish a temperature difference. He has thus decreased the entropy of the system without the performance of work. Some basic ideas of information theory are introduced in Appendix I.

## 9.2. THE SECOND EQUATION

It was seen in section 7.4 that the thermodynamic state of an $r$-co-ordinate $\chi$-component simple system can be specified by $1 + r + \chi$ independent thermodynamic variables. We shall consider the case of $r = 1$, $\chi = 1$ as the generalizations are obvious. In this case one has three independent variables, which can all be taken to be extensive variables: volume $v$, energy level $u$, and number of particles $n$, for example. One may replace one or two of these variables by intensive variables, but, of these, at most two are independent, as was seen in equation (7.6). It clearly does not matter which of the possible choices of two independent variables is made in the *thermodynamic* description of the system.

The situation is different in *statistical mechanics* where it is recognized explicitly that if an intensive variable is included in the thermodynamic specification of a state of a system that system must be in contact with a reservoir much larger than itself. For example, the specification of a temperature $T$ implies thermal contact between the system and a heat reservoir whose temperature is kept at $T$ and which has such a large heat capacity that heat exchange with the system of interest (which is possible because of the required thermal contact) does not affect the value of its temperature. Again, if a pressure $p$ is stipulated for the system of interest, it must be mechanically coupled, for instance by a smoothly moving and closely fitting piston, to a very large pressure reservoir in which the pressure has the required value. In the first case the energy of the system undergoes fluctuations; in the second case the volume does. A similar situation occurs if the chemical potential $\mu$ of the particles is specified, when the number of particles in the system can fluctuate. In this case a particle reservoir is involved.

One normally makes the identification that the mean quantities of statistical mechanics, as averaged over an appropriate probability distribution, correspond to the thermodynamic quantities. Thus when $1 + r + \chi$ variables have given values for the specification of a thermodynamic equilibrium state of the system, this situation is specified in statistical mechanics by the same variables, together with an equilibrium probability distribution (over the quantum states of the system) which is compatible with these $1 + r + \chi$ values. For example, if $n,v,T$ are given for a simple system, the internal energy $u$ can fluctuate, but if the probability distribution is determined by the nature of the system and the given values of $n$, $v$, and $T$, the mean energy $\equiv U$ and the higher moments $\overline{u^m}$, are determined. Thus, in passing from a specification involving $n$, $v$, and $U$ to one involving $n$, $v$, and $T$, we replace in effect the specification of a definite internal energy for the system by a specification of the probability distribution for the internal energy $u$. What is the form of this distribution?

For its determination it is sufficient to suppose that a specification of variables such as $n,v$, and $T$ implies a definite value for the mean energy $\overline{u}$. It turns out that with plausible assumptions this implies the form of the probability distribution, and hence the calculation of the higher moments $\overline{u^m}$ also becomes possible. One need not confine attention to one extensive variable $u$, which fluctuates if its canonically conjugate variable $T$ is given. If several independent intensive variables are given, their canonically conjugate extensive variables can fluctuate, and we now give the general argument.

For a given state of the system we may suppose that the variable $X_1$ can have any one of a set of discrete values $\{X_{1i_1}\}$, where $i_1$ labels one value of the set. Similarly $X_2$ has one of the values $\{X_{2i_2}\}$, etc., up to $X_t$. A state of the system is then specified by a set of numbers (these are the quantum numbers of quantum mechanics)

$$(i_1, i_2, \ldots, i_t) \equiv \mathbf{i}$$

which may be written as a $t$-component vector $\mathbf{i}$. In this state the fluctuating variables have the values

$$(X_{1i_1}, X_{2i_2}, \ldots, X_{ti_t}) \ .$$

Appropriate modifications, which involve the replacement of sums by integrals, have to be made if a variable has a continuous spectrum. To find the distribution law of the $p_{\mathbf{i}}$ for an equilibrium state, we must maximize the entropy $\overline{S}$ with respect to each $p_{\mathbf{i}}$ subject to the normalization $\Sigma p_{\mathbf{i}} = 1$, making a total of $t$+1 subsidiary conditions. Only in this way can one be sure that in the theoretical treatment the system has in equilibrium the largest entropy value compatible with the given experimental situation. Using undetermined multipliers $\alpha$, $\beta_1, \ldots, \beta_t$ one must therefore maximize the function (see Appendix II)

$$F \equiv -k \sum_{\mathbf{i}} p_{\mathbf{i}} (\alpha + \sum_{j=1}^{t} \beta_j X_{ji_j} + \ln p_{\mathbf{i}})$$

with respect to each $p_{\mathbf{i}}$, the subsidiary conditions now being taken into account automatically. It is found that

$$\alpha + \sum_{j=1}^{t} \beta_j X_{ji_j} + \ln p_{\mathbf{i}} + 1 = 0 \ ,$$

that is

$$p_{\mathbf{i}} = Q^{-1} \exp\,(- \sum_{j=1}^{t} \beta_j X_{ji_j}) \qquad (i_j = 1, 2, \ldots) \qquad (9.10)$$

where $Q$ is a normalizing factor

$$Q = \sum_{j_1} \sum_{j_2} \ldots \sum_{j_t} \exp(-\beta_1 X_{1j_1} - \beta_2 X_{2j_2} \ldots -\beta_t X_{tj_t}). \quad (9.11)$$

The sums are over all values $X_{1j_1}, \ldots, X_{tj_t}$ which the variables $X_1, \ldots, X_t$ can assume.

Instead of giving the probability $p_{\mathbf{i}}$ for quantum state $\mathbf{i}$ of the whole system, it is sometimes more convenient to regard the $X_1, \ldots, X_t$ as continuous variables, and write the probability law

$$p(X_1,\ldots,X_t) \propto \exp(-\sum_{j=1}^{t} \beta_j X_j) \tag{9.10a}$$

Equations (9.10) and (9.10*a*) represent the required distribution law.

Suppose $t = 1$ in (9.11) and the summation over $X_1$ is over the energy levels of the system. Suppose also that the volume is one of the specified thermodynamic variables to be denoted by $x_1,\ldots,x_s$. Then clearly the value of $Q$ will depend on the volume of the system since the energy levels do. It follows by similar arguments that the function $Q$ has a value which depends on the variables indicated

$$Q = Q(x_1,\ldots,x_s,\beta_1,\ldots,\beta_t) \ . \tag{9.12}$$

What can be said about these new physical quantities $\beta_1,\ldots,\beta_t$ and $Q$?

To answer this question, evaluate the entropy, using (9.10), to find

$$\overline{S} = k \sum_i \{p_i \sum_{j=1}^{t} \beta_j X_{ji_j})\} + k \sum_i p_i \ln Q$$

that is

$$\overline{S} = k \sum_{j=1}^{t} \beta_j \overline{X}_j + k \ln Q \ . \tag{9.13}$$

Since $\overline{S}$ and the $\overline{X}_j$ are extensive variables, it follows that $\ln Q$ is also an extensive variable, while the $\beta$'s are intensive variables. It is clear that $Q$, the $\beta_j$, and all mean values calculated from (9.10) must ultimately depend on the specified values of the $s$ extensive and the $t$ intensive variables.

The identification of the $\beta_j$ and of $\ln Q$ may be carried out by comparing (9.13) with an appropriate thermodynamic relation, subject to the assumption that the statistical mechanical $\overline{S}$ is equal to the thermodynamic entropy $S$. A similar assumption for the $\overline{X}_j$ is also required.

## 9.3. SOME SPECIAL DISTRIBUTIONS

We shall consider only systems which contain one type of particle since this is the most important case and simplifies the exposition. For our purposes the most important distribution is the *grand canonical distribution* in which two intensive variables ($\mu$ and $T$) are given. This means equivalently that the internal energy $u$ and the number of particles $n$ are only specified on average and can otherwise vary accordingly to the probability law (9.10) as applied to this situation. Hence in (9.10)

$$t = 2; \quad \overline{X}_1 = \overline{u}, \quad \overline{X}_2 = \overline{n} \quad \text{(given)}. \tag{9.14}$$

Possible values of internal energy and number of particles are

$$X_{1i_1} \to E_i, \quad X_{2i_2} \to n_j$$

and a state of the system is specified by (9.10) which, written fully, is

$$P(v, \mu, T; i, j) = \Xi^{-1} \exp(-\beta_1 E_i - \beta_2 n_j) . \tag{9.15}$$

The first group of variables in parenthesis on the left are those which are the same for *all* states of the system. The second group specify the state which is subject to the probability distribution. The normalizing constant in (9.15) can be evaluated by summing that equation over all $i$ and $j$ :

$$1 = \Xi^{-1} \left\{ \sum_i \exp(-\beta_1 E_i) \right\} \left\{ \sum_j \exp(-\beta_2 n_j) \right\}$$

as in (9.11). It follows that

$$P(v, \mu, T; i, j) = \frac{\exp(-\beta_1 E_i - \beta_2 n_j)}{\sum_{k,l} \exp(-\beta_1 E_k - \beta_2 n_l)} . \tag{9.16}$$

Each of the averages in (9.14) has led to an undetermined multiplier being found in (9.16). Equation (9.12) becomes

$$\Xi = \Xi(v, \mu, T; \beta_1, \beta_2)$$

and (9.13) is

$$\bar{S} = k\,\beta_1\,\bar{u} + \beta_2\,\bar{n} + k\ln\Xi \quad . \tag{9.17}$$

Comparing this with equation (7.4) ,

$$S = \frac{1}{T}\,(U - \mu N + pv)\ ,$$

one finds

$$\beta_1 = \frac{1}{kT}\ , \qquad \beta_2 = -\,\frac{\mu}{kT}\ , \qquad \ln\Xi = \frac{pv}{kT}\ . \tag{9.18}$$

The grand canonical distribution is accordingly

$$P(v, \mu, T; E_i, n) = \Xi^{-1}\exp\left(\frac{\mu n - E_i}{kT}\right) \tag{9.19}$$

$$\Xi = \exp\left(\frac{pv}{kT}\right) = \sum_{n,i}\exp\left(\frac{\mu n - E_i}{kT}\right). \tag{9.20}$$

The quantity (9.20) is called the *grand canonical partition function.*

If one replaces the specification of the *average* total number of particles $n$ in (9.11) by the precise total number, then (9.15) changes as follows

$P(v, \mu, T;\ \ , j\ )$ becomes $P(v, n, T; i\ )$

$\mu$ being replaced by $n$. The constraint which was taken care of by the Lagrangian multiplier $\beta_2$ is removed and one can simply put $\beta_2$ equal to zero. The energy $E_i$ is then the only variable quantity. Hence one finds instead of (9.19) and (9.20)

$$P(v, n, T; i) = Z^{-1}\exp\left(-\,\frac{E_i}{kT}\right) \tag{9.21}$$

$$Z = \exp\left(-\,\frac{F}{kT}\right) = \sum_i \exp\left(-\,\frac{E_i}{kT}\right) \tag{9.22}$$

where $F$ is the Helmholtz free energy. Its occurrence is justified by comparing

$$\bar{S} = k\,\beta\,\bar{u} + k \ln Z$$

and

$$S = T^{-1}(U + pv - \mu N) = T^{-1}(U - F)$$

noting that $pv - \mu N = TS - U = -F$. Expression (9.21) is called the *canonical distribution* and $Z$ is the *canonical* partition function. This distribution applies to a system of a fixed number of particles whose temperature and volume are also given. A reservoir of given temperature $T$ is still in contact with the system, but the particle reservoir, which impresses its chemical potential $\mu$ on the system, has now been removed.

The word 'partition function' expresses vaguely what is clear from the expressions for $\Xi$ and $Z$, that these quantities can be 'partitioned' into sums of a certain type. In German the word *Zustandssumme* is used, which emphasizes that one has to sum over states to obtain it.

The process of passing from the grand canonical to the canonical distribution by replacing a given intensive variable $\mu$ by giving the corresponding extensive variable $n$ instead can be carried further. In this step one replaces $T$ by $E$ thus fixing the total energy not only on average but exactly. The resulting distribution describes a system which is *not in contact with any reservoir*. It is thus an isolated system. The constraint for the average energy has been removed, and one may put $\beta_1 = \beta$ equal to zero. Only the normalizing constant is now left and one finds

$$\boxed{p(v,\, E,\, n) = 1/N}$$

where $N$ is the number of states of energy $E$ accessible to the system. This distribution is called *microcanonical*. It corresponds to the entropy

$$S = -k \sum_{i=1}^{N} \frac{1}{N} \log \frac{1}{N} = k \log N$$

as already obtained in equation (9.6). It is this formula, in the form $S = k \log W$, which has been engraved on Boltzmann's tombstone in recognition of his long struggle to clarify the statistical interpretation of the thermodynamic entropy. It has here been seen that the even more fundamental formula $S = -k \sum_i p_i \log p_i$ yields a framework for statistical mechanics.

## *9.4. REMARKS ON THE FOUNDATIONS OF STATISTICAL MECHANICS

The two basic equations of statistical mechanics have been obtained in a very simple manner. This simplicity should, however, not mislead the reader into believing that in this approach one can successfully avoid the well-known difficulties concerning the foundations of statistical mechanics.

This may be seen most clearly by taking stock of the assumptions which have been made. A system has been considered, the qualitative structure of which - for instance the nature of the particles contained in it and the nature of the container - is supposed given. Its macroscopic (thermodynamic) state, and the mode of description of this state, were also supposed given. The following assumptions have then been made:

(1) The system can be in any one of a finite number of quantum states.

(2) With each of these states one can associate a probability $p_i$ of finding the system in this state (we have reverted to a single subscript, $i$, for $p$ in order to simplify the notation).

(3) A statistical entropy function $S(p_1, p_2, \ldots)$ exists which can be expressed in terms of these probabilities only.

(4) In an equilibrium state this entropy function attains a maximum value with respect to each $p_i$.

At the present time all these stipulations do, in fact, seem very reasonable. Yet a link with the celebrated problems which arise in the foundations of statistical mechanics can be established by scrutinizing these assumptions more closely.

Take assumption (2) for example. Given the system and its macroscopic equilibrium state, what do we mean by assigning probabilities to the quantum states compatible with the given specification. The most reasonable answer is that we observe the system for a very long time $T$ and note the fraction of this time which the system spends in quantum state $i$; in principle one can estimate this fraction for all quantum states $i$. This furnishes a definition of the quantities $p_i$. However, the system may be started off in any one of the quantum states $i$ under consideration, and it is by no means certain that the set of probabilities $p_i$ will be the same, whatever the initial quantum state (compatible, of course, with the macroscopic specification) in which the system finds itself. Nor is it obvious that if $N$ quantum states are accessible to the system (and, of course, compatible with the macroscopic specification) when it is started off in quantum state $a$, then the same $N$ states remain accessible with lapse of time, or equivalently that the same $N$ states are accessible to the system if it is started off in a different quantum state $b$. The assumption which has in fact been made is therefore: the probabilities exist, and their values $p_1, p_2, \ldots, p_N$ together with the value of $N$ as established by observation over a long time from $t = 0$ to $t = T$, are independent of the initial state of the systems. The assumption just formulated implies that in the course of time the system passes through all its accessible quantum states, and that for a long time interval $T$ the fraction $p_i$ of it which is spent in the $i$th state is the same wherever on the time co-ordinate the interval of duration $T$ is chosen, and that this is valid for all states $i$. This assumption is essentially the *ergodic hypothesis*, which is incidentally not universally valid.

If one were to make a list of the main thermodynamic properties of the entropy, one would find more items than were used to establish the uniqueness theorem. Thus the establishment of this theorem (p. 128) is really not sufficient. One should take the remaining properties of the entropy, which were not used in the proof of this theorem, and show that they are a consequence of the functional form which was adopted for the statistical analogue of the entropy. Only when this has been done can one regard assumption (3) as really satisfactory. The most important property of the thermodynamic entropy, which was not used, is its increase with time for an adiabatically isolated system. The deduction of this property, or some approximation to it, from $S = -k \sum_i p_i \ln p_i$ is the content of the *H-theorem* (see section 9.5).

In the thermodynamics of equilibrium situations, however, one merely needs a somewhat more restricted form of the H-theorem. One considers the entropy $S_1$ of a system in an equilibrium situation. One then removes a constraint which allows a new equilibrium situation to establish itself and one determines the new entropy $S_2$. Examples for the removal of a constraint are as follows: two systems previously adiabatically enclosed are placed into thermal contact, or two kinds of liquids are allowed to diffuse into each other, or a chemical reaction is allowed to proceed. Thermodynamics asserts that $S_1 \leq S_2$, but not how the change from $S_1$ to $S_2$ takes place in time. Now it is clear that in our way of establishing the basic equations of statistical mechanics this result must necessarily hold. For, by assumption (4), in order to identify the $p_i$ for the second equilibrium situation, the statistical entropy is maximized with respect to the $p_i$. However, it is maximized subject to a smaller number of subsidiary conditions (or constraints) than it was in the first equilibrium situation, so that the condition $S_2 \geq S_1$ is necessarily satisfied by the statistical analogue of this entropy. Now many equilibrium situations in statistical mechanics may be discussed on the basis of assumptions (1) to (4). In such cases the procedure used here is satisfactory, and the restricted form of the H-theorem is necessarily satisfied.

It is clear, therefore, that the principles adopted here enable one to pass from thermodynamics to statistical mechanics by a single procedure which incorporates (or, if one prefers this way of speaking, ensures the validity of) some form of the ergodic theorem and of the H-theorem. This route to statistical mechanics is, however, not the only one. The alternative route of deducing thermodynamics from the principles of mechanics via statistical mechanics is also of great interest, but will not be discussed here.

The interpretation of the probabilities $p_i$ in terms of observations on a system, which are extended over a long time $T$, enables one to give a definite meaning to the equations derived in this section, even if the system considered is small. Consider, for instance, one labelled molecule in a gas of weakly interacting molecules. If the whole system is kept at constant temperature and volume, then the labelled molecule can be regarded as being in contact with a 'heat bath', which consists in fact of all the remaining molecules. In the course of time the labelled molecule will pass through all its accessible states, and the probability of finding it in any one of its states is given by the canonical distribution. Hence one can give a very simple proof of the equipartition theorem of classical statistical mechanics (see section 11.7), which asserts, in its simplest form, that with every degree of freedom of a labelled molecule in a gas there is associated a mean energy of $\frac{1}{2}\, kT$. The average may here be interpreted to be a time average.

## *9.5. THE FINE-GRAINED ENTROPY IS CONSTANT IN TIME

Surprisingly, a wide variety of views exist as to why entropy increases and on the causes of irreversibility. We therefore seek to clarify at least *one* aspect of this question in the present section, which gives the present author's preferred approach.

### 9.5.1. *Basic quantum mechanics*

Consider a typical wavefunction depending on time $t$ and position $\mathbf{r}$

$$\phi(t,\mathbf{r}) = \sum_j f_j(t,\mathbf{r})\ \varepsilon_j\ (\mathbf{r}) \ . \tag{9.23}$$

The $\mathbf{r}$-dependencies will be suppressed in the sequel, and often the time dependencies as well. The $\varepsilon_i$ are an orthonormal basis in the space and the $f_j$ are complex numbers. A scalar product exists such that, with $\psi = \Sigma_i p_i\ \varepsilon_i$,

$$(\psi,\phi) \equiv (\sum_i p_i^*\ \varepsilon_i\ ,\ \sum_j f_j\ \varepsilon_j)$$

$$= \sum_{i,j} p_i^*\ f_j(\varepsilon_i,\varepsilon_j) = \sum_i p_i^*\ f_i \ . \tag{9.24}$$

Orthonormality $(\varepsilon_i,\varepsilon_j) = \delta_{ij}$ has here been used. The asterisk denotes the complex conjugate number. Let $Y$ be an operator. Then $Y\phi$ is also of the form $\Sigma_i\ c_i\ \varepsilon_i$ :

$$Y\phi = \sum_i\ (\sum_j y_{ij}\ f_j)\ \varepsilon_i \ . \tag{9.25}$$

The term in parentheses arises from the multiplication of a matrix by a vector. In particular if $f_j = \delta_{jk}$ then $\phi$ is $\varepsilon_k$, and for this case one can obtain an expression for the matrix element of $Y$ from (9.25)

$$(\varepsilon_l,\ Y\varepsilon_k) = \sum_i\ (\varepsilon_l,\ y_{ik}\ \varepsilon_i) = y_{lk} \ . \tag{9.26}$$

Also the square of the 'length' of our wavefunction is, with $f_i^*\ f_i \equiv |f_i|^2$,

$$(\phi,\phi) = \sum_{i,j}\ (f_i^*\ \varepsilon_i\ ,\ f_j\ \varepsilon_j) = \sum_i\ |f_i|^2 \ .$$

We shall assume that this quantity is normalized to unity. Thus $|f_i|^2$ is the quantum mechanical probability of finding the system whose wavefunction is $\phi$ in the state $\varepsilon_i$.

We now need two basic concepts.

(*a*) The hermitian conjugate $Y^+$ of an operator or matrix is defined by its elements

$$(Y^+)_{ji} = (Y)^*_{ij} \quad .$$

It follows that

$$(Y\phi,\psi) = \sum_{i,j,k} (y^*_{ij}\, f^*_j\, \varepsilon_i,\, p_k\varepsilon_k)$$

$$= \sum_{i,j} y^*_{ij}\, f^*_j\, p_i = \sum_{i,j,k} (f^*_j\, \varepsilon_j, (Y^+)_{ki}\, p_i\varepsilon_k) \quad (9.27)$$

whence

$$(Y\phi,\psi) = (\phi, Y^+\psi) \quad (9.28)$$

(*b*) A unitary operator $U$ is defined by $U^+U = I$, where $I$ is the identity operator and has the property $I\phi = \phi$ for all $\phi$. A unitary operator has the property of transforming an orthonormal basis $\{\varepsilon_j\}$ to another one $\{\eta_j\} = \{U\,\varepsilon_j\}$ :

$$(\eta_i,\, \eta_j) = (\varepsilon_i,\, U^+U\,\varepsilon_j) = (\varepsilon_i,\, \varepsilon_j) = \delta_{ij} \quad . \quad (9.29)$$

Also by (9.28) and (9.29)

$$(\varepsilon_l,\, Y\varepsilon_k) = (U^{-1}\eta_l,\, YU^{-1}\eta_k) = (\eta_l,\, UYU^{-1}\eta_k) \quad . \quad (9.30)$$

Thus in the new basis the matrix elements of $Y$ are equal to those of $Y_u \equiv UYU^{-1}$ and in that sense $Y$ is transformed to $Y_u$. It is easily verified that the trace of a matrix

$$\mathrm{Tr}\ Y = \sum_i y_{ii} \quad (9.31)$$

is invariant under unitary transformation:

$$\mathrm{Tr}\ (Y_u) = \mathrm{Tr}\ (UYU^{-1}) = \mathrm{Tr}\ Y \quad . \quad (9.32)$$

In the last step use has been made of

$$\mathrm{Tr}\ (AB) = \mathrm{Tr}\ (BA) \tag{9.33}$$

with $A \equiv UY$ and $B \equiv U^{-1}$.

*9.5.2. Density matrix and entropy*

We now have the tools to imagine an ensemble of systems $u = 1,2,\ldots,M$ in states $\phi^{(u)} = \Sigma_i f_i^{(u)} \varepsilon_i$ with expectation values of the variable $y$ defined by

$$y^{(u)} \equiv (\phi^{(u)},\ Y\phi^{(u)}) = \sum_{i,j} f_i^{(u)*}\ f_j^{(u)}\ y_{ij}$$

where (9.27) has been used. The ensemble average of $y^{(u)}$ is

$$\bar{y} \equiv \frac{1}{M} \sum_{u=1}^{M} y^{(u)} = \sum_{i,j} \rho_{ji}\ y_{ij} = \mathrm{Tr}\ (\rho Y) \tag{9.34}$$

where the density matrix $\rho$ is defined by its matrix elements

$$\rho_{ji} \equiv M^{-1} \sum_{u=1}^{M} f_i^{(u)*}\ f_j^{(u)}. \tag{9.35}$$

It depends on time $t$ via the $f_j$'s. Since $\rho$ transforms like $Y$ under a unitary transformation, $\bar{y}$ is invariant under such a transformation:

$$\mathrm{Tr}\ \{(U\rho U^{-1})(UYU^{-1})\} = \mathrm{Tr}\ (U\rho Y U^{-1}) = \mathrm{Tr}\ (\rho Y)\ .$$

Note that

$$\sum_i \rho_{ii} = M^{-1} \sum_{u=1}^{M} \sum_i |f_i^{(u)}|^2 = M^{-1} \sum_{u=1}^{M} 1 = 1$$

so that $\rho_{ii}$ is the probability that a system chosen arbitrarily from the emsemble will be in state $i$.

If $\rho$ is diagonal we can write for any power series $f(x)$ which is convergent in a certain domain an operator $R$ which is defined by its matrix elements in that representation

$$(R)_{ii} = f(\rho_{ii})\ .$$

The operator $R$ can then be the interpretation of the operator $f(\rho)$ .

In this way one can define an operator $\ln\rho$ and hence

$$S \equiv -k \,\mathrm{Tr}\,(\rho \ln\rho) \,. \tag{9.36}$$

The right-hand side is, for a diagonal density matrix,

$$-k \sum_{i,j} \rho_{ij} (\ln\rho)_{ji} = -k \sum_i \rho_{ii} \ln\rho_{ii}$$

and is the entropy. Definition (9.36) extends it to an arbitrary basis.

### 9.5.3. *Time dependence*

If $H$ be the Hamiltonian of the system, assumed not to depend on the time explicitly, then the time dependence is given by the Schrödinger equation

$$H\psi + \frac{\hbar}{i}\frac{\partial\psi}{\partial t} = 0$$

into which (9.23) has to be substituted:

$$\sum_s f_s H\varepsilon_s + \frac{\hbar}{i}\frac{\partial f_s}{\partial t}\varepsilon_s = 0 \,.$$

Taking the scalar product with $\varepsilon_j$,

$$\frac{\partial f_j}{\partial t} = -\frac{i}{\hbar}\sum_s f_s h_{js}$$

$$\frac{\partial f_k^*}{\partial t} = \frac{i}{\hbar}\sum_s f_s^* h_{ks}^* \,.$$

Hence, using the fact that $H$ is hermitian,

$$\frac{\partial f_j f_k^*}{\partial t} = -\frac{i}{\hbar}\sum_s [h_{js} f_s f_k^* - f_j f_s^* h_{sk}] \,.$$

Supplying the superscripts $(u)$, averaging over $u$, and using (9.35)

$$\frac{\partial\rho}{\partial t} = -\frac{i}{\hbar}(H\rho - \rho H) \,. \tag{9.37}$$

This is the quantum mechanical form of *Liouville's theorem*. It has the formal solution

$$\rho_t = U(t)^{-1}\rho_0 U(t), \quad U(t) \equiv \exp\left(\frac{iHt}{\hbar}\right) , \tag{9.38}$$

as may be verified by differentiation. The operator $U$ is defined by the exponential series. It is seen to be unitary: $U^{+} = U^{-1}$.

One can now see how a typical average (9.34) depends on time. We have

$$\frac{\partial \bar{y}}{\partial t} = \frac{\partial}{\partial t} \operatorname{Tr} (\rho Y) = \operatorname{Tr} \left(\frac{\partial \rho}{\partial t} Y\right)$$

$$= -\frac{i}{\hbar} \operatorname{Tr} (H\rho Y - \rho H Y) = \frac{i}{\hbar} \operatorname{Tr} (\rho [H,Y]) \tag{9.39}$$

where $[H,Y] = HY - YH$ and (9.33) has been used in the first term. One could define an operator $\dot{Y}$ by the condition

$$\frac{\partial}{\partial t} \bar{y} = \operatorname{Tr} (\rho \dot{Y}) \tag{9.40}$$

so that it is the operator which yields the time derivative of the ensemble average of the expectation value. One then finds from (9.39) and (9.40) a standard result of quantum mechanics, namely

$$\dot{Y} = \frac{i}{\hbar} (HY - YH) . \tag{9.41}$$

*9.5.4. The increase of entropy as due to coarse-graining.*

The entropy (9.36) is called the fine-grained entropy as it makes full use of quantum mechanics and presumes the states $\phi^{(u)}$ to be known for all systems $u$. The ensemble introduced for the definition of the density matrix is then really not needed. In that case the entropy is in fact independent of time. For one finds from (9.38) and (9.33) for any positive integral power $j$ of $\rho_t$

$$\operatorname{Tr} (\rho_t^j) = \operatorname{Tr} (U^{-1} \rho_0 U \ldots U^{-1} \rho_0 U) = \operatorname{Tr} (\rho_0^j) .$$

Since $\ln \rho_t$ can be expanded as a power series in $\rho_t$ it follows that

$$\operatorname{Tr} (\rho_t \ln \rho_t) = \operatorname{Tr} (\rho_0 \ln \rho_0) = \text{independent of time.} \tag{9.42}$$

This means that full quantum mechanical knowledge of the systems of an ensemble does not leave one with a mechanism for entropy increase. The statistical mechanical entropy increase must therefore depend critically on the incomplete information normally available about physical systems. For example, one may know merely that the $N$ states in a given system energy range $(E, E+\delta E)$ are accessible to the system of interest. The appropriate ensemble of systems which represent the system of interest must then contain systems in all $N$ states which are compatible with this energy range in such a way as to give each such state equal weight in the ensemble. The density matrix is in this case expressible in terms of the energy, and here states are lumped together without altering their number.

$$S_{\text{coarse}} > S_{\text{fine}},\ \rho_{ji} = \begin{cases} N^{-1}\ \delta_{ij} \text{ if } E_i \text{ is in } (E, E+\delta E) \\ 0 \qquad \text{otherwise} . \end{cases} \tag{9.43}$$

Thus $\rho$ commutes with the Hamiltonian and is by (9.37) time-independent. This example corresponds to the microcanonical equilibrium distribution and represents the simplest way of coarse-graining

*9.5.5. Reduced coarseness as leading to a greater entropy*

At the end of section 9.1 it was shown that

$$S(\text{AB}) = S(\text{A}) + S_{\text{A}}(\text{B}) \geq S(\text{A}) . \tag{9.44}$$

We can interpret $S(\text{A})$ as a coarse-grained entropy which recognizes states $A_1, A_2, \ldots, A_n$, but not states $B_1, B_2, \ldots, B_m$. A less coarse-grained analysis recognizes that the use of one state of the set A *and* one state of the set B is an improved (less coarse or finer) description of the system. One sees that the finer entropy exceeds the less fine entropy because the number of distinguishable states has been reduced.

$S_{\text{fine}} > S_{\text{coarse}}$, i.e. the micro-entropy exceeds the macro-entropy. (9.45)

In the limit when the coarse description recognizes only one state, then the system has to be in that state and the entropy is zero. One sees that entropy has a meaning only against a background of experimental measurement - at least this is the present author's view point.

A coarse-grained entropy is obtainable as

$$S = -k \sum_i P_i \ln P_i$$

from the fine-grained density matrix by averaging over groups of $g_i$ states which are considered indistinguishable:

$$P_i = \frac{1}{g_i} \sum_{j=1}^{g_i} \rho_{jj} \; .$$

One can then show that this (coarse-grained) entropy increases with time for an adiabatically isolated system. This result, known as the H-theorem, will not be proved here, but it is of great importance.

PROBLEMS

9.1. The probability of a system being in its ith state is $p_i (i=1,2,\ldots N)$ where $\Sigma p_i = 1$. Suppose that an extensive variable $x$ takes on the value $x_i$ when the system is in its ith state. Search for the maximum entropy distribution if the average value of $x$ ( $= \Sigma_{i=1}^{N} p_i x_i$) is known to have the given value $x_0$. Show that for $j=1,2,\ldots N$ and an undetermined multipler $\beta$, one finds

$$p_i = \frac{1}{Z(x)} \exp(-\beta x_i)$$

where $Z(x) \equiv \Sigma \exp(-\beta x_i)$ . Also show that

$$x_0 = -\left\{\frac{\partial \ln Z(x)}{\partial \beta}\right\}_{x_1, x_2, \ldots}$$

(see p. 189 for an application). Show also that

$$S = k\beta x_0 + k \ln Z \;(\equiv S_2) \; .$$

9.2. The extensive variables $x$ and $y$ take on values $\{x_j\}$ and $\{y_k\}$ respectively when a system is in a definite state. If the average values of $x$ and $y$ are fixed at $x_0$ and $y_0$ for the system, show that for the maximum entropy of the distribution function it has the form

$$p(x_j, y_k) = \frac{1}{\Xi} \exp(-\beta x_j - \gamma y_k) \qquad (j,k = 1,2,\ldots)$$

where $\Xi = \Sigma_{j,k} \exp(-\beta x_j - \gamma y_k)$, and $\beta$ and $\gamma$ are undetermined multipliers. Show also that

$$x_0 = -\left(\frac{\partial \ln \Xi}{\partial \beta}\right)_{x_1,\ldots y_N,\gamma}, \qquad y_0 = -\left(\frac{\partial \ln \Xi}{\partial \gamma}\right)_{x_1,\ldots y_N,\beta}$$

and

$$S = k\beta x_0 + k\gamma y_0 + k \ln \Xi \quad (\equiv S_3) .$$

9.3. A collection of copies of the system which are at any given time distributed over their states in proportion to the probabilities $p_j$ obtained in the preceding three problems is called an ensemble. Problem 9.1 described a canonical ensemble if $x$ is the internal energy of the system, and problem 9.2 a grand canonical ensemble if $x$ is the internal energy and $y$ the number of (identical) particles in the system. $Z$ and $\Xi$ are called partition functions. The term grand ensembles is sometimes used if only the mean total number of particles is fixed. In petit ensembles the total number of particles itself is fixed.

With this interpretation, verify that the following identifications are consistent with thermodynamics:

$$\beta \to \frac{1}{kT}, \quad \gamma \to -\frac{\mu}{kT}, \quad -kT \ln Z \to F, \quad kT \ln \Xi \to pv$$

where $F$ is the Helmholtz free energy, provided only that the statistical entropy and the thermodynamic entropy are identical.

9.4. A one-dimensional normal distribution of zero mean and standard deviation $\sigma$ is given by

$$\boxed{p(x) = (2\pi\sigma^2)^{-\frac{1}{2}} \exp\left(-\frac{x^2}{2\sigma^2}\right)} \quad (-\infty < x < \infty)$$

(*a*) Show that its entropy is $\frac{1}{2}k \ln (2\pi e\sigma^2)$, where e is the base of the natural logarithms.

(*b*) Show that for given $\boxed{\int_{-\infty}^{\infty} x^2 p(x)\,dx \equiv \sigma^2}$ the normalized probability distribution having the largest entropy is the one-dimensional normal distribution.

(Note that $p(x)$ is of dimension $[x^{-1}]$ and is called a *probability distribution*, while $p(x)dx$ is the *probability* of finding the variable in the range $(x, x+dx)$.)

9.5. The states of a quantum mechanical system are labelled by a complete set of quantum numbers. Suppose one of these determines its energy $E_j$ and that the corresponding energy degeneracy is $g_j (j = 1,2,\ldots,)$. Consider an ensemble of $N$ copies of this system in the sense of problem 9.3 . Let one State of such an ensemble be specified by the numbers $(n_1, n_2, \ldots)$, where $n_j$ is the number of systems with energy $E_j$. The capital S is a reminder that a State of an ensemble is considered.

(*a*) Prove that the number of ways of realizing this State is

$$G = \frac{N! g_1^{n_1} g_2^{n_2} \ldots}{n_1! n_2! \ldots} .$$

(*b*) If $n$ is large enough, check from a book on special functions that Stirling's approximation holds:

$$n! = \Gamma(1+n) = n^n e^{-n} (2\pi n)^{\frac{1}{2}}$$

where $\Gamma$ is the gamma function and $n$ is a positive integer.

(*c*) Make the continuity assumption with regard to $n$, and define Gauss's $\psi$ function by

$$\psi(n) \equiv \frac{d}{dn} [\ln \Gamma(1+n)] .$$

Prove from (*b*) that

$$\psi(n) \approx \ln(n+\tfrac{1}{2}) .$$

Verify by the use of tables of $\psi(n)$, that for $n \geq 3$ this approximation holds with an error of less than 0·26 per cent.

(*d*) Make an assumption of equal *a priori* probabilities of different ways of realizing a State, and assume that the energy of the ensemble is given. Hence show, using part (*c*), that for the most probable State of the ensemble the probabilities $n_j/N$ are given by

$$\frac{n_j}{N} = \frac{n_j^*}{N} \quad \frac{g_j \exp(-\beta E_j)}{\sum_j g_j \exp(-\beta E_j) + O(1/N)}$$

where $\beta$ is a Lagrangian multiplier.

(*e*) Discuss the relation between this result and that of problem 9.1.

9.6. Develop the ideas of problem 9.5 for an ensemble of $N$ identical systems each of which has two non-degenerate states. Take $N$ to be even, and specify a State of the ensemble by the number $n$ (which is even) giving the difference between the number of systems in the lower energy state and the number in the upper energy state. Assume that $N$, but not the total energy, of the ensemble is given.

(*a*) For $n \ll N$ show that the number of distinct ways of realizing a State $n$ of the ensemble is, within the Stirling approximation,

$$G(n) = 2^N \left(\frac{2}{\pi N}\right)^{\frac{1}{2}} \exp\left(-\frac{n^2}{2N}\right) .$$

(*b*) Assuming that distinct ways of realizing a State of the ensemble are equiprobable, show that the probability of a State $n(\ll N)$ is

$$P(n) = A\left(\frac{2}{\pi N}\right)^{\frac{1}{2}} \exp\left(-\frac{n^2}{2N}\right)$$

where $A$ is a normalization constant. Determine its value, and hence verify that $P(n)$ is a one-dimensional normal distribution of zero mean and standard deviation $\sqrt{N}$ (defined in problem 9.4) .

(*c*) Compare the mean value of $n$ and the most probable value of $n$.

(*d*) Compare the total number, $G_T$, of States of the ensemble with the number of ways $G(0)$ of realizing the most probable State as $N$ becomes very large. Repeat this process for the corresponding entropies, and use the result to discuss the key properties of the method of the most probable distribution for this example.

9.7. (*a*) A system is specified by variables $x_1, x_2, \ldots, x_N$ which have independent normalized probability distributions $p_1(x_1), p_2(x_2), \ldots, p_N(x_N)$. Show that the entropy may be written as

$$S = -k \sum_{i=1}^{N} \left\{ \int p_i(x_i) \ln p_i(x_i) \, dx_i \right\} .$$

(*b*) The $x_i$ are interpreted as the three Cartesian velocity components $V_i$ of a particle of a gas with point interactions (i.e., the particles are points and interact only if they are at the same point) in equilibrium at temperature $T$. Assuming that the mean kinetic energy associated with each component is $\frac{1}{2}kT$, show that for a state of maximum entropy

$$p_i(V_i) = \left(\frac{m}{2\pi kT}\right)^{\frac{1}{2}} \exp\left(-\frac{mV_i^2}{2kT}\right)$$

where $m$ is the mass of a molecule.

(*c*) Derive from (*b*) that the probability for a molecule in this gas to have a speed in the range $(V, V + dV)$ is $p(V)dV$, where $V$ can have any value from 0 to $\infty$ and

$$p(V) = 4\pi V^2 \left(\frac{m}{2\pi kT}\right)^{3/2} \exp\left(-\frac{mV^2}{2kT}\right) .$$

(This is the Maxwell velocity distribution.)

(*d*) Discuss the range of validity of these results by inspecting the assumptions needed to obtain them.

(*e*) Show that the mean kinetic energy of a particle is $\frac{3}{2}$ kT.

9.8. The 'fluctuation' of $X_k$ can be defined by

$$\Delta X_k \equiv X_k - \overline{X}_k$$

where the bar indicates an average over the equilibrium distribution (9.10). Partial derivatives $\partial/\partial\beta_r$ will be taken with the other $\beta$'s kept constant. Hence establish the following results:

(*a*) $$\frac{\partial \ln Q}{\partial \beta_r} = -\overline{X}_r$$

(*b*) $$\frac{\partial^2 \ln Q}{\partial \beta_r \partial \beta_s} = \overline{X_r X_s} - \overline{X}_r \overline{X}_s = \overline{\Delta X_r \, \Delta X_s} = -\frac{\partial \overline{X}_r}{\partial \beta_s} \ .$$

(*c*) The 'information' $-\ln p_{\mathbf{i}}$ of state **i** satisfies

$$\overline{-\ln p_{\mathbf{i}}} = \frac{\overline{S}}{k} \ .$$

(*d*) Its fluctuation is

$$\Delta(-\ln p_{\mathbf{i}}) = \sum_r \beta_r \, \Delta X_r \ .$$

(*e*) The mean square fluctuation in the information is

$$\overline{\Delta(-\ln p_{\mathbf{i}})\Delta(-\ln p_{\mathbf{i}})} = -\sum_{r,s} \beta_r \beta_s \frac{\partial \overline{X}_r}{\partial \beta_s} \ .$$

(See Chapter 16 for an elementary discussion of fluctuations) .

9.9. Let intensive variables $\gamma_r$ ($r = 1,\dots$) be used as follows:

$$\beta_1 \equiv \gamma_1 \equiv \frac{1}{kT} \ , \qquad \gamma_r \equiv -\frac{\beta_r}{\gamma_1} \qquad (r = 2,3,\dots)$$

in generalization of equation (9.18). By (9.13) the entropy $\overline{S}$ can be regarded as a function of the $\overline{X}_r$ ($r = 1,\dots$), $\overline{X}_1$ being the mean energy of the system. Let $(\ )_{\overline{X}}$ mean that the 'other' $\overline{X}$'s are kept constant. Show that

(*a*) $$\left(\frac{\partial \overline{S}}{\partial \overline{X}_s}\right)_{\overline{X}} = k\beta_s$$

(*b*) $$\left(\frac{\partial \overline{S}}{\partial \beta_s}\right) = k \sum_r \beta_r \left(\frac{\partial \overline{X}_r}{\partial \beta_s}\right)_\beta \ .$$

(*c*) $$C_\gamma \equiv T\left(\frac{\partial \overline{S}}{\partial T}\right)_\gamma = -\gamma_1 \left(\frac{\partial \overline{S}}{\partial \gamma_1}\right)_\gamma$$

$$\gamma_1 \left(\frac{\partial \beta_r}{\partial \gamma_1}\right)_\gamma = \beta_r \qquad (r = 1,2,\dots) \ .$$

(*d*) Show that the heat capacity $C_\gamma$ at fixed $\gamma_2, \gamma_3, \dots$ is equal to the mean square fluctuation of the 'information' $-\ln p_i$ multiplied by Boltzmann's constant.

9.10. A distribution function ${}_0p(x)$ satisfies

$$\int_{-\infty}^{\infty} p \, dx = 1, \quad \int_{-\infty}^{\infty} xp dx = a$$

and is constrained so that its $\nu$th moment is given:

$$M_\nu^\nu = \int_{-\infty}^{\infty} [x-a]^\nu p \, dx \ .$$

Show that the distribution of maximum entropy satisfies

$${}_0p(x) = B_\nu^{-1} \exp\left[-\frac{|x-a|^\nu}{\nu M_\nu^\nu}\right]$$

where $B_\nu \equiv 2\nu^{1/\nu} M_\nu \Gamma(1+1/\nu)$. Show also that the entropy $S$ of any other distribution of given $M_\nu$ satisfies

$$\exp\left(\frac{S}{k} - \frac{1}{\nu}\right) \le B_\nu \ .$$

(An interpretation in terms of $\nu$th-law rectification and a comparison between entropy and other measures of dispersion is possible. The case $\nu=2$ leads back to problem 9.4.)

# 10. Mean occupation numbers and negative temperatures

## 10.1. MEAN OCCUPATION NUMBERS

The grand canonical distribution (9.19) applies to a system without restriction as to its structure. In this section we consider the particles which constitute the system, and assume that the system consists of *identical* particles.

The hypothesis of *weak interactions* among the particles of the gas will be adopted, and has the advantage that it enables one to express any energy $E_i$ of a system as a sum $\Sigma_j\, n_{ij}\, e_j$. This sum extends over all single-particle quantum states $j$, $e_j$ being the corresponding single-particle energy, and $n_{ij}$ being the number of particles in state $j$ when the system is in its $i$th state. It follows that instead of specifying a state of the system (of given $T$, $v$, and $\mu$) by $E_i$ and $n$, it may be specified more completely by the independently variable occupation numbers $n_{i1}, n_{i2}$ ... or, dropping the subscript $i$, by $n_1, n_2$, ... . Each set of numbers $n_1, n_2$ ... clearly implies a definite system energy $\Sigma n_j e_j$ and a definite number of particles $\Sigma n_j$. It is also assumed here that, in accordance with quantum mechanics, identical particles are *indistinguishable*. If the particles were distinguishable, a state of the system would have to be specified by stating in which single-particle state each (labelled) particle is to be found. Equations (9.19) and (9.20) may now be written

$$P(v,\mu,T;E_i,n) = \exp\left(-\frac{pv}{kT}\right)\exp\left\{\frac{1}{kT}\left(\mu\sum_j n_j - \sum_j n_j\, e_j\right)\right\}.$$

The weak interaction hypothesis is crucial for the *statistical mechanical* (as against the thermodynamic) definition of an *ideal gas*. How weak must it be? Clearly if it is too weak the particles will move through each other and equilibrium cannot establish itself. We therefore adopt the fiction of point particles and *point interactions*, so that particles interact only on collision, but can then interact strongly.

Passing to a state specification by *single-particle state occupation numbers*, instead of total energy and total number of particles,

$$P(v,\mu,T;n_1,n_2,\ldots) = \exp\left(-\frac{pv}{kT}\right) \exp\left\{\frac{1}{kT}\sum_j (\mu-e_j)n_j\right\} .$$

It follows that

$$P(n_1,n_2,\ldots) = (t_1^{n_1}t_2^{n_2}\ldots)\exp\left(-\frac{pv}{kT}\right) \tag{10.1}$$

where

$$t_j \equiv \exp\left\{\frac{\mu-e_j}{kT}\right\} .$$

Equation (10.1) gives the probability of finding the system in a state in which the single-particle quantum states 1,2,... have the definite occupation numbers $n_1,n_2,\ldots$ respectively. The factor $\exp(-pv/kT)$ in (10.1) is a normalizing constant and may therefore be written in the form of a multiple sum over all admissible values of $n_1,n_2,\ldots$ . The convergence of such sums is assumed, and it will in fact be possible to estimate their values in all the cases of interest here. The multiple sum breaks up into a product of sums:

$$\boxed{\Xi \equiv \exp\frac{pv}{kT} = \left(\sum_{n_1=0} t_1^{n_1}\right)\left(\sum_{n_2=0} t_2^{n_2}\right)\ \cdots} \tag{10.2}$$

This is again the grand partition function, and the probability law (10.1) is said to specify the grand canonical distribution. It is sometimes convenient to think of a very large number of copies of the given system of fixed $T$, $v$, and $\mu$ as constituting a grand canonical ensemble. The systems in the ensemble are distinguished by different energies $E_i$ and particle numbers $n$ the distribution law of which is given by (10.1). It is probably more in accord with modern ideas about probabilities to use the distribution rather than the ensemble language, and we shall normally do so.

Only two cases are of interest. In the first a quantum state can contain either 0 or 1 particle. This is the case of *fermions* which are subject to the Pauli exclusion principles which is discussed in any book on quantum mechanics. It may be stated for our purposes as follows: no single-particle quantum state can be occupied by more than one particle. (In a more general statement, the principle asserts the antisymmetry of an $N$-particle wavefunction if the variables (co-ordinates and spin) of two particles are interchanged.) The other case refers to *bosons* for which the wavefunction is symmetrical in the particles. For our purposes it is sufficient to suppose that any number of bosons may be in the same quantum state. Hence

$$\begin{aligned} n_j &= 0 \text{ or } 1 \quad \text{(fermions)} \\ n_j &= 0, 1, 2, \ldots \quad \text{(bosons)} \end{aligned} \tag{10.3}$$

Hence (10.2) yields

$$\Xi = \Pi (1 \pm t_j)^{\pm 1} \tag{10.4}$$

where the top signs refer to fermions and the bottom signs to bosons.

We now establish the interesting general formula for an arbitrary system described by a grand canonical distribution

$$\boxed{-\frac{\partial \ln \Xi}{\partial \eta_j} = \bar{n}_j \,, \quad \eta_j \equiv \frac{e_j}{kT} \,.} \tag{10.5}$$

Here $\bar{n}_j$ is the mean single-particle occupation number of state $j$, whose energy is $e_j$. This formula is independent of the special result (10.3); it holds for *any* system for which the energy can be written as $\Sigma n_j e_j$.

The left-hand side of (10.5) is by (10.2)

$$\begin{aligned} -\frac{1}{\Xi}\left(\frac{\partial \Xi}{\partial \eta_j}\right)_{\mu,T} &= \frac{1}{\Xi} \sum_{n_1} \sum_{n_2} \cdots \sum \exp\left(\frac{\mu \Sigma_i n_i}{kT}\right) n_j \exp(-\sum_i n_i \eta_i) \\ &= \sum_{n_1} \sum_{n_2} \cdots \sum n_j P(n_1, n_2, \ldots) \qquad \text{(by (10.1))} \\ &= \sum_{n_j} n_j P_j(n_j) = \bar{n}_j \end{aligned}$$

in agreement with (10.5). We have here used the fact that the probability of finding an occupation number $n_j$ of state $j$ is

$$\sum_{n_1}\sum_{n_2}\cdots\sum_{n_{j-1}}\sum_{n_{j+1}}\cdots P(n_1,n_2,\ldots) \equiv P_j(n_j). \qquad (10.6)$$

The average value of $n_j$ is then clearly $\Sigma_{n_j} n_j P_j(n_j)$.

Equation (10.6) can be illustrated by considering the probability $P(\mathrm{A}_j,\mathrm{B}_i)$ that an $N_\mathrm{A}$-sided and an $N_\mathrm{B}$-sided die each comes up with any one particular score, say $\mathrm{A}_j$ and $\mathrm{B}_i$. This probability is $1/N_\mathrm{A}N_\mathrm{B}$. It follows that the probability of finding $\mathrm{A}_j$, whatever the B score, is clearly

$$P(\mathrm{A}_j) = \sum_{\nu=1}^{N_\mathrm{B}} \frac{1}{N_\mathrm{A}N_\mathrm{B}} = N_\mathrm{B}\frac{1}{N_\mathrm{A}N_\mathrm{B}} = \frac{1}{N_\mathrm{A}} \qquad (10.7)$$

which is a reasonable conclusion satisfying $\Sigma_j P(\mathrm{A}_j) = 1$. Equation (10.7) illustrates the result (10.6) in the form

$$\sum_i P(\mathrm{A}_j,\mathrm{B}_i) = P(\mathrm{A}_j) \ .$$

Applying the result (10.5) to the partition function (10.4), we find the Fermi-Dirac and Bose-Einstein distribution laws:

$$\boxed{\bar{n}_j = \mp\frac{\partial}{\partial\eta_j}\sum_i \ln(1\pm t_i) = \frac{t_j}{1\pm t_j} = \frac{1}{t_j^{-1}\pm 1}} \ . \qquad (10.8)$$

Note that the result (10.8) is exact within the framework of the grand canonical distribution. Most other derivations - for example those based on the canonical distribution function or those based on the microcanonical distribution - involve mathematical approximations such as Stirling's approximation for factorials or the approximations involved in the method of steepest descents.

If the discrete set of values $e_j$ is replaced by a continuous variable $e$ which gives the energy of a single-particle state, (10.8) states that the mean occupation number of this state is

$$\bar{n} = \begin{cases} \dfrac{1}{(\lambda^{-1} \exp \eta + 1)} & \text{(FD)} \\ \dfrac{1}{(\lambda^{-1} \exp \eta - 1)} & \text{(BE)} \end{cases} \tag{10.9}$$

where

$$\lambda \equiv \exp\left(\frac{\mu}{kT}\right), \quad \eta \equiv \frac{e}{kT} \quad . \tag{10.10}$$

The Fermi-Dirac expression drops from the value unity at the lowest negative energies to the value ½ at $e = \mu$, and it then goes exponentially to zero at high energies. The Bose-Einstein expression has the value infinity at $e = \mu$, and also drops exponentially to zero at high energies. No energy level can lie below the value $\mu$ in this case.

## 10.2. LOW- AND HIGH-TEMPERATURE LIMITS

The chemical properties of elements are largely dominated by the outer electron shells so that an atom with two outer electrons is called a helium atom whether the nucleus consists of two protons and two neutrons or of two protons and one neutron; in both cases two electrons are needed to make the atom neutral. These two forms of helium are called *isotopes* of helium, and they are denoted by the symbols $^4$He and $^3$He respectively. Now a nucleus with an even number of nucleons (protons or neutrons) satisfies Bose statistics, while it satisfies Fermi statistics if it has an odd number of nucleons. One sees from this observation and (10.9) that these two isotopes will have very different *low-temperature* properties. The atoms of $^4$He are expected to crowd into the lowest single-nucleon energy level, while those of $^3$He will spread themselves over a wide range of levels. It is believed that some of the peculiar characteristics (e.g. superfluidity) of $^4$He below 2.17K are due to the Bose statistics; $^3$He does not exhibit them at these temperatures. Note that $^3$He and $^4$He are both liquid under their own saturated vapour pressure at the lowest temperatures yet achieved (see also p. 317).

At *high temperatures* the particles are found from (10.9) to be spread more uniformly over the available energy levels than at low temperatures. This leads one to expect (see Chapter 9) an increase of entropy with temperature when the volume is kept constant - a purely thermodynamic characteristic already encountered in problem 6.1 . The interpretation of this principle in terms of our statistical mechanical model presumes of course that an infinity of single-particle levels are available. One would expect this result to be modified if the number of such levels is finite or if there is an upper bound to the single-particle energies, for as more and more particles tend to move into the highest part of the single-particle energy spectrum one would then expect the entropy to fall as the temperature is increased further. This situation occurs in certain magnetic systems (see section 10.3).

Suppose the volume and the chemical potential (or the mean number of particles $\bar{N}$) to be fixed. Then, as the temperature is increased, one would expect an increasing fraction of the particles to be in 'high-energy states'. It is simplest to characterize such states by the requirement that the exponent $(e-\mu)/kT$ is large enough for the additive terms $\pm 1$ in (10.9) to be negligible. This neglect leads to a high temperature, or better high energy, approximation for the non-relativistic ideal quantum gas. The states covered by this approximation lie in the high-energy exponential tail of the distributions (10.9).

The above approximation has, however, another significance. Before the discovery of quantum statistics, the *Maxwell-Boltzmann distribution law* $\bar{n}(e) = C\exp(-e/kT)$ was normally used, $C$ being independent of energy. This shows that at sufficiently high temperatures there will be a tendency for an ideal quantum gas to exhibit the properties of a system described by the classical statistics. Given volume and temperature, one may obtain this *classical approximation* by thinking of the chemical potential as a large, negative quantity.

This discussion shows that, while high temperatures tend to lead to classical behaviour, there is a tendency for low temperatures to encourage the emergence of quantum phenomena. All the results of this section incorporate the various assumptions made before. These are as follows:

(i) Volume, temperature, and chemical potential of the whole system are given, so that the grand canonical distribution may be used.

(ii) The system consists of indistinguishable particles of one type only and with point interactions, so that states of the system can be specified by occupation numbers $n_1, n_2, \ldots$ .

## 10.3. OCCUPATION NUMBERS FOR LOCALISED SPINS

Consider magnetic atoms or dipoles or spins, each of magnetic moment $\mu$, localized on points in a finite crystal lattice of $N$ points. Let them be in equilibrium at a so-called spin temperature $T$. In the simplest case when the moment is capable of only two orientations the atoms will orientate themselves anti-parallel to an applied magnetic field $B$ (lower energy state 1) or parallel to the field (upper energy state 2). The energies of these states will be taken to be $\pm\mu B$, the energy gap opened up by the field being $2\mu B$. The energy zero has been taken as that of the two-fold degenerate level which arises if $B = 0$. Writing $x = \mu B/kT$ one finds, using the canonical distribution,

$$Z_N = (e^x + e^{-x})^N \qquad (10.11)$$

for the partition function in the absence of spin-spin interaction. The occupation probability of the states is

$$P_1 = \frac{\exp x}{Z_1}\,, \qquad P_2 = \frac{\exp(-x)}{Z_1}\,. \qquad (10.12)$$

The mean occupation numbers of these states are $NP_1$ and $NP_2$. The internal energy is

$$U = N\mu B(-P_1 + P_2) = -\frac{\mu NB}{Z_1}(e^x - e^{-x})\,.$$

The magnetic dipole moment $M$ is due to the moments arising from anti-parallel spins, reduced by that due to the parallel spins, and is given by

$$M = N\mu\ (P_1 - P_2) \tag{10.13}$$

so that

$$MB = -U = \mu NB \tanh\left(\frac{\mu B}{kT}\right) . \tag{10.14}$$

This is the ideal quantum gas relation $pv = gU$ again (with $g = 1$), provided, as noted in connection with equation (7.29), one replaces $pv$ by $-MB$. The entropy is obtained in the usual way as

$$\begin{aligned} S &= -Nk(P_1 \ln P_1 + P_2 \ln P_2) \\ &= Nk \ln Z_1 - \frac{N\mu B}{T}\,\frac{e^x - e^{-x}}{e^x + e^{-x}} . \end{aligned} \tag{10.15}$$

We now make two observations about this model:

(i) For small fields (10.14) yields the *Curie law* for paramagnetic solids

$$\frac{M}{B} = \frac{C}{T} \qquad C \equiv \frac{N\mu^2}{k} \tag{10.16}$$

and it yields an isothermal susceptibility $\chi_T = C/T$. Note that paramagnets have $\chi_T > 0$ and are distinguished from diamagnetic materials for which $\chi_T < 0$.

(ii) In the model of stronger (ferromagnetic) magnetization, domains of magnetization exist and are due to strong interaction between the elementary magnets. If the ferromagnet, for example iron, is not magnetized it consists of these domains but they are in random directions so that the average is zero. The magnetization of the material is the result of lining up these domains. A strong molecular field proportional to $M$ is in fact supposed to act on each individual magnet thus providing a tendency towards alignment which is additional to that provided by the external field $B$:

$$B_T = B + D(P_1 - P_2) = B + \frac{DM}{\mu N} \tag{10.17}$$

where $D$ is called the molecular field constant.

The theory embodied in equations (10.11) to (10.16) applies if $B$ is replaced by (10.17). For example, (10.14) becomes

$$M = \mu N \tanh\left(\frac{\mu BN + DM}{NkT}\right) . \tag{10.18}$$

The value of $M$ which solves (10.18) is given by the intersection of two curves, both plotted against $y$,

$$z_1 = \tanh y , \quad z_2 = \frac{y}{\alpha} - \frac{B}{D} \tag{10.19}$$

where

$$y \equiv \frac{\mu NB + DM}{NkT} , \quad \alpha \equiv \frac{\mu D}{kT} \equiv \frac{T_C}{T} . \tag{10.20}$$

The *Curie temperature* $T_C$ introduced in (10.20) receives a simple interpretation for the special case when the molecular field dominates over the applied field, but the energy $DM/N$ of an individual magnet due to this field is still small compared with the energy $kT$ of thermal agitation:

$$\frac{\mu B}{kT} \ll \frac{DM}{NkT} \ll 1 . \tag{10.21}$$

In that case (10.19) yields $y - \frac{1}{3} y^3 = (y/\alpha) - (B/D)$, and this is

$$B = (T - T_C) \frac{D^2}{NkTT_C} M + \frac{D}{3}\left(\frac{D}{NkT}\right)^3 M^3 . \tag{10.22}$$

This equation was noted at the end of section 7.9 as giving zero spontaneous magnetization for $T > T_C$ and non-zero magnetization for $T_1 > T_C$, so that at $T_C$ there is a phase transition. Reversal of $M$ for $T < T_C$ yields a first-order phase transition.

It is in fact true, even without assumption (10.21), that spontaneous magnetization requires $T < T_C$. One merely puts $B = 0$ and notes from (10.19) that

$$\frac{dz_1}{dy} = \frac{1}{\cosh^2 y} < 1$$

whereas

$$\frac{dz_2}{dy} = \frac{1}{\alpha} = \frac{T}{T_C}$$

so that the two curves rise from $y = 0$ and do not intersect for $M \neq 0$ unless $T < T_C$.

The above discussion presumes particles of spin $\hbar/2$ to be localized at the lattice points. Higher spins are possible; the above theory has then to be generalized to equally spaced energy levels, and this can be done. In some cases the spins interact with each other more strongly than they interact with the lattice. This makes it possible to examine the spin temperature $T$ (which alone has featured in the above argument) separately from the lattice temperature. The left-hand halves of Figs. 10.1 and 10.2 show some of the thermodynamic functions for the case $n = 4$.

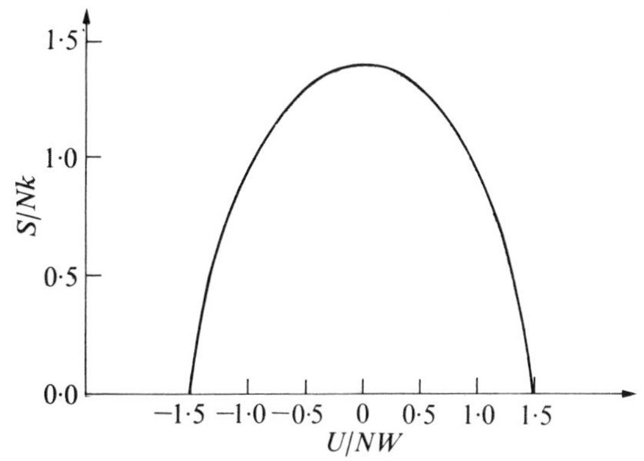


Fig.10.1. The entropy as a function of the internal energy for a system of N elements, each one having four equally spaced energy levels located at $-3/2W$, $-1/2W$, $1/2W$, $3/2W$. Here $W$ is energy-level spacing.

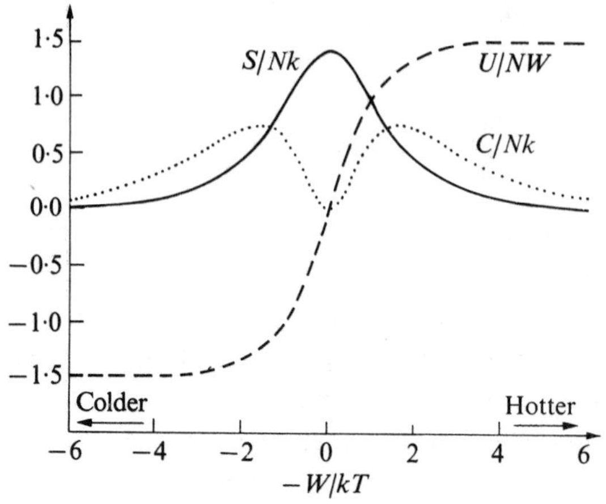


Fig.10.2. The internal energy, the entropy, and the specific heat are plotted as a function of $-W/kT$ for the same system as Fig.10.1; broken curve, internal energy $U$; solid curve, entropy $S$; dotted curve, specific heat $C$.

## 10.4. NEGATIVE TEMPERATURES

We shall now make a case for using the variable

$$\tau = -1/T \tag{10.23}$$

as one that measures 'hotness'. We first note that for the 'coldest' systems $\tau \to -\infty$, so that we would wish $\tau \to +\infty$ for the 'hottest' systems. For any system the addition of energy raises particles to higher levels, and it raises the temperature. Indeed (10.9) shows that there is a decrease in equilibrium occupation numbers as the energy of the quantum states increases, $\mu$ and $T$ being fixed. The hottest temperature on this basis is $T \to \infty$ or $\tau \to -0$ . If the energy spectrum has an upper bound, for example if the number of levels is finite, additional energy can be pumped into the system so as to take it beyond the stage of equiprobability, illustrated by $P_1 = P_2$ in (10.12) and $S = Nk\ln 2$ in (10.15). A *population inversion* must then be expected to correspond to a state hotter than $T = \infty$. The condition for this in the simple model of section 10.3 is $P_2 > P_1$, i.e.

$$\exp -\left(\frac{\mu B}{kT}\right) > \exp\left(\frac{\mu B}{kT}\right) , \quad \text{i.e. } T < 0 \quad \text{or} \quad \tau > 0 .$$

The maximum internal energy is obtained if $P_2 = 1$ and $P_1 = 0$, when the entropy is again zero. This corresponds to the limit as the negative temperatures tend to zero, $T \to -0$, and to $\tau \to +\infty$. This substantiates the claim made for (10.23).

The model of section 10.3 can be generalized to $n$ equally spaced levels, when the maximum entropy occurs at $T = \pm\infty$ or $\tau = 0$ and is $S = Nk \ln n$. In Figs. 10.1 and 10.2 negative temperatures occur on the right-hand halves.

Other generalizations include the following: (i) The third law stated in problem 6.7 as '$S(x,T)$ is finite, continuous, and unique as $T \to 0$' can be generalized to include the limits $T \to \pm 0, \pm \infty$. (ii) Instead of an upper bound to the energy, only the first moment $\int EN(E)dE$ of the density of states need be finite. (iii) More distant applications include the observation that the alignment of fish parallel or anti-parallel to the current, with turbulence in the water providing a randomizing effect, has statistical properties which may be expected to be analogous to those of a ferromagnet. Also the spreading of rumours, epidemics, and the role of agitators in society furnish further interesting examples.

Returning to Chapter 7, note that for negative temperatures the inequality (7.15) is reversed (problem 7.2), so that (7.16$a$) becomes $vK_S < 0$. Hence a simple fluid which can attain stable equilibrium at positive temperatures, which requires $vK_S > 0$, cannot do so at negative temperatures since neither $p$ nor $v$ can reverse sign. For magnetic systems $p \to B$ and $v \to -M$, where $B$ is the local magnetic field and $M$ is the magnetic moment, and inequality (7.15) becomes

$$(T_2 - T_1)(S_2 - S_1) + (B_2 - B_1)(M_2 - M_1) \geq 0 \quad (T > 0) .$$

Hence for stable equilibrium

$$\chi_T \propto \begin{cases} \left(\dfrac{\partial M}{\partial B}\right)_T > 0 & \text{for } T > 0 \\ \left(\dfrac{\partial M}{\partial B}\right)_T < 0 & \text{for } T < 0 . \end{cases} \tag{10.24}$$

Since the magnetic moment $M$ can reverse signs when the sign of $T$ is reversed (cf. equation (10.14)), a magnetic system can in principle be in stable equilibrium at positive *and* negative temperatures.

## *10.5. PHASE-SPACE STRUCTURE FOR A PARAMAGNET

Consider a paramagnetic material satisfying Curie's law and of internal energy

$$U(B,T) = -\frac{C(B^2+L^2)}{T} \quad \text{where} \quad M = \frac{CB}{T} . \tag{10.25}$$

Here $L$ is a constant which yields the energy for zero field

$$U(0,T) = -\frac{CL^2}{T} . \tag{10.26}$$

The entropy is given by

$$TdS(B,T) = \left\{\frac{\partial U(B,T)}{\partial T}\right\}_B dT = \frac{C(B^2+L^2)}{T^2} dT .$$

If $A$ be a constant of integration, it follows that (cf. Fig.10.3)

$$S(B,T) = A - \frac{C(B^2+L^2)}{2T^2} = A - \frac{M^2}{2C} - \frac{CL^2}{2T^2} . \tag{10.27}$$

This system is an ideal paramagnetic if $L = 0$. A non-zero value of $L$ can arise from the interaction between spins. The quasi-static adiabatic curves are shown in Figs. 10.4 and 10.5 and satisfy

$$B^2+L^2 = (2/C)(A-S)T^2 \quad \text{(Fig.10.4)}$$

$$(B^2+L^2)^{-1} = C\tau^2/2(A-S) \quad \text{(Fig.10.5)} .$$

We now deal with the sets $\gamma$, $F(\gamma)$, and $\beta$ for this system.

$\underline{\gamma}$

All boundary or frontier points $F(\beta)$ or $F(\gamma)$ of a proposed set of $\beta$ or $\gamma$ must be removed so as to leave one with an *open set* $\gamma$. Each point of $\gamma$ then has neighbourhoods lying entirely within $\gamma$. Hence the concept of an i-point can be defined, and an integrating factor for $d'Q$ can be deduced as seen in Chapter 5. In the present case $S(B,\infty) = A$ from (10.27)

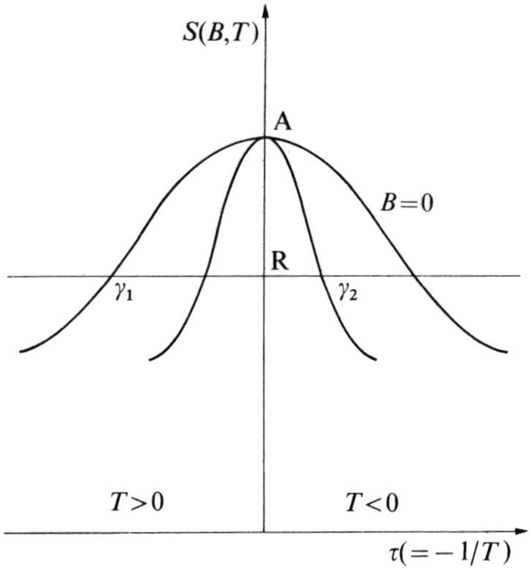


Fig.10.3. Thermodynamic phase space of spin system.

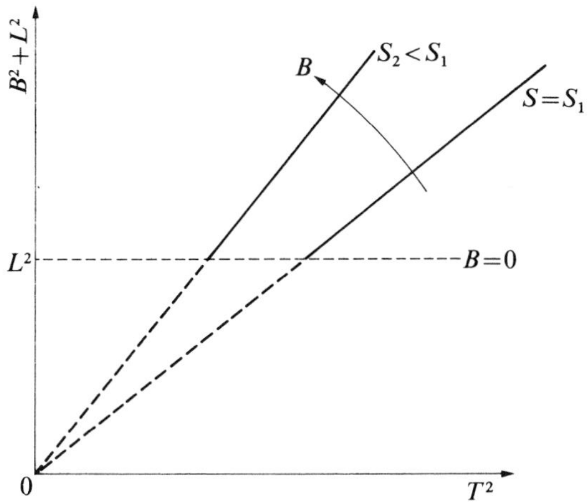


Fig.10.4. Alternative form of the phase space of Fig.10.3.

The point $A$ of Fig.10.3 is thus a member of $F(\gamma)$ as it can be reached only if $T \to \infty$. Similarly no point on the $\tau = 0$ axis can lie in $\gamma$, for these points, such as point R, cannot be reached adiabatically with finite fields. Fig.10.3 is thus decomposed into two sets $\gamma$, $\gamma_1$ and

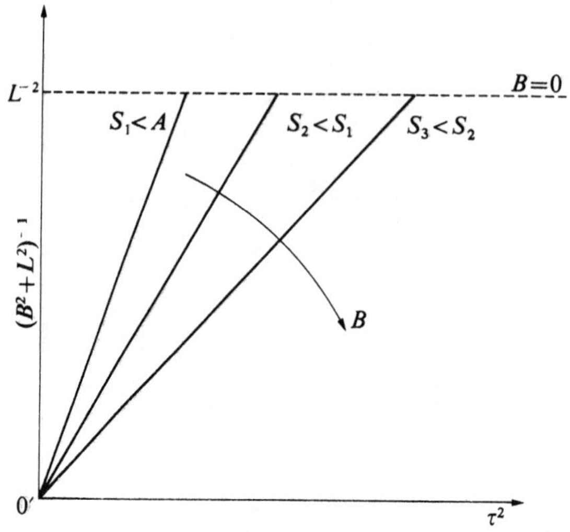


Fig.10.5. Alternative form of the phase space of Fig.10.4. The 'reduced' representation of this figure and of Fig.10.4 does not distinguish between positive and negative temperatures.

$\gamma_2$ say. Entropies can be associated with each point of $\gamma_1$ and $\gamma_2$, and quasi-static adiabatic changes are represented by curves in $\gamma_1$ and $\gamma_2$ as usual. However, these curves do not link $\gamma_1$ to $\gamma_2$ since the $\tau = 0$ axis is neither in $\gamma_1$ nor in $\gamma_2$ .

$F(\gamma)$

$F(\gamma)$ consists of the curve $B = 0$ in Fig.10.3, and part of the $(\tau=0)$-axis.

*Use of the third law*

In a generalized form the third law states that all points of a set $F(\gamma)$ may be counted as part of a set $\beta$, except for the points representing states at the absolute zero of temperature. This law enables us to regard part of the $(\tau=0)$-axis and the curve $B = 0$ as part of $\beta$. In this way $\gamma_1$ and $\gamma_2$ have been joined together at $\tau=0$ to form a single set $\beta$. The proper way to regard the third law is here illustrated; it determines the status of boundary points. It is then obvious that, having dealt with a second law in terms of sets $\gamma$, a third law is essential to remove all ambiguities.

$\underline{\beta}$
This set is now determined, except that nothing has been said about the mechanism whereby any two points of $\beta$ may be linked adiabatically. Such a mechanism must exist, otherwise we cannot claim to have a single set $\beta$. It cannot reside in quasi-static adiabatic linkage since this cannot operate between $\gamma_1$ and $\gamma_2$, as has been seen. The linkage between the sets *is* possible by means of non-static adiabatic processes. These have been studied in many experiments on spin systems using the 'adiabatic fast passage', described in the more specialized literature.

## *10.6. HEAT ENGINES AND HEAT PUMPS AT POSITIVE AND NEGATIVE ABSOLUTE TEMPERATURES

### *10.6.1. General considerations*

Following problems 5.5 and 5.6 , let constant temperature reservoirs h and c, at temperatures $T_h$ and $T_c$, *deliver* quantities of heat $Q_h$ and $Q_c$ during the isothermal parts of one cycle. Negative values of $Q_h$ or $Q_c$ mean that heat is given to the appropriate reservoir. The convention is made that h is 'hotter' than c in accordance with (10.23), so that $-1/T_h \geq -1/T_c$, i.e.

$$1/T_c - 1/T_h \geq 0 \quad . \quad . \tag{10.28}$$

However, equality $T_c = T_h$ is also allowed. The total system of reservoirs and medium is isolated so that its entropy cannot decrease. The working medium recovers its original state at the end of the cycle so as to be ready to execute the next cycle. The change of entropy is therefore that of the reservoirs and is in one cycle $-(Q_h/T_h + Q_c/T_c) \geq 0$ by the second law, so that

$$Q_h/T_h \leq -Q_c/T_c \ . \tag{10.29}$$

This cause of the inequality which resides in entropy-increasing dissipative processes depends on the spontaneous direction of heat flow. This direction is not obvious when the temperatures are not positive. However, it remains true in all cases that

$$\text{direction of spontaneous heat flow: h} \rightarrow \text{c} \ . \tag{10.30}$$

To see this, suppose some heat leaks spontaneously from h to c, subject to entropy increase, and determine its direction from (10.28) and (10.29). Hence we put $Q_h + Q_c = 0$ (heat lost by one reservoir is gained by the other). Now (10.29) shows that $Q_h/T_h \le Q_h/T_c$ so that

$$Q_h(1/T_c - 1/T_h) \ge 0 \ .$$

Using (10.28) it follows that $Q_h \ge 0$ and hence $Q_c \le 0$. Thus the hot reservoir gives up heat and the cold reservoir gains it. Note that nothing has been assumed here about the signs of $T_c$ and $T_h$, so that (10.30) is generally true. In fact *while the signs of* $Q_h, Q_c, T_h, T_c$ *are uncertain the inequalities must be manipulated with care*; the multiplication of an inequality by a negative number reverses the inequality sign!

If $l$ is the heat lost per cycle, for example by heat leakage from the reservoirs to the container walls, the work done by the medium per cycle is less than the heat $Q_h+Q_c$ given up by h and c:

$$W = Q_h + Q_c - l \qquad (l \ge 0) \ .$$

It will be assumed that $l$ does not exceed numerically any of the other three terms in this equation. In fact, to avoid algebraic complications it will be assumed to be always small enough so that its presence does not change inequalities that would hold if $l = 0$. Using (10.29), one finds

$$\frac{W+l}{T_c} \le \left(\frac{1}{T_c} - \frac{1}{T_h}\right)Q_h \tag{10.31}$$

$$-\frac{W+l}{T_h} \ge \left(\frac{1}{T_c} - \frac{1}{T_h}\right)Q_c \ . \tag{10.32}$$

A heat *engine* will be defined by the condition $W + l > 0$. It has an efficiency

$$\eta \equiv \frac{W}{(\text{all positive } Q)} \qquad (W + l > 0) \ . \tag{10.33}$$

If work has to be *supplied* to the medium ($W < 0$) to pump heat one has a heat *pump*, which is defined here by $W + l < 0$. It has a *coefficient of performance*

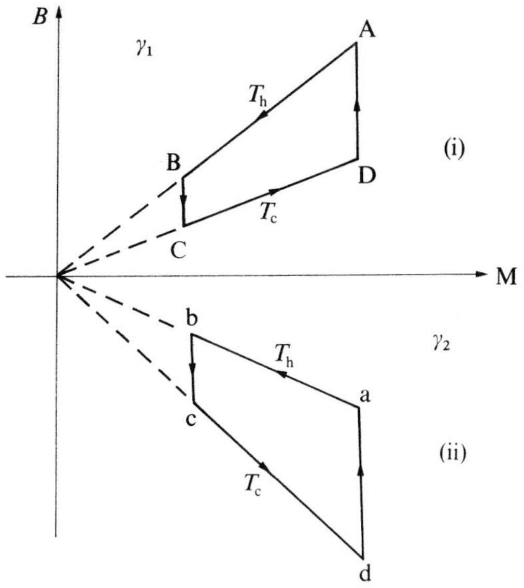


Fig.10.6. Work-delivering Carnot cycles for a Curie law paramagnet between (i) positive temperatures and (ii) negative temperatures.

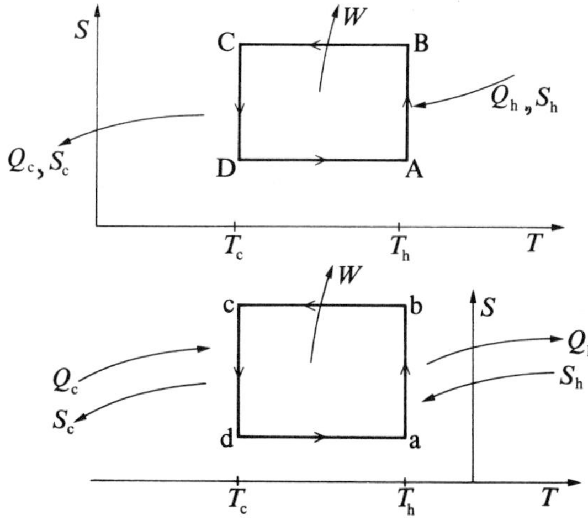


Fig.10.7. Entropy-temperature diagrams corresponding to Fig.10.6.

$$CP \equiv \frac{(\text{all negative } Q)}{|W|} \qquad (W + l < 0) \; . \tag{10.34}$$

Both $\eta$ and $CP$ are decreased by a heat leak, but whereas $\eta \leq 1$ the $CP$ can exceed unity, as is demonstrated in any case below.

Heat pumps include refrigerators (r) according to (10.34). The narrower definitions for $W + l < 0$

$$CP_r \equiv \frac{Q_c}{|W|} \; (Q_c > 0), \qquad CP_{hp} \equiv \frac{|Q_h|}{|W|} \; (Q_h < 0)$$

can of course be made, and lead for $Q_h < 0 < Q_c$ to

$$CP_{hp} - CP_r = \frac{-Q_h - Q_c}{-Q_h - Q_c + l} \leq 1 \; .$$

Of the heat pump situations covered by the general theory it turns out that none satisfies these narrower definitions, except for (10.36) below, which satisfies both. We therefore do not pursue these distinctions here.

Note that the cycles need not be quasi-static. They are largely arbitrary, except that the working fluid must recover its initial state at the end of a cycle. However, for an *ideal cycle* it is assumed to be quasi-static with $l = 0$.

### *10.6.2. Heat engine at positive temperatures*

On a $pv$-diagram the cycle proceeds in a clockwise sense and $W$ is the area enclosed since the work done by the working medium is $\int p dv$. In the magnetic case $pdv$ is replaced by $-Bd\mathcal{M}$ and on a $B\mathcal{M}$-diagram the cycle proceeds in an anticlockwise sense, the work done being again the area enclosed (Fig.10.6). By (10.31) and (10.32) we must have $Q_h \geq 0 \geq Q_c$, and the usual result $\eta \leq 1 - T_c/T_h$ is found. Entropy and heat are transferred from the hot to the cold reservoir.

For a Curie-law paramagnetic material (Fig.10.6) A → B is an isothermal demagnetization; B → C is an adiabatic reduction of the field. During this process $M$ is constant by virtue of equation (10.27) with $L = 0$. The temperature is thus lowered (Fig.10.7) from $T_h$ to $T_c$, where by (10.25)

$$B_B/T_h = B_C/T_c \; . \tag{10.35}$$

The cycle is completed by an isothermal magnetization C → D and an adiabatic increase of the field D → A. All processes shown are assumed to be quasi-static. This is, however, not a necessary restriction, as already noted.

A complete conversion of heat $Q_h$ into work $W$ can be achieved in an *ideal cycle* only if $Q_c = 0$, but this is ruled out by (10.32) as it would make $W$ negative. Hence there arises a famous formulation of the second law due to Kelvin and Planck which states that in an ideal cycle the heat withdrawn from one reservoir cannot be completely converted into work. It is violated by negative temperatures (see section 10.6.4) and has not been used in this book.

*10.6.3. Heat pumps at positive temperature*

The most important case in this subsection is that of a refrigerator, or a heat pump whch heats a dwelling at $T_h$ by using ambient heat at temperature $T_c$. In this first case

$$Q_h < 0 \le Q_c$$

where $Q_h = 0$ is ruled out to ensure that some heat is delivered and $W + l$ is negative. Since the work supplied can at positive temperatures be used to heat a fluid by irreversible stirring, the reservoir c is not needed, and all the work can be converted into heat ($CP = 1$) . In any case $Q_h < 0$ already ensures that entropy increases. Indeed after each cycle heat could be allowed to leak back from h to c, thus restoring c to its original state and reducing it to the role of a *catalyst* (which helps a process without being changed by it). A numerical example is

$$T_c = 20\text{ K},\ T_h = 40\text{ K},\ Q_c = 30\text{ J},\ Q_h = -70\text{ J},\ W = -40\text{ J},\ \Delta S = \frac{7}{4} - \frac{3}{2} = \frac{1}{4}$$

where $\Delta S$ is the entropy increase per cycle. By (10.31)

$$|W+l| \ge (1 - T_c/T_h)|Q_h|$$

so that

$$1 \le CP = \frac{|Q_h|}{|W|} = \left(1 - \frac{Q_c - l}{|Q_h|}\right)^{-1} \le \left|\frac{W+l}{W}\right| \frac{1}{1-T_c/T_h} . \tag{10.36}$$

If $T_c$ and $T_h$ lie close together little mechanical work is needed to transfer heat to the hotter region and the $CP$ rises. Broadly speaking, heat pumps become economically interesting for $CP \ge 4$.

*Secondly*, heat can be pumped into c:

$$Q_c < 0 \leq Q_h \ .$$

Here $Q_c = 0$ is ruled out again to ensure $W + l < 0$ . One finds (without the second law)

$$1 \leq CP = \left|\frac{Q_c}{W}\right| = \left(1 - \frac{Q_h - l}{Q_c}\right)^{-1} < \infty \ .$$

To ensure $W + l < 0$ one has $|Q_c| - Q_h + l > 0$ so that $CP$ must be finite. The complete conversion of work into heat is again possible, and the second law exercises no constraint.

*Thirdly*, $Q_c, Q_h \leq 0$, $Q_c + Q_h < 0$ is also possible.

To consider the use of heat pumps to heat the water d for a dwelling consider the system shown in Fig.10.8. The exhaust heat ($Q'_d$ per cycle) from a heat engine is used. The work $W$ per cycle from the engine is used to pump heat $Q_c$ from the (cool) ambient to the water at temperature $T_d$. For the heat engine part of the arrangement the efficiency is in an ideal

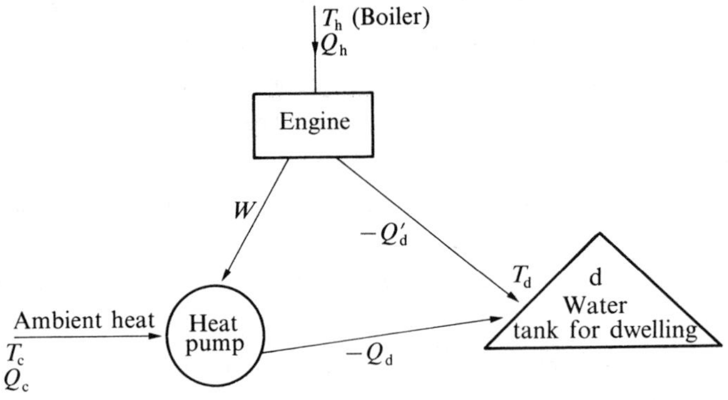


Fig.10.8. A coupled engine and heat pump for domestic water heating. The temperature of the condenser is $T_d$.

cycle

$$\eta = \frac{W}{Q_h} = 1 - \frac{T_d}{T_h} \ .$$

This yields an exhaust heat output

$$|Q'_d| = Q_h - W = \frac{T_d}{T_h} Q_h .$$

For the heat-pump arrangement the ideal coefficient of performance is, by (10.34),

$$CP = \left|\frac{Q_d}{W}\right| = \frac{1}{1-T_c/T_d} .$$

This yields an additional heat output

$$|Q_d| = \frac{W}{1-T_c/T_d} = \frac{(1-T_d/T_h)Q_h}{1-T_c/T_d} = \frac{T_d}{T_h}\frac{T_h-T_d}{T_d-T_c} Q_h .$$

It will be assumed that, typically,

$$T_h = 700 \text{ K}, \; T_d = 350 \text{ K}, \; T_c = 270 \text{ K}.$$

It follows that

$$|Q'_d|/Q_h = T_d/T_h \qquad \sim 0\cdot 5$$

$$|Q_d|/Q_h = \frac{T_d}{T_h}\frac{T_h-T_d}{T_d-T_c} \sim 2\cdot 19 .$$

Thus the heat pump is very effective and the performance of the whole system can be measured by the ratio of the heat energy delivered to the dwelling divided by the heat supplied by the boiler:

$$\frac{|Q_d| + |Q'_d|}{Q_h} \sim 2\cdot 7$$

*10.6.4. Heat engines at negative temperatures*

Three cases analogous to those noted in section 10.6.3 arise. In all of these heat is withdrawn from a negative-temperature reservoir, thus increasing its entropy (atoms drop back from their high-energy states). The second law does not then require the other reservoir, so that heat can be completely converted into work in an ideal cycle. The second-law inequalities require the efficiency to be *large* enough, since (10.32) yields

$$W + l \geq (1-T_h/T_c)Q_c \ .$$

Hence, *firstly*, if $Q_h \leq 0 < Q_c$ then

$$0 < 1 - \frac{T_h}{T_c} - \frac{l}{Q_c} \leq \frac{W}{Q_c} = \eta = 1 - \frac{|Q_h|+l}{Q_c} \leq 1 - \frac{|Q_h|}{Q_c} \ .$$

The value $\eta = 0$ is ruled out since for $W + l > 0$

$$W = Q_c - |Q_h| - l > 0 \ .$$

In an ideal cycle it is possible for $\eta = 1$ if $Q_h$ is zero. Alternatively, if the heat rejected into h is allowed to leak back to c after each cycle, then the effect is that $Q_c - |Q_h|$ is withdrawn from c and fully converted into work with $Q_h$ merely acting as a *catalyst*. A numerical example is

$$T_c = 40\ \text{K},\ T_h = 20\ \text{K},\ Q_c = 70\ \text{J},\ Q_h = -30\ \text{J},\ W = 40\ \text{J},\ \Delta S = \frac{7}{4} - \frac{3}{2} = \frac{1}{4} \ .$$

*Secondly*, for $Q_c \leq 0 < Q_h$,

$$0 < \eta = \frac{W}{Q_h} = 1 - \frac{|Q_c|-l}{Q_h} \leq 1 - \frac{Q_c}{Q_h} \leq 1 \ .$$

*Thirdly*, for $0 \leq Q_c,\ Q_h$ with $0 < Q_c + Q_h$,

$$0 < \eta < \frac{W}{Q_c+Q_h} = \frac{Q_c+Q_h-l}{Q_c+Q_h} \leq 1 \ .$$

In each case the Kelvin-Planck statement, noted in section 10.6.2, fails. It is also easily verified that it fails for heat engines working between temperatures of opposite sign. Fig.10.9 gives an example for a paramagnetic working material. This case, the analogous heat pump case, and heat pumps at negative temperatures will not be further discussed here.

### *10.6.5. Second Law formulations*

It is reasonable to define a perpetuum mobile *of the third kind* as one which withdraws heat from a negative-temperature reservoir and converts it completely into work. In the second kind of perpetuum mobile a positive-temperature reservoir is used, and this device cannot exist. From

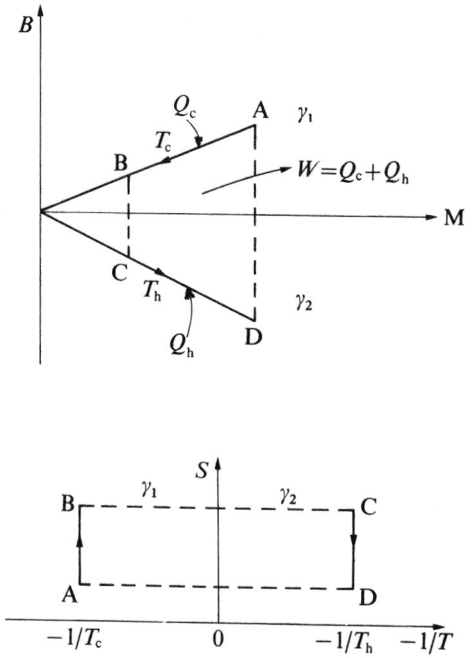


Fig.10.9. Magnetic field B, magnetic moment, entropy, and temperature for an engine operating between temperatures of opposite sign.

what has been learnt in section 10.6.4 one sees that an easily understood modification of the Kelvin-Planck formulation of the second law is: Of the three kinds of perpetuum mobile only the third kind is possible. Or, more explicitly: *Heat can be completely converted into work by taking a medium through a cyclic process if, and only if, that heat is withdrawn from a negative-temperature reservoir.* Unfortunately this third kind of perpetuum mobile has not as yet been turned to use owing to the limited life of negative-temperature states.

PROBLEMS

10.1. Without using equation (10.5), obtain the distributions (10.8) directly from the formula (10.1) by first proving the formula

$$\bar{n}_k = \frac{\left(\Sigma_{n_1} t_1^{\,n_1}\right)\left(\Sigma_{n_2} t_2^{\,n_2}\right) \cdots \left(\Sigma_{n_k} n_k t_k^{\,n_k}\right) \cdots}{\left(\Sigma_{n_1} t_1^{\,n_1}\right)\left(\Sigma_{n_2} t_2^{\,n_2}\right) \cdots \left(\Sigma_{n_k} t_k^{\,n_k}\right) \cdots} .$$

10.2. Writing $\bar{n}(e)$ for the mean Fermi-Dirac occupation number of a state of energy $e$, show that the line joining the values $\bar{n}(\mu-2kT)$ and $\bar{n}(\mu+2kT)$ when plotted against energies $e$ has the slope $-0{\cdot}19/kT$.

10.3. Prove that for fermions and for arbitrary energy $e$

$$n(e) + n(e') = 1$$

provided $e' = 2\mu - e$.

10.4. The *principle of detailed balance* asserts that in thermal equilibrium a process and its inverse occur with equal frequency. Apply this idea to a transition in which two fermions (or bosons) make transitions $1 \to 1'$, $2 \to 2'$ in such a way that the energies of these four states satisfy $e_1 + e_2 = e_{1'} + e_{2'}$ (so that this transition is energetically permitted). Hence show from the reaction

$$1,2 \rightleftarrows 1',2'$$

that the following factors contribute to the reaction rate, as discussed in Chapter 8, $n$ being the mean occupation number of a state.

TABLE 10.1

| | Fermion | Boson |
|---|---|---|
| From an initial state | $n$ | $n$ |
| From a final state | $1-n$ | $1+n$ |

and that the Bose distribution implies in addition that

$$(A_{IJ})_0 = (B_{IJ})_0 \ .$$

(The 'new' term is the induced or stimulated emission introduced by Einstein in 1917. The coefficients are called the Einstein $A$-$B$-coefficients, although they are sometimes defined a little differently.)

(The principle of detailed balancing is very important as can b
seen from the application of it which is explained in problem 1

10.5. Transitions with the emission of photons of energy $h\nu = E_I - E_J$ place by fermions dropping from quantum states $I$ (in some group states $i$) of energy $E_I$ to quantum states $J$ (in some group of st $j$) of energy $E_J$. The upward and downward rates for given $I$ and are respectively $p_J\, S_{JI}\, q_I$ and $p_I\, S_{IJ}\, q_J$ where $S_{IJ}$ is the trans probability $I \to J$ per unit time, $p_I$ is the occupation probabili of the single-fermion state $I$, and $q_I$ is the probability of sta $I$ being empty. Show that in thermal equilibrium at temperature denoted by subscript 0,

$$\left(\frac{S_{JI}}{S_{IJ}}\right)_0 = \exp\left(\frac{E_J - E_I}{kT}\right) .$$

(Assume either the Maxwell-Boltzmann distribution or the Fermi-Dirac distribution, $q_I = 1 - p_I$, and apply the principle of det balance.)

10.6. Let $A_{IJ}$, $B_{IJ}$ be quantum mechanical probabilities per unit time are independent of temperature and the mean number $N_\nu$ of photo of frequency $\nu$ present in the material. For absorption, set $S_{I}$ proportional to $N_\nu$ so that

$$S_{IJ} = A_{IJ} , \quad S_{JI} = B_{JI} N_\nu .$$

Show, by considering the limit $T \to \infty$ and using the results of t preceding problem, that this analysis of radiative transitions runs into trouble on physical grounds.

10.7. Show that the difficulty encountered in the preceding problem r be overcome by the introduction of an additional emission rate, that we have now

$$S_{IJ} = A_{IJ} + B_{IJ} N_\nu, \quad S_{JI} = B_{JI} N_\nu .$$

Hence show that

$$(B_{IJ})_0 = (B_{JI})_0$$

# 11. The continuous spectrum approximation

## 11.1. THE GENERAL PROCEDURE

We shall wish to consider more closely the expressions

$$N \equiv \sum_j \bar{n}_j = \sum_j \frac{1}{t_j^{-1} \pm 1} \tag{11.1}$$

$$F = \mu N - pv = \mu N \mp kT \sum_j \ln(1 \pm t_j) \tag{11.2}$$

for the mean of the total number of particles ($N$) and for the free energy ($F$) respectively.

In order to obtain expressions in closed form for the thermodynamic functions from (11.1) and (11.2), it is convenient to replace summations by integrations. In order to do so it will be assumed that the number of single-particle quantum states lying in the energy range ($e$, $e + \mathrm{d}e$) can be approximated by the function $Ae^s \mathrm{d}e$. Here $A$ is independent of the energy, and it will be assumed to be independent of temperature and proportional to volume; $s$ is a constant. Some density of state functions will be investigated later, when it will be seen that these assumptions are indeed fulfilled in many important cases, and in most of the situations envisaged in the preceding sections.

The above two equations now become

$$N = A(kT)^{s+1} \int_0^\infty \frac{x^s \mathrm{d}x}{\exp(x-\mu/kT) \pm 1} \tag{11.3}$$

$$F = \mu N \mp (kT)^{s+2} A \int_0^\infty x^s \ln[1 \pm \exp\{(\mu/kT) - x\}] \mathrm{d}x \; . \tag{11.4}$$

It is convenient to introduce the following integrals:

$$\boxed{I(c,s,\pm) \equiv \frac{1}{\Gamma(s+1)} \int_0^\infty \frac{x^s \mathrm{d}x}{\exp(x-c) \pm 1} \; .} \tag{11.5}$$

Using these, one finds that provided $s > -1$

$$N = A(kT)^{s+1}\Gamma(s+1)I\left(\frac{\mu}{kT},s,\pm\right) \tag{11.6}$$

$$F = \mu N - A(kT)^{s+2}\Gamma(s+1)I\left(\frac{\mu}{kT},s+1,\pm\right) \quad . \tag{11.7}$$

In these equations use has been made of the gamma function (p.436)

$$\Gamma(s+1) \equiv \int_0^\infty x^s \exp(-x)\mathrm{d}x \ , \tag{11.8}$$

whose properties are discussed in any book on the calculus. The derivation of (11.6) is immediate, and equation (11.7) is obtained by a partial integration from (11.4). This yields

$$F = \mu N \mp (kT)^{s+2}A\left\{\left|\frac{x^{s+1}}{s+1} \ln\{1 \pm \exp(\gamma - x)\}\right|_0^\infty - \frac{1}{s+1}\int_0^\infty x^{s+1} \frac{\mp\exp(\gamma-x)}{1\pm\exp(\gamma-x)} \mathrm{d}x\right\}$$

$$= \mu N - \frac{A}{s+1}(kT)^{s+2}\int_0^\infty \frac{x^{s+1}\mathrm{d}x}{\exp(x-\gamma)\pm 1}$$

where $\gamma \equiv \frac{\mu}{kT}$ as required.

It will be clear that the thermodynamic properties of the ideal gas can now be derived by differentiation. One needs no more than (11.6) and (11.7) together with the standard thermodynamic relations. This will be demonstrated on p. 185. The integrals (11.5) have been tabulated for the more important values.

## 10.2. SIGNIFICANCE OF THE MEAN TOTAL NUMBER OF PARTICLES: THE CANONICAL DISTRIBUTION

Equation (11.6) makes it clear that for a physical system at a given volume at a given temperature, and having a given set of single-particle quantum states, the chemical potential determines the mean number of particles $N = \bar{n}$ in the system (averaged over a grand canonical distribution). Conversely $N$ determines $\mu$. One may therefore suppose $v$, $T$, and the average number $N = \bar{n}$ to be given. If the treatment by the grand canonical distribution gives rise to small fluctuations in the total number $n$ of particles, the situation described approximates closely to one for which $v$, $T$, and $\mu = \mu_1$ are not given, but for which $v$, $T$, and the total number of particles $n = n_0$, say, are given, provided that $\mu_1$ is a value which gives rise to a mean number $N = \bar{n}$ which lies very close to $n_0$. Under these conditions one may expect a grand canonical distribution to represent to a high approximation a closed physical system. The small fluctuation in the total number of particles may then be regarded as brought about by the sticking of particles, or the formation of adsorbed films, on the surfaces of the container.

Suppose one goes a step further, and supposes one's physical system is such that the total number of particles in it (and free of the container walls) is rigorously constant. The probability law (9.21) then applies. This is the *canonical probability distribution.*

In expression (9.13) for the entropy the summation over $j$ is reduced to just one term. Hence one has for a canonical distribution:

$$\left.\begin{aligned} P(E_i) &= \frac{\exp(-E_i/kT)}{Z}, \qquad Z = \sum_i \exp(-E_i/kT), \\ S &= -k \sum_i P(E_i) \ln P(E_i) . \end{aligned}\right\} \quad (11.9)$$

The question of how the thermodynamic properties of a general system may be derived from a canonical distribution can be dealt with by a method which is given on page 137. This close analogy is, however, somewhat misleading because the exact theory of the ideal quantum gas is in fact mathematically simpler if it is developed on the basis of the grand canonical ensemble (see section 10.1).

10.3. PROPERTIES OF THE FERMI AND BOSE INTEGRALS $I(c,s,\pm)$

Consider the integral

$$\frac{1}{\Gamma(s+1)} \int_0^\infty x^s \left\{ \frac{1}{\exp(x-c)-1} - \frac{2}{\exp\{2(x-c)\}-1} \right\} dx . \qquad (11.10)$$

Since $1/(a-1) - 2/(a^2-1) = 1/(a+1)$ this is a Fermi integral. In contrast, by changing the variable in the second integral from $x$ to $y = 2x$, it can be converted to a Bose integral, whence it is seen that

$$I(c, s, -) - 2^{-s} I(2c, s, -) = I(c, s, +) . \qquad (11.11)$$

It is convenient to note here two other properties of the these integrals:

(*a*) In the classical approximation

$$I(c, s, \pm) = \exp c \qquad \text{for all } s \qquad (11.12)$$

(*b*) $$\boxed{\frac{\partial}{\partial c} I(c, s+1, \pm) = I(c, s, \pm) \quad \text{for} \quad s > -1 \ .} \qquad (11.13)$$

Result (*a*) follows immediately from (11.5) and (11.8). Result (*b*) is obtained by differentiating with respect to $c$ under the integral sign, writing the result in the form

$$\frac{\mathrm{d}I(c,s+1,+)}{\mathrm{d}c} = \frac{1}{\Gamma(s+2)} \int_0^\infty x^{s+1} \frac{\mathrm{d}}{\mathrm{d}x} \left\{\frac{-1}{\exp(x-c)\ 1}\right\} \mathrm{d}x \qquad (11.14)$$

and integrating partially.

**10.4**. VERIFICATION THAT THE STATISTICAL MODEL SPECIFIES A CLASS OF IDEAL QUANTUM GASES

One can now return to equations (11.6) and (11.7) for the free energy of the statistical mechanical system under investigation, and obtain the thermodynamic functions. For simplicity the *reduced chemical potential*

$$\gamma \equiv \mu/kT \qquad (11.15)$$

will be used.

It is helpful to note that upon differentiating (11.6) first at constant volume and at constant $N$ and secondly at constant temperature and at constant $N$, one finds simple relations for the corresponding derivatives of the reduced chemical potential, use being made of the fact that $(\partial N/\partial x)_N = 0$. One finds, by (11.13)

$$\left(\frac{\partial\gamma}{\partial T}\right)_{v,N} = -\frac{(s+1)I(\gamma,s,\pm)}{TI(\gamma,s-1,\pm)} \tag{11.16}$$

$$\left(\frac{\partial\gamma}{\partial v}\right)_{T,N} = -\frac{I(\gamma,s,\pm)}{vI(\gamma,s-1,\pm)} \tag{11.17}$$

provided $s > 0$. The entropy $-(\partial F/\partial T)_{v,N}$ is found from (11.6) and (11.7) for $s > 0$ as

$$S = -N\left(\frac{\partial\mu}{\partial T}\right)_{v,N} + (s+2)kN\frac{I(\gamma,s+1,\pm)}{I(\gamma,s,\pm)} + A(kT)^{s+2}\Gamma(s+1)I(\gamma,s,\pm)\left(\frac{\partial\gamma}{\partial T}\right)_{v,N}$$

$$= kN\left\{(s+2)\frac{I(\gamma,s+1,\pm)}{I(\gamma,s,\pm)} - \gamma\right\} . \tag{11.18}$$

More interesting in the present context is that the pressure exerted by the gas is

$$p = -\left(\frac{\partial F}{\partial v}\right)_{T,N} = \frac{NkT}{v}\frac{I(\gamma,s+1,\pm)}{I(\gamma,s,\pm)} . \tag{11.19}$$

The internal energy is found to be

$$U = F + TS = (s+1)NkT\frac{I(\gamma,s+1,\ )}{I(\gamma,s,\ )} . \tag{11.20}$$

Note the remarkable fact that we have in (11.19) a proof of Boyle's law if the classical approximation is used. Furthermore, the defining equation for the ideal quantum gas, $\boxed{pv = gU,}$ is satisfied by the system under investigation, provided

$$\boxed{g = \frac{1}{s+1} .} \tag{11.21}$$

It is seen therefore that the Grüneisen constant $\Gamma = g$ of the ideal quantum gases receives a very simple microscopic interpretation for the model investigated in this section. As $g$ increases to 1 the equation of state $p = g(U/v)$ is said to become 'stiffer'.

Since the general thermodynamic theory of problem 6.6 must apply to the present statistical model, it is desirable to verify some of its general relations. Note in the first place that by (11.6) $\mu/kT$ is a function only of $vT^{s+1}$, since

$N$ is to be regarded as a constant and $A$ is proportional to the volume. Now from (11.20) $U/T$ is a function only of $\gamma$ and therefore only of $vT^{s+1}$. Hence $Uv^{1/(s+1)}$ is also a function only of $Tv^{s+1}$, as required by the thermodynamic result $Uv^g = f(Tv^g)$ .

## 11.5. THE HEAT CAPACITIES

These must be calculated rather carefully. $C_v$ may be obtained from the internal energy (11.20), using (11.16) to find

$$C_v = \frac{U}{T}\left\{s+2-(s+1)\,\frac{I^2(\gamma,s,\pm)}{I(\gamma,s+1,\pm)I(\gamma,s-1,\pm)}\right\} . \qquad (11.22)$$

Note that the last term has to be omitted for systems for which the chemical potential is identically zero.

For the heat capacity at constant pressure, it is convenient to note from (11.16) that

$$\left(\frac{\partial\gamma}{\partial T}\right)_{p,N} = -\frac{I(\gamma,s,\pm)}{I(\gamma,s-1,\pm)}\left\{\frac{1}{v}\left(\frac{\partial v}{\partial T}\right)_{p,N} + \frac{s+1}{T}\right\} . \qquad (11.23)$$

Also the enthalpy is

$$H = U + pv = (s+2)A(kT)^{s+2}\Gamma(s+1)I(\gamma,s+1,\pm) . \qquad (11.24)$$

$C_p$ is now readily obtained:

$$C_p = \left(\frac{\partial H}{\partial T}\right)_{p,N} = \frac{s+2}{s+1}C_v$$

$$+ \left\{1 - \frac{I^2(\gamma,s,\pm)}{I(\gamma,s+1,\pm)I(\gamma,s-1,\pm)}\right\}\frac{s+2}{s+1}\frac{U}{v}\left(\frac{\partial v}{\partial T}\right)_{p,N} . \qquad (11.25)$$

Note that $(\partial v/\partial T)_{p,N}$ diverges for systems for which the chemical potential is identically zero. With this observation, equations (11.23) to (11.25) hold also for systems for which $\mu \equiv 0$.

## 11.6. THE CLASSICAL APPROXIMATION

It has been seen (Chapter 10) that the classical approximation is equivalent to the high-temperature approximation for ideal quantum gases. It follows that for the ideal gases, and systems resembling them, low temperatures will favour departures from classical behaviour. Now equation (11.12) shows that the extent of the departure from classical behaviour is governed by the value of the chemical potential divided by $kT$. Equation (11.6) teaches that, as far as its effect on the value of $\mu/kT$ is concerned, an increased particle density should have the same effect as a lowered temperature. Hence, *departures from classical behaviour should also be favoured by greater densities.*

We next use (11.12) in (11.6) and (11.7), to observe that in the classical approximation

$$\exp\left(-\frac{F}{kT}\right) = N^{-N}[\mathrm{e}A(kT)^{s+1}\Gamma(s+1)]^{N} \tag{11.26}$$

where e is the base of natural logarithms. Let $F_N$ denote the free energy of the system on the basis of the grand canonical ensemble when the mean total number of particles is $N$ and the volume and temperature are $v$ and $T$ respectively. Assuming the fluctuations in the total number of particles in the grand canonical ensemble to be small, equation (9.22) in the form $Z_N = -kT\ln F_N$ shows that (11.26) is the classical approximation, $Z_N$, of the correct quantum statistical formula for the canonical partition functions for an ideal system of $N$ indistinguishable weakly interacting particles.

The older classical statistical mechanics gave an incorrect formula of the type (11.26). This can be written in our notation as

$$Z_{N,\mathrm{cl}} = \exp\frac{-F_{N,\mathrm{cl}}}{kT} = [A(kT)^{s+1}\Gamma(s+1)]^{N} \tag{11.27}$$

where the subscript cl indicates the older classical theory. That it is in error can be seen at once, since

$$F_{N,\mathrm{cl}} = -N\ kT\ \ln[A(kT)^{s+1}\Gamma(s+1)]$$

is not extensive in character; on doubling the volume and the number of particles $A$ is doubled, as well as $N$, but not $F_{N,\mathrm{cl}}$. To find a precise relation between our theory and the older classical theory, note from (11.26) and (11.27) that

$$Z_N = Z_{N,\mathrm{cl}}\left(\frac{e}{N}\right)^N = \frac{(Z_{1,\mathrm{cl}})^N}{N!} . \qquad (11.28)$$

We have here used Stirling's approximation for large $N$:

$$N! \sim (N/e)^N .$$

Thus one sees from (11.28) that

$$F_N = F_{N,\mathrm{cl}} + N\,kT(\ln N-1)$$

We shall see in section 11.8 how formulae such as (11.27) follow from the older classical statistical mechanics.

The result (11.28) can be made basic in discussions of statistical mechanics which do not aim at taking full account of quantum statistics. In this section, however, quantum statistics has been taken as basic, so that (11.28) appears as merely an approximate result which is valid under certain conditions. If statistical mechanics is not developed in this way, the factor $(N!)^{-1}$ appears as a correction to a theory of a gas of particles which, in the absence of this factor, would be treated as distinguishable. If one starts in this way, it is then impossible to discuss the properties of the gas when the classical approximation does not apply. The reason for this is clear from the foregoing discussion: equation (11.28) implies the validity of the classical approximation.

In order to apply the general theory developed in this section to particular cases, one must be able to answer the following questions in each instance: (i) Are Fermi or Bose statistics applicable? (ii) What are the values of $A$ and $s$? In addition, one has to make the assumption that the particles have point interactions, and that the density of states law $N(e) = Ae^s$ applies. Lastly, one will need to have a table giving values of the integrals (11.5) or, alternatively, one needs a mathematical theory which enables one to compute their values.

## 11.7. CLASSICAL STATISTICAL MECHANICS: THE EQUIPARTITION THEOREM

In classical statistical mechanics it is adequate for our purposes to replace sums over single-particle quantum states by integrals over momenta and co-ordinates. Thus the canonical partition function for a one-dimensional harmonic oscillator of frequency $\nu$, or angular frequency $\omega = 2\pi\nu$, is

$$\begin{aligned} Z &= z \iint_{-\infty}^{\infty} \exp\left\{-\frac{E(p,q)}{kT}\right\} \mathrm{d}p\,\mathrm{d}q \\ &= z \iint \exp\left\{-\frac{ap^2+bq^2}{kT}\right\} \mathrm{d}p\,\mathrm{d}q \; . \end{aligned} \tag{11.29}$$

Here $z$ is a constant which makes $Z$ still dimensionless, and $a$ and $b$ are independent of $p$ and $q$. This expression for the energy is easily obtained from the differential equation for the harmonic oscillator:

$$\ddot{q} = -\omega^2 q$$

where $\omega$ is a constant. It follows by multiplying by $2\dot{q}$ and integrating that

$$\dot{q}^2 = -\omega^2 q^2 + C$$

where $C$ is a constant. The total energy is essentially this constant:

$$\tfrac{1}{2}m\dot{q}^2 + \tfrac{1}{2}m\omega^2 q^2 = \tfrac{1}{2}mC = ap^2+bq^2 \qquad \left(a \equiv \frac{1}{2m},\; b \equiv \frac{m\omega^2}{2}\right).$$

Reverting to the partition function, and putting $x = \sqrt{(a/kT)}p$ and $y = \sqrt{(b/kT)}q$ ,

$$Z = z\sqrt{\left(\frac{kT}{a}\frac{kT}{b}\right)} \int_{-\infty}^{\infty} \exp(-x^2)\,\mathrm{d}x \int_{-\infty}^{\infty} \exp(-y^2)\,\mathrm{d}y = \frac{kT}{\nu}\, z\, . \tag{11.30}$$

Writing $kT = 1/\beta$ one sees that the average energy is, using a result of problem 9.1,

$$U = -\frac{\partial \ln Z}{\partial \beta} = -\frac{\partial}{\partial \beta}\left(\ln \frac{1}{\beta} + \text{constant}\right) = \frac{1}{\beta} = kT \; .$$

Thus each squared term (or *degree of freedom*) in the expression for $E(p,q)$ contributes $\frac{1}{2}kT$ to the average energy in classical statistical mechanics. This is the classical theorem of the equal distribution of the average energy of a system over all its degrees of freedom, and is called the *equipartition theorem*. It was once believed to be generally valid. However, turning to equation (11.20), it is seen that an $N$-particle ideal gas has an internal energy of the form $U \propto NkT$ only in the special case when the classical approximation (11.12) can be used. One then finds for the model of this section an equipartition type of result

$$U = (s+1)NkT \; . \tag{11.31}$$

For non-relativistic particles of non-zero rest mass in a field-free box $s = \frac{1}{2}$ and $U = \frac{3}{2}\,NkT$. This is the expected result since for each such particle the energy $E(p,q)$ is purely kinetic, and is the sum of three squared terms:

$$\frac{1}{2m}\left(p_x^2 + p_y^2 + p_z^2\right) \; .$$

For $N$ particles the equipartition theorem does indeed yield $3N\,\frac{1}{2}kT$. The full form (11.20) of $U$ shows that equipartition is normally invalid.

A more general theorem is obtained from the result (11.29) in a form which applies to a system of $f$ degrees of freedom $q_1,\ldots,q_f,\; p_1,\ldots,p_f$:

$$1 = \frac{z}{Z}\underbrace{\int\cdots\int}_{2f} \exp\left(-\frac{E}{kT}\right)\, dq_1 \ldots dp_f \; .$$

Integrating partially with respect to $q_1$,

$$1 = \frac{z}{Z}\underbrace{\int\cdots\int}_{2f-1}\left|q_1 \exp\left(-\frac{E}{kT}\right)\right| dq_2\ldots dp_f + \frac{z}{ZkT}\underbrace{\int\cdots\int}_{2f} q_1\,\frac{\partial E}{\partial q_1}\exp\left(-\frac{E}{kT}\right) dq_1\ldots dp_f .$$

If at both limits $q_1 = 0$ or $E = \infty$

$$\frac{z}{Z}\int\cdots\int q_1\frac{\partial E}{\partial q_1}\exp\left(-\frac{E}{kT}\right)dq_1\ldots dp_f = \langle q_1\frac{\partial E}{\partial q_1}\rangle = kT\ . \qquad (11.32)$$

The angular brackets imply an average over the distribution. If $E(q_1,q_2,\ldots,p_f)$ involves $q_1$ in the following way

$$E = aq_1^2 + f(q_2,q_3,\ldots,p_f)$$

then the simple result

$$\langle aq_1^2\rangle = \tfrac{1}{2}kT \qquad (11.33)$$

is recovered. Results $\langle p_j\ \partial E/\partial p_j\rangle = kT$ hold similarly.

### *11.8. AN INTERPRETATION OF THE FORMULA $Z_N = Z_1^N/N!$

It was emphasized in section 11.6 that the formula (11.28) takes approximate account of the *indistinguishability* of the particles in a gas consisting of one type of particle. It is instructive to approach this result from the opposite direction by considering first a gas of $N$ *distinguishable* particles and making a correction for indistinguishability afterwards.

Consider a gas of $N$ distinguishable particles with point interactions in a volume $v$ on the basis of a canonical distribution. By (11.30) the classical partition function has the form

$$Z_N = z^N\int\cdots\int\exp\left[\frac{K_1+K_2+\ldots+K_N}{kT}\right](d\mathbf{p}_1\,d\mathbf{q}_1)(d\mathbf{p}_2\,d\mathbf{q}_2)\ldots(d\mathbf{p}_N\,d\mathbf{q}_N) \qquad (11.34)$$

where $K_i$ is the kinetic energy of the $i$th particle. A field-free box has been assumed so that the potential energy of each particle can be taken to be zero. For $N$=2 there are separate contributions to this integral from the following two states:

(i) Particle 1 in a volume element $(\Delta\mathbf{p}\ \Delta\mathbf{q})_A$ at a point A in phase space; particle 2 in a volume element $(\Delta\mathbf{p}\ \Delta\mathbf{q})_B$ at a point B in phase space.

(ii) The positions of particles 1 and 2 are reversed.

This checks that the integral treats particles as distinguishable, since for indistinguishable particles there should be only one contribution to the integral namely

(iii) One particle is near A and one particle is near B in phase space.

One sees at once that

$$Z_N = z^N v^N \left\{ \int \exp\left(-\frac{K_1}{kT}\right) d\mathbf{p}_1 \right\}^N = Z_1^N \tag{11.35}$$

for distinguishable particles. This can now be 'corrected' for indistinguishability, if necessary, by dividing by the number of permutations of the particles among themselves. This yields a division by $N!$

Similarly for one particle in a field-free container of volume $v$ and wall temperature $T$ the classical statistical mechanics yields

$$Z_{1,\mathrm{cl}} = z^3 \int d\mathbf{q} \int \exp\left(-\frac{p^2}{2mkT}\right) d\mathbf{p} \ .$$

The integration can be converted to one over the kinetic energy $e$ of the particle. If one assumes our general density of states law $Ae^s$ where the constant $A$ includes the volume and the coefficient $z^3$,

$$\begin{aligned} Z_{1,\mathrm{cl}} &= \int_0^\infty Ae^s \exp\left(-\frac{e}{kT}\right) de \\ &= A(kT)^{s+1} \int_0^\infty x^s e^{-x}\, dx \\ &= A(kT)^{s+1}\, \Gamma(s+1) \end{aligned}$$

where the formula for the gamma function has been used. The older classical statistical mechanics then assumed that for $N$ particles with point interaction $Z_{N,\mathrm{cl}} = (Z_1)^N$, and that does indeed yield (11.27). Incidentally, if one takes the simple kinetic energy relation

$$e = p^2/2m, \qquad de = (p/m)\,dp$$

one can identify both $A$ and $s$. One merely notes that

$$d\mathbf{p} = 4\pi p^2 dp = 4\pi\times 2me \times \frac{m}{(2me)^{\frac{1}{2}}}\, de = 4\pi m(2me)^{\frac{1}{2}}\, de \ .$$

Hence $s = \frac{1}{2}$ and $A = 4\pi z^3 vm(2m)^{\frac{1}{2}}$. This also identifies $z$ as the reciprocal of Planck's constant $h$. It foreshadows a formula of Table 12.1, which will be obtained more systematically later.

PROBLEMS

11.1. Obtain the internal energy directly from its expression as a sum in the form

$$\boxed{U = A(kT)^{s+2}\Gamma(s+2)I(\gamma,s+1,\pm)\ .}$$

11.2. Obtain from the above expression for $U$ and from the free energy (11.7) the entropy in the form

$$S = k\{(s+2)A(kT)^{s+1}\Gamma(s+1)I(\gamma,s+1,\pm) - \gamma N\}\ .$$

11.3. Obtain from the above expressions for $U$ and $S$ an expression for $p$ in the form

$$p = \frac{A}{v}(kT)^{s+2}\Gamma(s+1)I(\gamma,s+1,\pm)\ .$$

11.4. Prove (11.16), (11.17), and (11.23) in detail.

11.5. Check in detail the calculations leading via (11.6) and (11.7) to

$$Z_N = Z_1^N/N!$$

11.6. Obtain the formula (11.22) for $C_v$.

11.7. A classical gas at temperature $T$ is in an infinitely high vertical cylinder which is at rest in a constant gravitational field of acceleration $g$. If $m$ is the mass of a molecule show that the one-molecule partition function is proportional to $(kT)^{5/2}$. Hence show that the internal energy is that of a classical gas with five degrees of freedom.

11.8. A classical gas at temperature $T$ has particles of rest mass $m_0$ moving with relativistic speeds in a field-free box so that for one particle

$$E(q_1,q_2,q_3,p_1,p_2,p_3) = c(p_1^2 + p_2^2 + p_3^2 + m_0^2c^2)^{1/2}$$

where $c$ is the velocity of light. Show that for $j = 1,2,3$ and with

$m = m_0/(1-v^2/c^2)^{\frac{1}{2}}$ the equipartition theorem yields

$$\langle \frac{c^2 p_j^2}{2E} \rangle = \langle \frac{p_j^2}{2m} \rangle = \langle \frac{m\dot{q}_j^2}{2} \rangle = \tfrac{1}{2}kT$$

that is

$$\langle mv^2 \rangle = 3kT \ .$$

The speed of the particle is denoted by $v$.

# 12. The density of states functions $Ae^s$

In order to obtain the density of states function, for example for radiation, consider a cavity with its walls at temperature $T$. (We could allow reflecting walls, provided only that some matter at temperature $T$ is in contact with the radiation.) We shall fit various modes of vibration into the cavity and attempt to count how many different types we have available.

In order to achieve generality we shall allow the cavity to be filled with an isotropic medium of refractive index $\mu(\nu)$. It can depend on the frequency $\nu$ (in which case it is called a *dispersive medium*). All essential results are, however, still obtainable by putting $\mu(\nu) = 1$ for all frequencies $\nu$. The velocity of the radiation, which in an anisotropic medium depends on direction, has for our isotropic medium the same value for all directions of propagation.

For a three-dimensional continuum one can imagine fitting various wave shapes into the box containing the medium. If the walls of the box are rigid, they do not participate in the vibrations and the disturbance must have nodes at the surfaces. For a prism of rectangular cross-section and sides of lengths $D_1$, $D_2$, $D_3$ these six surfaces are

$$x_1 = 0, \quad x_2 = 0, \quad x_3 = 0,$$

$$x_1 = D_1, \quad x_2 = D_2, \quad x_3 = D_3 \,. \tag{12.1}$$

The possible modes of vibrations are therefore disturbances whose magnitude at position $\mathbf{r} = (x_1, x_2, x_3)$ are

$$u_{\mathbf{k}}(\mathbf{r}) = A_{\mathbf{k}} \sin 2\pi k_1 x_1 \sin 2\pi k_2 x_2 \sin 2\pi k_3 x_3 \,. \tag{12.2}$$

Here $A_{\mathbf{k}}$ is the amplitude, and $\mathbf{k} = (k_1, k_2, k_3)$ specifies the direction of propagation of the wave. This implies the existence of integers $n_j$ so that $2k_j D_j = n_j$ and (12.1) is satisfied.

Thus the allowed modes **k** are, in a sense, quantized to discrete values:

$$\mathbf{k} = \frac{1}{2}\left(\frac{n_1}{D_1}, \frac{n_2}{D_2}, \frac{n_3}{D_3}\right) . \qquad (12.3)$$

Displacements in a two-dimensional 'box' can be visualized as in Fig.12.1, and the three-dimensional case is a simple

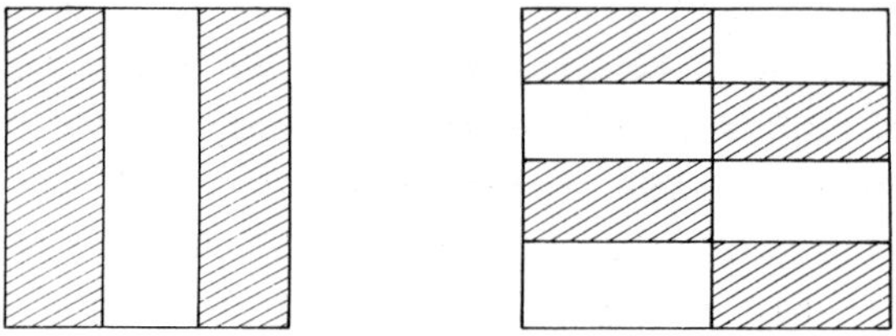

$(n_1,n_2) = (3,1)$ $(n_1,n_2) = (2,4)$

Fig.12.1. The modes $(n_1,n_2)$ for a two-dimensional box. The shaded and unshaded parts have displacements of opposite sense.

generalization. This static pattern does not suggest any single wavelength which can be associated with a mode $(n_1,n_2,n_3)$. To achieve this a time-dependent part, appropriate to a frequency $\nu$ of the disturbance, is required:

$$\psi_{\mathbf{k}}(\mathbf{r},t) = u_{\mathbf{k}}(\mathbf{r}) \sin 2\pi\nu t .$$

The wave equation

$$\nabla^2\psi = c^{-2} \frac{\partial^2\psi}{\partial t^2}$$

then yields (after cancelling $-4\pi^2\psi$)

$$k^2 = \nu^2/c^2 = 1/\lambda^2 .$$

Using the de Broglie relation for the momentum $p = h/\lambda$, where $h$ is Planck's constant, one finds

$$p = hk = h/\lambda \ . \tag{12.4}$$

For a one-dimensional pattern a wavelength can be defined without appeal to the time dependence of the disturbance by requiring, with $k = |\mathbf{k}|$,

$$\sin(2\pi\ \mathbf{k}.\mathbf{r}) = \sin\left\{2\pi\mathbf{k}.\left(\mathbf{r} + \frac{\lambda\mathbf{k}}{k}\right)\right\}$$

whence $\lambda k = 1$, but this procedure fails when applied to equation (12.2).

A further interpretation of the *wave vector* **k** is obtained by passing from position **r** to position **r'** by one wavelength along its direction of propagation. Then the new position vector is subject to

$$2\pi\mathbf{k}.\mathbf{r}' = 2\pi\mathbf{k}.(\mathbf{r} + \lambda\hat{\mathbf{k}})$$

and the disturbance must be the same in all respects at **r'** so that

$$2\pi kr' = 2\pi kr + 2\pi\lambda k = 2\pi kr + 2\pi \ .$$

It follows that, although sometimes (as on p. 261) $k$ is defined to be $2\pi/\lambda$, one has here

$$k = 1/\lambda \ .$$

Each mode of vibration can be specified by a set of integers $(n_1, n_2, n_3)$ such that, by (12.3) and (12.4),

$$\sum_{j=1}^{3}\left(\frac{n_j}{a_j}\right)^2 = 1, \quad a_j = \frac{2D_j}{\lambda} \quad (j = 1,2,3) \ . \tag{12.5}$$

This is the equation of an ellipsoid with semi-axes $a_j$ in a space in which the integers $n_j$ are plotted along the three axes.

Each mode of vibration is represented by a point $(n_1, n_2, n_3)$ with integer co-ordinates. Confining attention to modes of negligible $n_2$ and $n_3$, the modes with wavelengths lying between $\lambda_0$ and $\infty$ satisfy, by (12.5),

$$\frac{1}{\lambda_0^2} \geq \frac{n_1^2}{4D_1^2} = \frac{1}{\lambda^2} \geq 0 \ .$$

Thus values of $n_1 = 1, 2, \ldots$ up to a maximum are admissible. In general, the admissible points satisfy

$$\sum_{j=1}^{3} \left(\frac{n_j\lambda_0}{2D_j}\right)^2 \leq 1$$

and they therefore lie in an ellipsoid with semi-axes $a_{j0} = 2D_j/\lambda_0$. Since each point can be visualized as the centre of a cube with sides of length unity, the volume of this ellipsoid also yields the number of modes. This is

$$N(\lambda_0) = \frac{1}{8}\,\frac{4\pi}{3}\,a_{10}a_{20}a_{30} = \frac{4\pi}{3}\,\frac{v}{\lambda_0^3} \tag{12.6}$$

where $v = D_1\,D_2\,D_3$ is the volume of the box. The division by the factor 8 occurs because one is interested only in the independent modes. Now in passing from $n_1$ to $-n_1$ one simply passes from $u_{\mathbf{k}}(\mathbf{r})$ to $-u_{\mathbf{k}}(\mathbf{r})$ and this is not an independent function. Hence for two dimensions one requires to confine the integral to one quadrant of $(n_1,n_2)$ space. In three dimensions the octant in which $n_1$, $n_2$ and $n_3$ are positive will suffice.

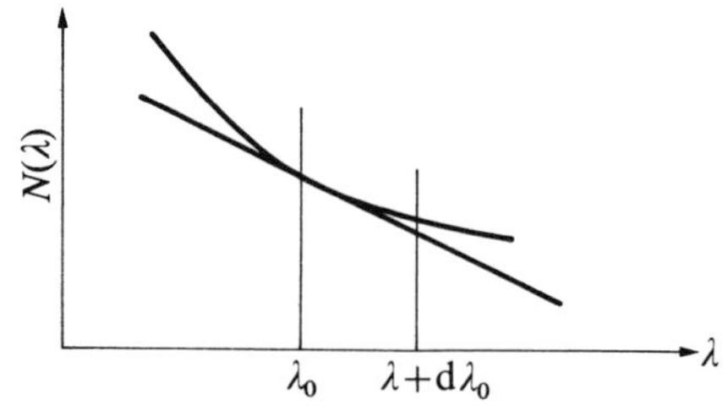


Fig.12.2. Counting normal modes.

Fig.12.2 illustrates that the number of normal modes whose wavelengths lie in the range $(\lambda_0,\lambda_0 + d\lambda_0)$ is $d\lambda_0$ multiplied by the slope of the curve of $N(\lambda)$ against $\lambda$ at the value $\lambda = \lambda_0$. The required expression is, apart from sign,

$$dN(\lambda_0) = \left\{\frac{dN(\lambda)}{d\lambda}\right\}_{\lambda=\lambda_0} d\lambda_0 = 4\pi v\, \lambda_0^{-4}\, d\lambda_0 \;. \tag{12.7}$$

Because the surface of the ellipsoid cuts the line between neighbouring lattice points at various positions, this result is not accurate. The accuracy improves as the surface/volume ratio of the ellipsoid is made smaller, i.e. as $v$ is increased for given $\lambda_0$. The condition for accuracy is $v \gg \lambda_0^3$. It can be shown that (12.7) remains val d for containers of arbitrary shape in the limit in which $v/\lambda^3 \to \infty$ (see problem 13.11 ).

We now obtain similar expressions for a frequency range $d\nu$ and for an energy range $de$. For this purpose we shall write

$$\nu\lambda = c_0/\mu \tag{12.8}$$

for the phase velocity of the disturbance in the enclosure. Here $c_0$ is the speed of light *in vacuo*, $\nu$ is the frequency of the disturbance, and the factor $\mu$ scales the speed down appropriately. In optics or electromagnetic theory $\mu$ is called the refractive index. The disturbance is called light, i.e. electromagnetic radiation, or a gas of states of excitation called *photons*. From (12.8)

$$\lambda^{-4}\, d\lambda = -\left(\frac{\mu\nu}{c_0}\right)^4 \frac{c_0}{\mu\nu^2}\, d\nu = -\frac{\mu^3\nu^2}{c_0^3}\, d\nu \quad .$$

Writing $dN(\lambda) = dN(\nu)$ (in an obvious notation), one finds

$$dN(\nu) = \frac{4\pi v\mu^3}{c_0^3}\, \nu^2\, d\nu \;. \tag{12.9}$$

One finds similarly

$$dN(e) = \frac{4\pi v\mu^3}{h^3 c_0^3}\, e^2\, de \tag{12.10}$$

where the Planck relation

$$e = h\nu \tag{12.10'}$$

has been used between the energy and a frequency of a mode. In (12.10) we have a first formula for the number of quantum states in a range $de$ which has the form

$$\frac{dN(e)}{de} = Ae^s$$

where $A$ is proportional to volume and $s$ is a constant.

In some cases of so-called dispersion, $\mu$ can itself depend on frequency. In that case the argument leading to (12.9) is replaced by

$$\lambda^{-4}\, d\lambda = \left(\frac{\mu\nu}{c_0}\right)^4 \frac{c_0}{(\mu\nu)^2} \frac{d(\mu\nu)}{d\nu}\, d\nu = \frac{\mu^2\nu^2}{c_0^3} \frac{d(\mu\nu)}{d\nu}\, d\nu$$

$$= \frac{\mu^3\nu^2}{c_0^3} \frac{d\{\ln(\mu\nu)\}}{d(\ln\nu)}\, d\nu$$

$$dN(\nu) = 4\pi v\, \mu^3\nu^2 c_0^{-3}\, [d\{\ln(\mu\nu)\}/d(\ln\nu)]d\nu \,.$$

This way of writing the result shows that in the absence of dispersion the differential coefficient is simply

$$\frac{d(\ln\mu)}{d(\ln\nu)} + \frac{d(\ln\nu)}{d(\ln\nu)} = 1$$

and (12.9) is recovered. The derivation is preferable in terms of wavelengths, because equation (12.7) is free of complications due to dispersion.

The main applications of the result just derived are indicated in Table 12.1. The phase velocity of longitudinal waves has been denoted by $c_l$ and the phase velocity of transverse waves by $c_t$. The above argument applies to each class of waves having a given direction of polarization. If there are two independent directions of polarization, the number of distinct normal modes is, of course, multiplied by 2. This gives rise to the first three rows of the table.

The importance of (12.7) rather than (12.9) or (12.10) can be illustrated by our next computation. We wish to derive the number of quantum states of a structureless particle

TABLE 12.1

*Main applications of equation (12.7)*

| Medium | Number of modes per unit volume in range $(\lambda, \lambda + d\lambda)$ | $(\nu, \nu + d\nu)$ | $(e, e + de)$ | Used in |
|---|---|---|---|---|
| Normal modes of vibration of gas transmitting sound waves only (Bose statistics) | $4\pi\lambda^{-4}d\lambda$ | $4\pi c_\ell^{-3}\nu^2 d\nu$ | | Classical theory of sound |
| Normal modes of vacuum or ether (Bose statistics) | $8\pi\lambda^{-4}d\lambda$ | $8\pi c_t^{-3}\nu^2 d\nu$ | $8\pi h^{-3}c_t^{-3}e^2 de$ | Planck's and Rayleigh's theory of radiation |
| Normal modes of an elastic solid (Bose statistics) | $12\pi\lambda^{-4}d\lambda$ | $4\pi(2c_t^{-3} + c_\ell^{-3})\nu^2 d\nu$ | $4\pi h^{-3}(2c_t^{-3} + c_\ell^{-3})e^2 de$ | Debye's theory of specific heats |
| Quantum states of free particle, of non-zero rest mass, allowing for a $g$-fold degeneracy due to spin (Bose or Fermi statistics) | $4\pi g\lambda^{-4}d\lambda$ | | $4\pi mgh^{-3}(2me)^{\frac{1}{2}}de$ | Theory of ideal gas of bosons or fermions having non-zero rest mass |
| Quantum states of free particle of arbitrary rest mass (zero or non-zero), allowing for a $g$-fold degeneracy due to spin (Bose or Fermi statistics) | $4\pi g\lambda^{-4}d\lambda$ | | $4\pi g\, m^2 c h^{-3}\ (x+1)\times (x+2)^{\frac{1}{2}}x^{\frac{1}{2}}de$, $x \equiv e/mc^2$ | Theory of relativistic gas of fermions or bosons |

The expressions in terms of frequency $\nu$ and single-particle energy $e$ are valid only for non-dispersive media, as explained in connection with equation (12.9). The velocities are the phase velocities, so that in the second row one may write $c_t = c_0/\mu$ where $c_0$ is the velocity of light *in vacuo* and $\mu$ is the refractive index of the medium.

confined to a three-dimensional box by infinitely high potential walls. The box and the co-ordinate system are chosen as before. The wavefunction of the particle vanishes at the planes $x = 0$, $y = 0$, $z = 0$ of the box, and (12.2) gives a suitable form of the wavefunction of the particle. Indeed, the whole calculation up to (12.7) goes through unchanged. However, while the relation $e = h\nu$ remains valid, the relation between the single-particle energy $e$ and its wavelength $\lambda$ must be determined from the Schrödinger wave equation which the wavefunction must satisfy

$$\sum_1^3 \left(\frac{\partial^2 u}{\partial x_j^2}\right) + 8\pi^2 h^{-2} m(e-\phi)u = 0$$

where $\phi(x_1, x_2, x_3)$ is the potential energy of the particle at a point in the box. Hence

$$e - \phi = \frac{h^2k^2}{2m} = \frac{h^2}{2m\lambda^2}.$$

(One often writes alternatively $\hbar k$ instead of $hk$. In that case $k$ is taken as $2\pi/\lambda$ and $\hbar$ denotes Planck's constant divided by $2\pi$.) By (12.8)

$$\frac{c_0}{\mu} = \nu\lambda = \frac{h\nu}{\{2m(e-\phi)\}^{\frac{1}{2}}} \tag{12.11}$$

Thus, solving for $\mu$, the *matter waves of frequency* $\nu$ are seen to behave as if they moved in a medium of refractive index $\{2m(e-\phi)\}^{\frac{1}{2}}(c_0/h\nu)$, which varies continuously with the potential $\phi$. Substituting (12.11) in (12.7) yields for the density of quantum states

$$\mathrm{d}N(e) = 4\pi v m h^{-3}(2me)^{\frac{1}{2}}\mathrm{d}e \tag{12.12}$$

in the case of a free and structureless particle. For particles having spin $\frac{1}{2}$, each quantum state counted in (12.7) gives rise to two states which differ only in the spin quantum number. The generalization to $g$-fold spin degeneracy is immediate, and this gives rise to the fourth row of Table 12.1. One should note that the change in sign which occurs upon passing from $\mathrm{d}N(\lambda)$ to $\mathrm{d}N(e)$ has been ignored. This merely reflects the fact that $N(\lambda)$ decreases as $\lambda$ increases, while $N(e)$ increases as $e$ increases. This is clearly irrelevant to a count of quantum states.

It is at first sight curious that we have been able to treat quantum mechanical entities of zero and non-zero rest mass on the basis of the same formula, equation (12.7), provided we confined interest to the distribution of states of given wavelength and subsequently introduced the energy in terms of the wavelength. There should exist an argument of equal generality which uses the kinetic energy directly as a variable. This will now be given for the case of a vacuum.

This last application of (12.7) generalizes (12.12) to the case of a gas whose particles move with velocities great enough to bring relativistic effects into play. First note that the energy $e$ in the preceding paragraph may be interpreted as wholly kinetic, since the particle moves in a field-free vacuous box. Writing $e = p^2/2m$, where $\mathbf{p}$ is the linear momentum of the particle, and $p = |\mathbf{p}|$, one recovers

$$p = h/\lambda \, , \tag{12.13}$$

the de Broglie relation which is elucidated in treatments of quantum mechanics, and can be seen to follow from (12.11). It holds for photons and particles of a non-zero rest mass. According to the special theory of relativity the linear momentum of a free particle of rest mass $m$ is related to its total energy $E$ by

$$\frac{E^2}{c_0^2} = p^2 + m^2 c_0^2 \, . \tag{12.14}$$

Here $c_0$ is the velocity of light *in vacuo* and $E \equiv e + mc_0^2$ includes the energy due to the rest mass. By (12.7), (12.13), and (12.14) one finds for the number of single-particle quantum states in the energy range $(E, E + dE)$

$$\begin{aligned} dN(E) &= \frac{4\pi g v}{\lambda^2}\, d\left(\frac{1}{\lambda}\right) = \frac{4\pi g v}{\lambda^2}\, \frac{\lambda E\, dE}{h^2 c_0^2} \\ &= \frac{4\pi g v}{h^3 c_0^2} \left(\frac{E^2}{c_0^2} - m^2 c_0^2\right)^{\frac{1}{2}} E\, dE \, . \end{aligned} \tag{12.15}$$

Here again we have allowed for a $g$-fold degeneracy due to spin and have ignored a change of sign. It is instructive to change back to the kinetic energy variable, when one finds

$$\mathrm{d}N(e) = 4\pi g v m^2 c_0 h^{-3}(x+1)(x+2)^{\frac{1}{2}}x^{\frac{1}{2}}\,\mathrm{d}e \tag{12.16}$$

where $x \equiv e/mc_0^2$. On the basis of the formula one may develop a theory of the relativistic quantum gas of weakly interacting particles which is quite analogous to the theory developed in Chapter 11. However, the integrals which occur are more complicated.

The result (12.16) has the important property

$$\mathrm{d}N(e) = \begin{cases} 4\pi g v\, c_0^{-3}\, h^{-3}\, e^2\,\mathrm{d}e & (x \to \infty) \\ 4\pi g v m h^{-3}(2me)^{\frac{1}{2}}\mathrm{d}e & (x \to 0) \end{cases} \tag{12.17}$$

The case $x \to \infty$ corresponds to zero rest mass, or to non-zero rest mass particles in the extreme relativistic limit. This latter case is of importance in astrophysical applications (white dwarfs, neutron stars). In either case it gives rise to a special case of (12.10). The case $x \to 0$ corresponds to the non-relativistic case of particles of non-zero rest mass and gives rise to (12.12). It is seen therefore that (12.16) represents a general density of states expression in terms of the kinetic energy. Its implications are pursued in problems 12.5 to 12.7 .

Note that the expressions (12.10) and (12.12) or (12.17) are precisely of the form envisaged in Chapter 11 with $A/v$ a constant for a given system.

PROBLEMS

12.1. Show that the canonical partition function of one particle of mass $m$ with the density of states function (12.12) is

$$Z_1 = \left(\frac{2\pi m\, kT}{h^2}\right)^{3/2} gv$$

if the temperature is $T$, the volume is $v$, and the spin degeneracy is $g$.

12.2. By using equation (11.28) obtain the partition function $Z_N$ for indistinguishable particles in classical statistical mechanics given $Z_1$ in problem 12.1. Hence show, using problem 9.1 or otherwise, that this system satisfies the usual ideal classical gas equation

$$U_N = \frac{3}{2} NkT .$$

(This is the equipartition theorem).

12.3. Show that the system discussed in problem 12.2 satisfies Boyle's law.

12.4. Approach the statistical thermodynamics of the classical ideal gas from the full quantum statistics of Chapter 11 to show that, for general $A$ and $s$,

$$U_N = (s + \tfrac{1}{2})NkT \quad \text{and} \quad pv = NkT$$

in agreement with the results of problems 12.2 and 12.3. Show also that the free energy is given by

$$F_N = (\gamma_N - 1)NkT$$

where $\gamma_N$ is $1/kT$ times the chemical potential $\mu_N$ for $N$ particles.

12.5. (a) A gas of free relativistic particles of rest energy $\varepsilon_0 = mc_0^2$ and spin degeneracy $g$ is confined to a box of volume $v$ at temperature $T$ and chemical potential $\mu$. From (11.2) and (12.16) find the grand partition function in the continuous spectrum approximation in the form ($\eta_0 \equiv \varepsilon_0/kT$, $\gamma' \equiv (\mu-\varepsilon_0)/kT$, $x \equiv \varepsilon/\varepsilon_0$)

$$\Xi = \frac{4\pi g v m^4 c_0^5}{3h^3 kT} \int_0^\infty \frac{(x^2+2x)^{3/2}}{\exp(\eta_0 x-\gamma')\pm 1} \, dx .$$

(b) Show also directly from (12.16) that the mean total number of particles and the mean total energy (apart from the energy due to the rest masses) are

$$N = \frac{4\pi g v m^3 c_0^3}{h^3} \int_0^\infty \frac{(x^2+2x)^{\frac{1}{2}}(x+1)}{\exp(\eta_0 x-\gamma')\pm 1}\,dx$$

$$U = \frac{4\pi g v m^4 c_0^5}{h^3} \int_0^\infty \frac{x(x^2+2x)^{\frac{1}{2}}(x+1)}{\exp(\eta_0 x-\gamma')\pm 1}\,dx \quad .$$

(c) Verify that this system is not an ideal quantum gas in the sense that the relation $pv = gU$ is not satisfied (cf. problem 6.6 ).

(This is an example of the use of a density of states function which does not have the form $Ae^s$.)

12.6. Show that the entropy $S$ of the system introduced in problem 12.5 is given by

$$TS = \frac{4\pi g v m^4 c_0^5}{3h^3} \int_0^\infty \{4x^2 + 8x + 3 - \frac{3\mu}{\varepsilon_0}(x+1)\}(x^2+2x)^{\frac{1}{2}} f\,dx$$

where

$$f \equiv \{\exp(\eta_0 x-\gamma') \pm 1\}^{-1} \quad .$$

12.7. Although the systems discussed in problems 12.5 and 12.6 have a density of states function not covered by the formula $N(e) = Ae^s$ confirm by direct calculation from $N$ and $U$ that such systems are found in the following special cases and are ideal gases, $pv = gU$, with $\mu - \varepsilon_0$ taking the place of the chemical potential.

TABLE 12.2

| | $x$ | $s$ | $g$ |
|---|---|---|---|
| Non-relativistic limit | $\to 0$ | $\frac{1}{2}$ | $\frac{2}{3}$ |
| Extreme relativistic limit | $\to \infty$ | 2 | $\frac{1}{3}$ |

(In the extreme relativistic limit a gas of fermions and a gas of bosons each behave like a simple system of particles of zero rest mass. The most famous example of such a system is black-body radiation. This will be studied in Chapter 13.)

12.8. Independent particles of non-zero rest mass $m$ move in a field-free square 'box' of side $D$ and 'volume' $v = D^2$. Show that the density of states formula for this system has again the form $Ae^s$ with $A$ proportional to $v$, and show that for spin degeneracy $g$

$$A = \frac{2\pi mgv}{h^2}, \qquad s = 0 .$$

12.9. Let $\mu_F$ and $\mu_B$ be the chemical potentials of ideal non-relativistic fermion and boson gases of N particles of rest mass m each, in two dimensions, both at temperature T and in volume v.

(a) Show that

$$\int_0^\infty \frac{x\ dx}{\exp(x-\alpha_F)+1} - \int_0^\infty \frac{x\ dx}{\exp(x-\alpha_B)-1} = \frac{1}{2}\{\ln(1+\exp\alpha_F)\}^2$$

where

$$\alpha_F \equiv \mu_F/kT, \quad \alpha_B \equiv \mu_B/kT, \quad \exp(-\alpha_B) - \exp(-\alpha_F) = 1 .$$

(b) Hence prove that the heat capacities of the two systems are the same.

# 13. Bosons: black-body radiation

## 13.1. TEMPERATURE AND BLACK-BODY RADIATION

In order to elucidate the concept of black-body radiation suppose that isotropic radiation with an arbitrary frequency spectrum is trapped in an enclosure having perfectly reflecting walls and of linear dimensions large compared with the wavelengths considered. Such walls neither emit nor absorb radiation, and the system can be in equilibrium in the sense that there are no changes in the macroscopic properties of the system with lapse of time. Suppose now that we have two such enclosures separated by a perfectly reflecting wall and containing either similar or different radiation. In addition, one box contains a body A and the other a body B. These bodies are arbitrary, except that they are at the same temperature $T$. Because it is isotropic, the radiation can be specified by the two functions $u(\nu,T,A,x)$, $u(\nu,T,B,y)$ giving the energy in the radiation per unit volume per unit frequency range in the two enclosures. The $A$ indicates the possible dependence of the first frequency distribution on the nature of the body A and the symbol $x$ stands for other characteristics such as the shape of the container. Let the separating wall be made transparent to radiation lying in the frequency interval $\nu_1$, $\nu_1 + d\nu$ by a small filter of negligible heat capacity. Then energy will be radiated from one side to the other, unless $u(\nu_1,T,A,x) = u(\nu_1,T,B,y)$. If the bodies A and B can emit and absorb radiation of all frequencies (at least in a very broad band of frequencies), it follows that the body in the energy-gaining enclosure will share in this energy gain with a consequent rise in temperature. Thus heat is in effect transferred from one body to another hotter than itself without producing any changes outside the system under consideration. This has been seen to be impossible provided the absolute temperature $T$ is positive (section 10.6). It follows that $u(\nu,T,A,x) = u(\nu,T,B,y)$ for all $\nu$, so that the variables, $A$, $x$, $B$, $y$ are redundant. Hence the energy spectrum of isotropic radiation, in equilibrium with a body at temperature $T$, is independent of other factors. Thus a temperature $T$ (the temperature of the walls) can be assigned to any isotropic radiation in equilibrium with its walls. Also, by adding a (small) piece of matter, capable of absorbing and emitting all frequencies, to radiation in an enclosure with perfectly reflecting walls, this particle will by absorption and emission enable radiation to come into thermal equilibrium with it. The initially arbitrary radiation is thus converted to isotropic radiation corresponding to equilibrium with a body at some temperature $T$. The particle acts as a kind of catalyst for this transformation.

It is seen, therefore, that, given a body (not a perfect reflector) at a temperature $T$, radiation which is in equilibrium with it can have only one type of spectral distribution, whatever the nature of the body and whatever the nature of the radiation enclosure. The spectral composition of its energy density can be specified by the variables $\nu$, $T$ so that $U/v = \int u(\nu,T)d\nu$. The function $u(\nu,T)$ will be obtained later in equation (13.15). This radiation is called *black-body radiation* (or *complete radiation* or *temperature radiation*). We shall use this name as a convenient way of referring to isotropic radiation which, when isolated from its surroundings and at rest, is in equilibrium with a body at temperature $T$. The reason for this particular choice of name is that an enclosure (as described) with a small opening is a very good approximation to an ideally black body in the sense that radiation falling on to the opening is unlikely to be reflected out of the enclosure.

## 13.2. THE THERMODYNAMIC PROPERTIES AS OBTAINED FROM STATISTICAL MECHANICS

The formal theory of Chapter 11 holds for black-body radiation if Bose statistics is adopted with

$$\boxed{A = 4\pi g v\, c_0^{-3} h^{-3} \qquad s=2,\ g=2,\ \mu=0\ .} \tag{13.1}$$

The first three conditions have already been discussed. It is known that there is no conservation law for the total number of photons, so that the grand canonical constraint specifying the average number of particles can be removed. This is achieved very simply in the present formalism by putting $\mu=0$.

The consequences of this condition will be investigated first. It proves convenient here to introduce a constant $D$ which depends on the value of the power $s$ and is defined by

$$D = k^{s+2}\,\Gamma(s+1)I(0,\ s+1,\ \pm)\,\frac{A}{v}\ . \tag{13.2}$$

Since $A$ is itself proportional to volume, $D$ is in fact a constant. One then finds from (11.6)

$$\begin{aligned} N &= A(kT)^{s+1}\,\Gamma(s+1)I(\gamma,s,\pm) \\ &= \frac{D}{k}\,\frac{I(0,s,\pm)}{I(0,s+1,\pm)}\,T^{s+1}\,v\ . \end{aligned} \tag{13.3}$$

Also from problems 11.1 and 11.2

$$U = A(kT)^{s+2}\,\Gamma(s+2)I(\gamma,s+1,\pm) = (s+1)D\,T^{s+2}\,v \tag{13.4}$$

$$S = (s+2)D\,T^{s+1}\,v. \tag{13.5}$$

Also from problem 11.3

$$p = \frac{A}{v}\,(kT)^{s+2}\,\Gamma(s+1)I(\gamma,s+1,\pm) = D\,T^{s+2}\ . \tag{13.6}$$

Hence,

$$H = U + pv = (s+2)D\,T^{s+2}v\ . \tag{13.7}$$

The heat capacities are

$$C_v = \left(\frac{\partial U}{\partial T}\right)_v = (s+1)\ (s+2)D\,T^{s+1}v \tag{13.8}$$

and, using a result of problem 5.2 ,

$$C_p = \left(\frac{\partial H}{\partial T}\right)_p = \infty\ . \tag{13.9}$$

The last result is clear from (13.6). Constant pressure implies constant temperature. Thus while $H$ can be changed in

(13.9) by changing the volume, the denominator in (13.9) is fixed. These results can be summarized by

$$\boxed{U = (s+1)pv = \frac{s+1}{s+2}TS = \frac{1}{s+2}TC_v = k^{s+2}\Gamma(s+2)I(0,s+1,\pm)(A/v)T^{s+2}v}. \tag{13.10}$$

Using the rest of (13.1) in (13.10) one finds

$$U = 3pv = \tfrac{3}{4}TS = \tfrac{1}{4}TC_v = aT^4 v = 3kT\frac{\zeta(4)}{\zeta(3)}N \tag{13.11}$$

where $$\sigma = \frac{2\pi^5 k^4}{15h^3 c_0^2} \quad \text{(see also p. 230)}$$

and $$a \equiv k^4\Gamma(4)I(0,3,\pm)\left(\frac{4\pi g}{c_0^3 h^3}\right) \tag{13.12}$$

Using $g = 2$ and $I(0,3,-) = \pi^4/90$ (see solution of problem 13.2) one finds

$$D = \frac{8}{45}\frac{\pi^5 k^4}{h^3 c_0^3} = 2\cdot 55 \times 10^{-15}\ \text{erg cm}^{-3}\ \text{K}^{-4}$$

$$= 2\cdot 55 \times 10^{-30}\ \text{kg m}^{-3}\ \text{K}^{-4} \tag{13.13}$$

and $a = 3D$. Using the Riemann zeta function $\zeta(n) \equiv \Sigma_{j=1}^{\infty} j^{-n}$ $(n \geq 1)$, it can be shown by expansion and term-by-term integration that

$$I(0,s,-) = \zeta(s+1) \tag{13.12a}$$

The fact that $U/v \propto T^4$ is known as the Stefan-Boltzmann law. Josef Stefan (1879) discovered it experimentally and Boltzmann linked it to what was already known of the thermodynamics of radiation.

The results (13.11) also show that in a quasi-static adiabatic compression or expansion of black-body radiation the following are constants because they depend only on the entropy:

$$Tv^{1/3}, \quad pv^{4/3}, \quad p/T^4, \quad pv/T, \quad Uv^{1/3}. \tag{13.14}$$

For an ideal gas of extreme relativistic *fermions or bosons* the density of states law (12.17) is precisely (13.1), and one finds

$$N = 8\pi g v \left(\frac{kT}{hc_0}\right)^3 I(\gamma,2,\pm),$$

$$U = 3NkT\,\frac{I(\gamma,3,\pm)}{I(\gamma,2,\pm)} = 3pv \;,$$

$$S = kN\left[\frac{I(\gamma,3,\pm)}{I(\gamma,2,\pm)} - \gamma\right]$$

In quasistatic adiabatic processes the first three quantities (13.14) remain constant since these systems are ideal quantum gases with $g = 1/3$ and the results of problem 6.6(d) apply. These systems are therefore thermodynamically somewhat analogous to black-body radiation.

## 13.3. THE DISTRIBUTION LAW FOR PHOTONS

The amount of energy in black-body radiation at temperature $T$ and in a frequency interval $(\nu,\nu+d\nu)$ is readily obtained from the sum over all modes of vibration

$$U = \sum_j \bar{n}_j\, e_j = \sum_j \frac{e_j}{\exp\{(e_j-\mu)/kT\}-1}\;. \tag{13.15'}$$

Substituting $e_j = h\nu_j$ as in (12.10') and putting $\mu = 0$ by (13.1)

$$U = \sum_j \frac{h\nu_j}{\exp(h\nu_j/kT)-1}\;.$$

Summing over the groups of $8\pi v c_0^{-3}\nu_i^2 d\nu_i$ modes with frequencies in the range $(\nu_i,\ \nu_i+d\nu_i)$, the spectral distribution of energy per unit volume per unit frequency range $d\nu$ is

$$\boxed{u(\nu,T) = \frac{8\pi\, hc_0^{-3}\nu^3}{\exp(h\nu/kT)-1}\;.} \tag{13.15}$$

This may also be spoken of as the mean photon energy per unit volume per unit frequency range; the photons are the excitations which are the carriers of the energy of the radiation. The system studied can be said to be a *gas of photons*.

The expression (13.15) is the famous Planck radiation law. It has the classical limit (as $T \to \infty$)

$$u(\nu,T) = 8\pi c_0^{-3} \nu^2 kT . \tag{13.16}$$

This is the classical expression obtained by Rayleigh (1900) and Jeans. They treated each normal mode of vibration in a radiation field as a harmonic oscillator. Its mean energy was taken as $kT$, by the classical equipartition theorem (11.33). However, (13.16) diverges for large $\nu$ and so cannot hold in the ultraviolet (the 'ultraviolet catastrophe'). In contrast, the Planck formula (13.15) has no such divergence. It yields the expression (13.11) for $U$ on integration over frequencies.

The distribution (13.15) also satisfies another result known from classical considerations: one can infer from it that

$$u(\nu,T) = \nu^3 F(h\nu/kT) \tag{13.17}$$

where

$$F(x) \equiv \frac{8\pi h c_0^{-3}}{e^x - 1} .$$

Condition (13.17) is known as Wien's theorem (1894). It was obtained by classical methods, the form of the function $F$ being left unspecified. It will not be proved here, as we have in any case the full expression for $F$.

If one writes (13.17) in terms of the wavelength $\lambda$,

$$u(\nu,T)\mathrm{d}\nu = u(\lambda,T)\mathrm{d}\lambda$$

that is, in what we hope is an obvious notation,

$$u(\lambda,T) = u(\nu,T)\,\frac{\mathrm{d}\nu}{\mathrm{d}\lambda} = \frac{8\pi h c_0}{e^x - 1}\left(\frac{xkT}{hc_0}\right)^5$$

where $x = hc_0/\lambda kT$. It follows that

$$u(\lambda,T) = \frac{8\pi(kT)^5 x^5}{(hc_0)^4(e^x - 1)} .$$

Now the energy density $u(\lambda,T)$ has a maximum at a specific value of $\lambda$, obtainable from the condition

$$\left\{\frac{\partial u(\lambda,T)}{\partial\lambda}\right\}_T = 0 .$$

This yields

$$\frac{5x^4}{e^x - 1} = \frac{x^5 e^x}{(e^x - 1)^2}$$

that is

$$\left(1 - \frac{x}{5}\right)e^x = 1 .$$

This equation has the solution $x = 4{\cdot}965$ so that the maximum in $u(\lambda,T)$ when plotted against $\lambda$ occurs when

$$\frac{hc_0}{\lambda kT} = 4{\cdot}965 \quad \text{when} \quad \lambda T = \frac{hc_0}{4{\cdot}965k} .$$

This condition is, in units of m K:

$$\lambda T = 2{\cdot}9 \times 10^{-3} \text{ m K} .$$

Thus the maximum moves to lower wavelengths (higher energies) as the temperature of the radiation is raised. This rule for the displacement of the maximum is known as Wien's displacement law and it was already known as a consequence of (13.17) before Planck's radiation law was discovered.

## 13.4. MATTER AND RADIATION IN THE UNIVERSE

A rough estimate of the average density of matter in the universe can be made from astronomical observations, and is

$$\rho_{m0} \sim 4 \times 10^{-27} \text{ kg m}^{-3} = 4 \times 10^{-30} \text{ g cm}^{-3} .$$

Using the proton rest mass of $1{\cdot}67 \times 10^{-27}$ kg, the equivalent density in terms of nucleons is

$$n_{m0} \sim 2\cdot 4 \text{ m}^{-3},$$

where subscripts '0' denote present values.

There has also recently been discovered a black-body radiation at about $T_{r0} = 2\cdot 7$K which is believed to be uniformly distributed throughout the universe and to be a relic of the 'big bang' which gave rise to the present expansion. This leads to a mean photon number density $n_{r0}$ and a mean energy density $\rho_{r0}$ of radiation. These quantities may be estimated from equations (13.3) and (13.11) to (13.13) as follows:

$$n_{r0} = \frac{D}{k}\frac{I(0,2,-)}{I(0,3,-)}T_{r0}^3 \sim \frac{2\cdot 521 \times 10^{-15}}{1\cdot 381 \times 10^{-16}}\frac{\zeta(3)}{\zeta(4)}(2.7)^3 \times 10^6 \text{ m}^{-3}$$

$$= 4 \times 10^8 \text{ m}^{-3}$$

so that $n_{r0}/n_{m0} \sim 1\cdot 7 \times 10^8$. Also we have

$$\rho_{r0} = \frac{U_{r0}}{v} = 3DT_{r0}^4 = 3T_{r0}^4 \frac{kn_{r0}}{T_{r0}^3}\frac{\zeta(4)}{\zeta(3)} = \frac{3\pi^4}{1\cdot 202 \times 90} kT_{r0}\, n_{r0} .$$

Since $kT \sim 1/40$ eV at $T = 300$K and

$$1 \text{ eV} = 1\cdot 783 \times 10^{-36} \text{ kg}$$

one finds

$$\rho_{r0} \sim 2\cdot 70 \times \frac{2\cdot 7}{40\times 300} \times (4\times 10^8) = 2\cdot 4\times 10^5 \text{ eV m}^{-3} = 0\cdot 43\times 10^{-30}\text{kg m}^{-3} .$$

The present photon entropy per nucleon is, by (13.11), in units of Boltzmann's constant

$$\sigma_0 = \frac{4DT_{r0}}{n_{m0}k} = \frac{2\pi^4}{45\zeta(3)}\frac{n_{r0}}{n_{m0}} = 3\cdot 60\,\frac{n_{r0}}{n_{m0}} \sim 6\cdot 12 \times 10^8 .$$

The above figures are very approximate to a factor of 10 or 1/10. However, they show clearly that the energy resides at present in matter rather than in radiation. Nonetheless the reservoir of energy in the background radiation is such as to enable one to ionize all the atoms in the universe at the cost of only a negligible drop in background temperature. This may be shown by using 13.6 eV as the energy to ionize an atom. With about 1 atom $m^{-3}$ one requires radiation energy per $m^3$ given by

$$\delta U_r/v = 13{\cdot}6 \text{ eV m}^{-3} .$$

Now by (13.11) $U_r \propto T_r^4$ so that

$$\frac{\delta T_{r0}}{T_{r0}} = \frac{1}{4}\frac{\delta U_r/v}{U_{r0}/v} = \frac{1}{4}\frac{13{\cdot}6}{2{\cdot}4\times 10^5} = 1{\cdot}42 \times 10^{-5}$$

so that the drop in temperature resulting from the ionizations is less than $4 \times 10^{-5}$ K.

As one goes back in time we shall assume (i) that the mass of matter $M_m$ is constant and (ii) that the radiation changes adiabatically. These assumptions can be justified for a certain broad class of conditions. From (ii) and (13.11) one finds $S_r \propto vT_r^3 \propto (RT_r)^3$ is a constant, where the 'scale factor' $R(t)$ determines the extent of the expansion of the universe. If $A$ and $B$ are constants, we then have from (13.14) and (13.11) that

$$(RT_r)^3 = A \;, \quad \rho_r = BT_r^4 \quad . \tag{13.18}$$

Also

$$\rho_m = \frac{M_m}{v} \sim \frac{M_m T_r^3}{A} = \frac{M_m}{A}\left(\frac{\rho_r}{B}\right)^{3/4} . \tag{13.19}$$

Thus as one goes back in time $\rho_r$ rises more rapidly than $\rho_m$, and a time $t_c$ will be reached at which

$$\rho_r = \rho_m \;(\equiv \rho_c \text{ say}) \quad .$$

One can calculate the value of $\rho_c$ by noting from (13.19) that $\rho_m^4/\rho_r^3$ is a constant during the portion of the evolution of the universe for which assumptions (i) and (ii) hold. If these conditions include both the present time $t_0$ and the time $t_c$, then

$$\rho_c = \left(\frac{\rho_m^4}{\rho_r^3}\right)_{t=t_c} = \frac{\rho_{m0}^4}{\rho_{r0}^3} \sim (9\cdot 3 \times 10^3)^3 \times 4 \times 10^{-27} \text{ kg m}^{-3}$$

$$= 3216 \times 10^{-18} \text{ kg m}^{-3} . \qquad (13.20)$$

The temperature at which this density occurs is given, by virtue of (13.18), by

$$T_c^4 \rho_{rc}^{-1} = T_c^4 \rho_{r0}^{-1} \qquad (13.21)$$

so that

$$T_c = \left(\frac{\rho_c}{\rho_{r0}}\right)^{1/4} T_{r0} \sim \left(\frac{3216\times 10^{-18}}{0\cdot 43\times 10^{-30}}\right)^{1/4} \times 2\cdot 7 = 25\ 100 \text{ K}. (13.22)$$

Prior to the time at which (13.20) and (13.22) holds, radiation was presumably the dominant component in the contents of the universe. The point $(\rho_c, T_c)$ is shown in Fig.13.1. The names of particle groups which occur in the figure are explained in Table 13.1.

If one goes back even further, a time $t_b$ is reached at which a large number of positron-electron pairs can be created by the radiation. If $m_0$ is the rest mass of an electron, this condition is given by

$$m_0 c^2 \sim kT_b$$

whence

$$T_b \sim \frac{9\cdot 1 \times 10^{-28} \times 9 \times 10^{20}}{1\cdot 3 \times 10^{-16}} \sim 5 \times 10^9 \text{ K} .$$

The radiation density at this time is given by a formula analogous to (13.21), i.e.

$$\rho_{rb} = \left(\frac{T_b}{T_{r0}}\right)^4 \rho_{r0} \sim \left(\frac{5\times 10^9}{2\cdot 7}\right)^4 \times 0\cdot 43 \times 10^{-30} = 5\cdot 1 \times 10^6 \text{ kg m}^{-3}.$$

This radiation is clearly extremely dense, for 1 mm$^3$ of it has

TABLE 13.1

*The elementary particles*

| Particles with a continuing existence | Particles which are carriers of interactions |
|---|---|
| i.e. they are constant in number<br>They can be created (or destroyed) *only if* an appropriate *antiparticle* is simultaneously created (or destroyed). For example, the neutron n decays to a proton p, an electron $e^-$, and and an antineutrino $\bar{\nu}$: $n \to p + e^- + \bar{\nu}$. This conserves the number of baryons (one, before and after) and creates a lepton-antilepton pair: $e^- + \bar{\nu}$. Particles are counted positive and antiparticles negative, so the total number of both leptons and baryons remains constant.<br>They are all *fermions*. | i.e. they are required to carry forces between particles; they *may* also exist on their own.<br>Their number is not constant; they may be created (or destroyed) in interactions. For example, photons can be emitted (or absorbed) by atoms; the number of photons does not remain constant.<br>They are all *bosons* |

| Leptons | Baryons | Quarks | Gluons | Mesons | Intermediate vector boson | Photon | Graviton |
|---|---|---|---|---|---|---|---|
| | | | | (carriers of the strong interaction) | (carrier of the weak interaction) | (carrier of electro-magnetism) | (carrier of gravity) |
| mass: light | mass: heavy | | | mass: medium to heavy | mass: very heavy | mass: zero | |
| Electron e<br>Neutrino $\nu$ | nucleons: proton p, neutron n<br>Strange baryons: $\Delta, \Sigma$<br>Doubly-strange baryons: $\Xi$ | u,d,s,c<br>Conjectured as an interpretation of the observed hadrons: | | Pion $\pi$<br>Kaon<br>$\eta, \omega \ldots$<br>$\psi, \psi' \ldots$ | $(W^+)$<br>$(Z^\circ)$<br>$(W^-)$ | $\gamma$ | ? |

| | | | | | | |
|---|---|---|---|---|---|---|
| Muon $\mu^-$<br>Muon-neutrino $\nu_\mu$<br>(apparently $(e^-,\nu)$ and $(\mu^-,\nu_\mu)$ form Separate classes of leptons) | Triply-strange baryons: $\Omega^-$<br>Charmed baryons<br>Plus excited states | Baryon = 3 quarks bound by gluons<br>Meson = quark + anti-quark | plus excited states | (not yet observed) | | |
| | Hadrons: particles sensitive to the strong interaction | | | | | |
| ← Can be involved in the weak interaction → | | | | | | |
| ← Can be involved in the electromagnetic interaction → | | | | | | |
| ← Can be involved in the gravitational interaction → | | | | | | |

A similar table constructed in 1930 would have had only the entries in the three dotted boxes.

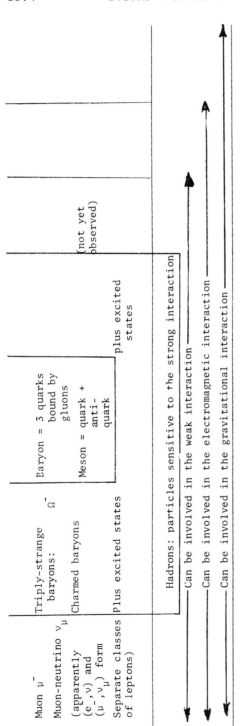

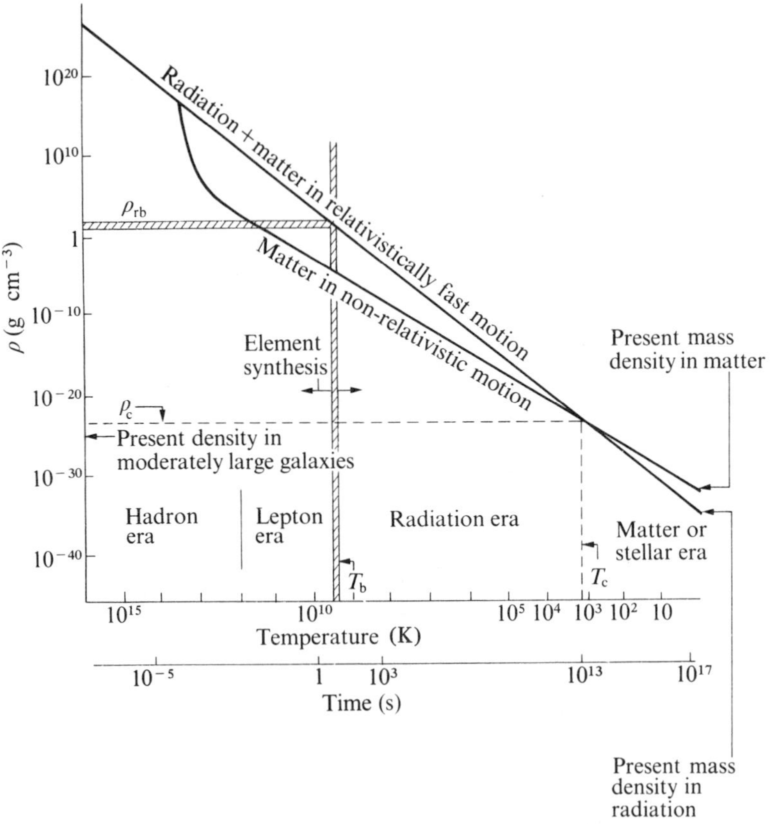


Fig.13.1. The composition of an expanding model universe changes as the temperature drops. In the hadron era strongly interacting particles predominate; the baryons annihilate and only 1 in $10^9$ of them survives as a nucleon. The lepton era ends when electrons are annihilated. The radiation era is over when the relative density of matter exceeds that of radiation (at about $10^6$ years). The present is in the stellar era.

an equivalent mass of about 5 g.

If we now go forward in time, the point $(\rho_b, T_b)$ indicates the end of the lepton era; it occurs a few minutes after the big bang and the temperature $T_b$ is low enough for electron-positron pairs to annihilate more rapidly than they are created. An even shorter time after the big bang is characterized by heavier particles, and this is the hadron era. It is not well understood, but it comes to an end when the temperature is low enough for hadron annihilation to dominate over hadron creation. Taking typically the pion mass ($m_h = 2\cdot5 \times 10^{-26}$Kg) as representative of the lighter hadrons, the hadron era comes to an end when the temperature is of order

$$m_h c^2/k \sim 1 \cdot 6 \times 10^{12} \text{ K}.$$

This also represents the beginning of the lepton era. The synthesis of helium and the lighter elements occurs close to the transition region from the lepton to the radiation era.

The above considerations are very rough and the numerical values indicate only orders of magnitude, but it is interesting how much insight one can gain by simple methods. Furthermore, since we have largely depended on thermodynamics, there is no time scale on these developments. One can call these considerations the 'statics of cosmology'. We shall now consider briefly the dynamics of cosmology.

## 13.5. THE EARLY UNIVERSE

A 'scale factor' $R(t)$ governs the expansion of a simple model of the universe whose uniform mass density is $\rho(t)$. Then $R(t)$ is small near the 'big bang' which is often taken to be the point in time beyond which we cannot investigate the behaviour of the universe. As the universe ages, $R(t)$ increases and the universe expands. It is then thought that the cosmological differential equation holds

$$\dot{R}^2 = \frac{8\pi}{3} G \, \rho(t) R^2(t) + \frac{\lambda}{3} R^2 - k c_0^2 .$$

Here $G$ is the gravitational constant, $\lambda$ is a constant governing a universal cosmological repulsion, $k = 0$, $+1$, or $-1$, and $c_0$ is the velocity of light *in vacuo*. The 'early' universe is such that radiation dominates among the contributions to $\rho(t)$. Assuming the expansion to be adiabatic, (13.14) suggests that $Uv^{1/3}$ is a constant. The matter density multiplied by $c_0^2$ is the energy density, so that

$$\rho c_0^2 \, v \times v^{1/3} \propto \rho c_0^2 R^4 \quad \text{is a constant, } A \text{ say.}$$

Accordingly

$$\dot{R}^2 = \frac{8\pi}{3c_0^2} G \, \frac{AR^2}{R^4} + \frac{\lambda}{3} R^2 - k c_0^2 .$$

For small $R$ the first term dominates and leads to

$$R\,\mathrm{d}R = \left(\frac{8\pi GA}{3c_0^2}\right)^{1/2} \mathrm{d}t \ .$$

Hence

$$\tfrac{1}{2}R^2 = \left(\frac{8\pi GA}{3c_0^2}\right)^{1/2} t$$

that is

$$R = \left(\frac{32\pi GA}{3c_0^2}\right)^{1/4} t^{1/2} .$$

By equation (13.14) $Tv^{1/3} \sim TR$ is another constant, $B$ say, so that

$$RT = B$$

so that

$$T(t) = \frac{B}{R(t)} = \left(\frac{3c_0^2}{32\pi Ga'}\right)^{1/4} t^{-1/2}$$

where

$$a' \equiv \frac{A}{B^4} = \frac{\rho c_0^2 R^4}{T^4 R^4} = \frac{U}{vT^4}$$

and this is just the expression for $a$ introduced in (13.11). Hence from (13.13),

$$a = 7{\cdot}56 \times 10^{-15}\ \mathrm{erg\ cm^{-3}\ K^{-4}} = 7{\cdot}56 \times 10^{-16}\ \mathrm{Jm^{-3}\ K^{4}} \ .$$

One finds therefore that the temperature of the early universe fell according to the law

$$T(t) \sim \frac{10^{10}}{\sqrt{t}}$$

where $T$ is in K and t is in s. This is of course very rough but gives some insight. Placing the beginning of the radiation era at $t$ = 10s and its end at $10^{13}$s, one sees that the temperature must have fallen in this period from $3 \times 10^{9}$ K to 3000 K.

If the time since the (last) big bang is inserted, say

$$t \sim 2 \times 10^{10} \text{ years} \sim 5 \times 10^{17} \text{ s}$$

one finds for the present average temperature of the universe 14 K, which is too high to account for the present temperature $T_{r0} \sim 2{\cdot}7$ K of the black-body background radiation mentioned in section 13.4. Considering that the formula was derived for the early universe, however, the agreement is not bad.

Contrary to one's expectations, the expanding universe seems to provide an example of the emergence of non-equilibrium states with $T_m \neq T_r$ from the near-equilibrium states with $T_m \sim T_r$ of the early universe when high density and temperature facilitated equilibrium. This phenomenon is easily understood in terms of two non-interacting ideal quantum gases which occupy the same volume $v$, have temperatures $T_1$ and $T_2$, and equations of state

$$v_1 = v_2 \qquad p_i v_i = g_i U_i \qquad (i = 1,2) .$$

If the volume is expanded adiabatically then problem 6.7*d* teaches one that $T_1 v^{g_1}$ and $T_2 v^{g_2}$ would each remain constant if the other gas were absent. If therefore the expansion is taking place too quickly for equilibrium to establish itself one would expect a temperature difference to develop from an initial equilibrium state. The gas with the larger $g$-value would be cooler. In the case of matter ($g_1 = 2/3$) and radiation ($g_1 = 1/3$) one would on this simple basis expect matter to be cooler while the expansion continues.

In the case of the universe, as in any other system, a return to equilibrium would depend on the interaction between matter and radiation which in turn depends on their densities. In fact it has been seen that for the early universe

$$T \propto t^{-1/2}, \rho \propto T^4 \propto t^{-2} .$$

As the total density decreases more rapidly than the temperature the re-establishment of equilibrium in an expanding universe may be long delayed.

As regards the paradox that 'disorder' seems to decrease in the expansion, it may well be that, because more states become distinguishable as a result of expansion, the disorder could be expected to decrease on theoretical grounds even though the entropy increases (see Appendix A, equation (A5)).

## 13.6. RADIATION IN EQUILIBRIUM WITH BLACK HOLES AND OTHER SYSTEMS

In this section we ask what the thermodynamic properties of a general system must be in order that black-body radiation can come into equilibrium with it. The analysis is purely thermodynamic in nature.

Suppose that the external parameters $X_1, X_2$ of two systems are kept constant and that the total energy $U_1 + U_2 = U$ is also kept constant. The total entropy is then

$$S = S_1(U_1, X_1) + S_2(U - U_i, X_2)$$

and depends on $U_1$ as the only variable. This superposition is the simplest assumption which can be made; note that it is artificial if gravitating systems are involved (as on p. 227). From a given entropy function one can derive the internal energy, the temperature, and the heat capacity $C_j$ as follows, $(j=1,2)$:

$$\left(\frac{\partial S_j}{\partial U_j}\right)_{X_j} = \frac{1}{T_j}\,, \quad U_j = \int^{T_j} \tau \left(\frac{\partial S_j}{\partial \tau}\right)_{X_j} \mathrm{d}\tau \tag{13.23}$$

$$C_j = \left(\frac{\partial U_j}{\partial T_j}\right)_{X_j} = T_j\left(\frac{\partial S_j}{\partial T_j}\right)_{X_j}$$

where $T\mathrm{d}S = C_X \mathrm{d}T + l_X \mathrm{d}X$ (= $\mathrm{d}U$ at constant $X$) has been used. For an entropy extremum, and using $\mathrm{d}U_2/\mathrm{d}U_1 = -1$,

$$\left(\frac{\partial S}{\partial U_1}\right)_{U, X_1, X_2} = \left(\frac{\partial S_1}{\partial U_1}\right)_{X_1} - \left(\frac{\partial S_2}{\partial U_2}\right)_{X_2} = \frac{1}{T_1} - \frac{1}{T_2} = 0\,,$$

so that $T_1 = T_2$ ($T$ say). Differentiating with respect to $U_1$ a second time, the condition for the extremum to be a maximum is

$$-\frac{1}{T_1^2}\left(\frac{\partial T_1}{\partial U_1}\right)_{X_1} + \frac{1}{T_2^2}\left(\frac{\partial T_2}{\partial U_2}\right)_{X_2}\frac{dU_2}{dU_1} = -\frac{1}{T^2}\left(\frac{1}{C_1}+\frac{1}{C_2}\right) < 0 \,.$$

Thus we have, as expected from (7.21), that for positive absolute temperatures

$$T_1 = T_2,\quad C_1^{-1} + C_2^{-1} > 0 \,. \tag{13.24}$$

The second condition is normally ($C_1, C_2 > 0$) satisfied. Only unstable equilibrium is possible if $C_1, C_2 < 0$ for the entropy is a minimum rather than a maximum at $T_1 = T_2$. An interesting case arises if only one system, say system 2, has a negative heat capacity. One then requires $|C_2| > C_1$, i.e. writing $S_j'$ for $(\partial S_j/\partial T_j)_{X_j}$ ,

$$C \equiv C_1 + C_2 = T(S_1' + S_2') < 0 \,. \tag{13.25}$$

The total heat capacity must be negative for the resulting equilibrium between the two systems to be stable.

An alternative argument to see that (13.25) must hold is as follows. Suppose heat $Q$ flows from system 2, of negative heat capacity $C_2$, to system 1, of positive heat capacity $C_1$, as a result of the inequality $T_2 > T_1$. Both systems become hotter, but for stability one requires that the hotter system shall be subject to a smaller rise in temperature as a result of the transaction and then

$$\frac{Q}{|C_2|} < \frac{Q}{C_1} \,.$$

This requires $|C_2| > C_1$, i.e. $C = C_1 + C_2 < 0$. This is again (13.25).

In particular one sees that a system of negative heat capacity $C_2$ cannot establish a stable equilibrium with a large heat reservoir (of positive temperature $T_1$ and positive heat capacity $C_1$) since then $C_1^{-1} + C_2^{-1} \sim C_2^{-1} < 0$, which violates (13.21). We therefore do not expect to give a statistical mechanical description of such a system in terms of a canonical or grand canonical ensemble, as these ensembles require contact with large heat reservoirs. This remark checks with the observation, following from equation (16.3), that $C_V > 0$ for any system described by a canonical ensemble.

Let us now consider equilibrium with black-body radiation by assuming, with $a$ an appropriate constant and $x = 3$,

$$S_1 = a\, T_1^x \,.$$

Hence

$$U_1 = \frac{x}{x+1}\, a\, T_1^{x+1} \,, \quad C_1 = a\, x\, T_1^x$$

where (13.23) has been used. Our conditions (13.24) for stable equilibrium are $[S_2' \equiv (\partial S_2/\partial T_2)_{X_2}]$

$$T_1 = T_2(\equiv T) \,, \quad (a\, x\, T_1^x)^{-1} + (T_2\, S_2')^{-1} > 0$$

*Case 1. Positive heat capacity: black-body radiation in contact with a special kind of ideal quantum gas*

Assume that

$$S_2 = (1+g)\, Bv\, T_2^{1/g} \,.$$

Then

$$U_2 = Bv\, T_2^{1+1/g} \,.$$

Here $v$ is the volume of the system and $B$ and $g$ are positive constants. The ideal quantum gases have been defined in problem 6.6 , and we here have the special case where

$$\text{in } U_2 v^g = f(T_2 v^g), \quad f(z) = Bz^{1+1/g}$$

and

$$\text{in } S_2 = h(T_2 v^g), \qquad h(z) = (1+g)Bz^{1/g} \,.$$

Thus $S_2' > 0$ and a stable equilibrium is possible by (13.24). Stable equilibrium between two kinds of black-body radiation is of course also possible, and is a special example for $g = 1/3$. By (13.24) it implies that the two black-body temperatures must then be equal.

*Case 2. Negative heat capacity: radiation in equilibrium with black holes*

We appeal to the following result of the general theory of relativity, as discussed in section 15.8. A non-rotating uncharged black hole of mass $M$ can be said to have an internal energy and an entropy given in terms of its temperature by

$$U_2 = \frac{\hbar c^5}{8\pi GkT_2} \qquad S_2 = \frac{\hbar c^5}{16\pi GkT_2^2}$$

where $k$ is Boltzmann's constant and $G$ the gravitational constant. Such a system has a negative heat capacity characteristic of gravitational systems (see section 18.1):

$$\frac{dU_2}{dT_2} = -\frac{\hbar c^5}{8\pi GkT_2^2} = -\frac{U_2}{T_2}$$

and therefore its temperature rises as it loses energy.

For the total system (13.24) can again be satisfied. The condition for this is $C_1 < |C_2|$, i.e. $4U_1/T < U_2/T$. Thus the energy which resides in the black holes must for stable equilibrium be in excess of four-fifths of the total energy. However, as $S_2 \propto T_2^{-2} \propto U_2^2 \propto M^2$, the fourth law of thermodynamics (p. 79) fails, and procedures using entropy maximization are therefore subject to some doubt, as noted on p. 369.

The simple-looking equation $T_1 = T_2$ will be discussed further for case 2. Using (13.11) and the entropy of a non-rotating uncharged black hole $S_2 = 4\pi kG\, U_2^2/\hbar c^5$, the equation $T_1^{-1} = T_2^{-1}$ is

$$\left\{\frac{8\pi^5 k^4}{15(8\pi^3\hbar^3)c^3}\,\frac{v}{U-U_2}\right\}^{1/4} = \frac{8\pi kG\, U_2}{\hbar c^5}$$

Raising this equation to the fourth power, and rearranging,

$$\left(\frac{U_2}{U}\right)^4 - \left(\frac{U_2}{U}\right)^5 = \frac{\hbar c^{17} v}{2^{12} 15\pi^2 G^4 U^5} \equiv \frac{2^8}{5^5} \frac{v}{v_h} .$$

The total energy $U$ of the system is a constant, so that $v_h$ is also a constant:

$$v_h = \frac{3\times 2^{20} \pi^2 G^4 U^5}{5^4 \hbar c^{17}} .$$

Let $U_2 = \lambda U$ and consider the function

$$y(\lambda) \equiv \lambda^4 (1-\lambda) - 2^8 v/5^5 v_h .$$

The solution of $y(\lambda) = 0$ gives the values of $\lambda$ for thermal equilibrium between the two systems. However, since $y(\lambda) < 0$ for small positive $\lambda$, there are no solutions, and hence the black hole will evaporate into radiation. In fact, to ensure that there is a solution of $y(\lambda) = 0$, observe that $y(0) = y(1)$ is negative. The maximum, lying between $\lambda = 0$ and $\lambda = 1$, occurs at $\lambda_m = 4/5$ and has the value

$$y(\lambda_m) = \left(\frac{4}{5}\right)^4 \times \frac{1}{5} - \frac{2^8 v}{5^5 v_h} = \frac{2^8}{5^5}\left(1 - \frac{v}{v_h}\right) .$$

Equilibrium is possible only if $y(\lambda_m) > 0$, i.e. if $v < v_h$: the enclosure must contain enough energy $U$.

## 13.7. THE NUMBER OF PHOTONS EMITTED BY A BLACK BODY PER UNIT TIME PER UNIT AREA

We now return to a classical topic. If radiation in a frequency range $(\nu, \nu + d\nu)$ falls on a surface, the fraction of the energy $dE$ falling on it per unit area per unit time to be absorbed is called the absorptive power $\alpha(\nu)$. Thus $\alpha(\nu)dE$ is the energy absorbed per unit area per unit time. The energy emitted by unit area in unit time by the surface in the range $(\nu, \nu + d\nu)$ is $e(\nu)d\nu$ and must clearly depend on the temperature of the surface. In equilibrium between the surface and the surroundings one must have

$$\alpha(\nu)\, dE = e(\nu)\, d\nu .$$

The incident energy in any frequency interval can always be taken from a black-body radiator so that $dE/d\nu$ is then known and is only a function of the temperature of the black body and of the frequency under investigation. It follows that $\alpha(\nu)/e(\nu)$ is a function of temperature and frequency only. This is *Kirchoff's law*. The argument given is rather rough since $\alpha$ and $e$ could, in principle, depend on the angles made by the incident radiation with the normal to the surface, as well as on the point of the surface on which the rays are incident. However, in an elementary account these possible complications are usually ignored.

It is shown in the kinetic theory of gases that, if a gas has a molecular concentration $n$ and the molecules have a mean velocity $c$, the number of collisions made by molecules with one side of a surface placed in the gas is $\frac{1}{4}nc$ per unit per unit time. Accordingly this number for photons of frequency range $(\nu, \nu + d\nu)$ in black-body radiation is by (13.15) (divided by $h\nu$ to obtain the photon concentration)

$$\frac{2\pi c_0^{-2}\ \nu^2\ d\nu}{\exp(h\nu/kT)-1} = \frac{2\pi(kT)^3 x^2 dx}{h^3 c_0^2 (e^x-1)} \quad . \tag{13.26}$$

The number is also a measure of the *emissive power of photons* for a black body of temperature $T$ in frequency range $(\nu, \nu + d\nu)$. This formula is used in many contexts (see problem 18.3).

The number of photons emitted per unit area per unit time by a black body at temperature $T$ in a frequency range $(\nu, \nu + d\nu)$ and into a solid angle $d\omega$ lying in a direction given by the angles $\theta$ and $\phi$ is

$$\frac{2\nu^2\ c_0^{-2}}{\exp(h\nu/kT)-1}\ \cos\theta\ \ d\nu\ d\omega \ . \tag{13.26a}$$

The factor $\cos\theta$ is characteristic of a whole class of emitters called Lambertian since they obey *Lambert's law*. This law states that the emission into various directions $(\theta,\phi)$ is proportional to the cosine of the angle $\theta$ made by the direction of emission with the normal to the surface. If one integrates over $d\omega$ one recovers (13.26) since

$$\int_0^{+1} \cos\theta \; d(\cos\theta) \int_0^{2\pi} d\phi = \pi .$$

The cosine of the angle occurs in the expression for the number of photons (or the energy) emitted per unit solid angle in the $(\theta,\phi)$-direction if it is assumed that the radiation from each surface element $da$ is proportional to the projection of the element on a plane which has the $(\theta,\phi)$-direction normal to it. An observer viewing $da$ at this angle will see the radiating element foreshortened by $\cos\theta$ and would thus regard the object as more luminous as $\theta$ increases. The factor $\cos\theta$ in (13.26a) just cancels out this increase. A luminous sphere thus appears uniformly luminous from all angles if it is Lambertian, the sun being a good example.

To obtain the energy emitted per unit area per unit time by a black body at temperature $T$ one has to multiply (13.26) by $h\nu$ and integrate to find an emissive power per unit area, usually denoted by $\sigma T^4$:

$$\sigma T^4 = \frac{2\pi (kT)^3}{h^3 c_0^2} \int_0^\infty \frac{x^2 \; h\nu \; dx}{e^x - 1} = \frac{2\pi (kT)^4}{h^3 c_0^2} \int_0^\infty \frac{x^3 \; dx}{e^x - 1} .$$

The integral has the value $\pi^4/15$ (see problem 13.2) so that

$$\sigma T^4 = \frac{2\pi^5 k^4 T^4}{15 h^3 c_0^2} . \tag{13.26b}$$

The numerical value of *Stefan's constant* $\sigma$ is

$$\sigma \sim 5{\cdot}67 \times 10^{-8} \text{ Wm}^{-2} \text{ K}^{-4}$$

The fact that the total emitted radiation is proportional to $T^4$ is *Stefan's law* (1879). We shall now apply this result to solar radiation.

The *solar variable* (often called *solar constant*) $f$ is defined as the energy received per unit time per unit area (perpendicular to the radiation) at the earth's mean distance $R$ from the sum. Assuming the sun to be a black body sphere of radius $R_s$ and temperature $T_s$, $f$ can be calculated from the equation

$$\sigma T_s^4 \times 4\pi R_s^2 = f \times 4\pi R^2 .$$

The left-hand side gives the rate of energy flow across the sphere of radius $R_s$, the right-hand side gives the flow across the sphere of radius $R$, the sun being the centre of the sphere. Hence

$$f = (R_s/R)^2 \sigma T_s^4 .$$

Taking $T_s = 6000\text{K}$, $R = 1{\cdot}5 \times 10^8$ km, and $R_s = 6{\cdot}4 \times 10^5$ km as an *effective* radiative solar disc radius one finds

$$f = \left(\frac{6{\cdot}4}{1{\cdot}5\times10^3}\right)^2 \times (5{\cdot}67\times10^{-8})(6000)^4 = 1353 \text{ Wm}^{-2} ,$$

in agreement with the accepted value. Alternatively, one can chose an *effective* temperature $T_s = 5760$ K, and the known outer radius $R_s = 6{\cdot}95 \times 10^5$ km of the sun. One then finds a slightly low value

$$f = 1349 \text{ Wm}^{-2} .$$

The use of effective values is needed since the sun's radiation intensity drops to about 75% of its disc-centre-value at the edge. Hence either a temperature reduced from the surface temperature of 6000 K or a reduced solar radius is needed in this simple calculation. Apart from small variations in $f$ with time a systematic increase in $f$ is expected (by a few per cent in $10^9$ years) from our knowledge of the evolution of stars. The current solar emission rate is

$$\sigma T^4 4\pi R_s^2 \sim 5{\cdot}67 \times 10^{-8} (6000)^4 4\pi(6{\cdot}4\times10^8)^2 \sim 4 \times 10^{26}\text{W}.$$

It leads to solar power incident on earth equivalent to 396 000 60 Watt bulbs burning per person. World power consumption comes to only 33 such bulbs per person, food consumption accounting for two of them ( see p.438).

## *13.8. THERMODYNAMICS OF ENERGY CONVERSION

Thermodynamics provides the natural tools for the investigation of problems of energy conversion. In this section a simple framework will be given for this purpose. Consider a device which absorbs energy from some 'pumping' system and emits energy in the form of light while at the same time being in contact with a heat reservoir at temperature $T$. The actual processes going on in the device will be of no interest for the purposes of the general theory, so that the device is essentially a 'black box'. We shall describe as 'heat' both electromagnetic radiation and the thermal motion of particles. The term 'light' will mean electromagnetic radiation in excess of black-body radiation at the temperature $T$. That electromagnetic radiation can act as both heat and light will cause no problem. It will be assumed that a *single* temperature $T_p$ can be assigned to the pumping system, which can give up energy $E_p$ and entropy $S_p$ to the black box. The pump is the part of the device which forces it to depart from thermal equilibrium. Thus a laser which is activated by incident radiation is covered by the analysis, provided only an effective temperature can be associated with the incident radiation. If a laser is pumped electrically the pumping temperature will in general be that of the electrical circuit used, which will be the ambient temperature, and in general also that of the black box:

$$T_p = T. \tag{13.27}$$

If it is pumped electrically *and* optically, then there are *two* temperatures to be associated with the pumping system, and such cases are here excluded for simplicity.

It will be convenient to denote the energy and entropy of the box by $E$ and $S$ respectively, and the rate at which entropy is generated in the box by $\dot{S}_g$. Let $\dot{E}_s$ and $\dot{S}_s$ be the rates at which the radiation emitted by the box carries away energy and entropy. More generally, energy and entropy are given up to a *sink*.

In order to explain the origin of the entropy generation inside the box, suppose that the pumping system consists of incident radiation with some arbitrary frequency spectrum. The rate $\dot{S}_p$ at which this pump gives up entropy to the box is a function of the randomness in this spectrum, that is of the extent to which it is not monochromatic, and it also depends on the directions of the photons. This randomness is reproduced in the box, and may here be increased. For example, if the radiation excites electrons from a valence band of a semiconductor, the electrons can then lose energy to the lattice in the form of heat (i.e. of phonons, see Chapter 15) as they make transitions to new states. This heating up of the lattice is responsible for the term $\dot{S}_g$. Thus $\dot{S}_g \geq 0$ in this case, and this inequality is true generally. It is an expression, in the thermodynamics of non-equilibrium states, of the second law of thermodynamics. The first and second law of thermodynamics can now be expressed by

$$\dot{E} = \dot{E}_p - \dot{Q} - \dot{E}_s - \dot{W} \tag{13.28}$$

$$\dot{S} = \dot{S}_p - \frac{\dot{Q}}{T} - \dot{S}_s + \dot{S}_g, \quad \dot{S}_g \geq 0 \ . \tag{13.29}$$

Here the surroundings of the device are also assumed to be at temperature $T$ and the surroundings are assumed to take up heat from the box at the rate $\dot{Q}$ while the box or device yields useful work at the rate $\dot{W}$. Mechanical work, being 'ordered' energy, has no entropy term associated with it. We shall neglect this entropy term also if $W$ denotes energy deposited in the form of a new chemical material, or 'biomass' etc. Multiplying equation (13.29) by $T$ and subtracting from (13.28) gives

$$\dot{W} + \dot{E}_s - T\dot{S}_s = \dot{E}_p - T\dot{S}_p - (\dot{E} - T\dot{S}) - T\dot{S}_g \ . \tag{13.30}$$

Eliminating the entropies by $S_i = (E_i - F_i)/T_i$, where $F_i$ is the free energy and the temperatures $T_i$ are independent of time,

$$\dot{W} + \left(1 - \frac{T}{T_s}\right)\dot{E}_s = \left(1 - \frac{T}{T_p}\right)\dot{E}_p + \frac{T}{T_p}\dot{F}_p - \frac{T}{T_s}\dot{F}_s - \dot{F} - T\dot{S}_g \ . \tag{13.31}$$

Two efficiencies are convenient. The first measures the energy transformation from pump energy into light emission or, more generally, into sink energy, using (13.31),

$$\eta_s = \frac{\dot{E}_s}{\dot{E}_p} = \frac{1-T/T_p}{1-T/T_s} + \frac{(T/T_p)\dot{F}_p - (T/T_s)\dot{F}_s - (\dot{F}+T\dot{S}_g+\dot{W})}{(1-T/T_s)\dot{E}_p} \ . \tag{13.32}$$

The second efficiency measures the energy transformation into useful work, using (13.30),

$$\eta_W \equiv \frac{\dot{W}}{\dot{E}_p} = 1 - \frac{T\dot{S}_p}{\dot{E}_p} - \left(1 - \frac{T\dot{S}_s}{\dot{E}_s}\right)\frac{\dot{E}_s}{\dot{E}_p} - \frac{\dot{E}-T\dot{S}+T\dot{S}_g}{\dot{E}_p} \ . \tag{13.33}$$

Lastly, it is convenient to have an expression for $\dot{Q}$. This is obtained from (13.29) as

$$-\dot{Q} = T\dot{S} - \frac{T}{T_p}(\dot{E}_p - \dot{F}_p) + \frac{T}{T_s}(\dot{E}_s - \dot{F}_s) - T\dot{S}_g \tag{13.34}$$

that is

$$\left.\begin{aligned} -\dot{Q} &= \frac{T}{T_s}\dot{E}_s - \frac{T}{T_p}\dot{E}_p + \frac{T}{T_p}\dot{F}_p - \left(\frac{T}{T_s}\dot{F}_s - T\dot{S} + T\dot{S}_g\right) \\ &= \dot{E}_s - \dot{E}_p + \dot{E} + \dot{W} \,. \end{aligned}\right\} \tag{13.35}$$

An important observation one can make at once is that in the case of continuous light emission (without work being delivered) in a steady state of the device, i.e. when $\dot{E} = 0$ and $\dot{W} = 0$, by (13.28) $\dot{Q} = \dot{E}_p - \dot{E}_s$. This shows that $\eta_s > 1$ and $\dot{Q} < 0$ are then equivalent statements. This means that when the efficiency of energy conversion into light is in excess of unity, then the extra energy which has been converted comes from the surroundings. In fact the device acts as a refrigerator. (Incidentally, in the steady state $\dot{S} = 0$ as well, but we did not need to use (13.29).)

One can obtain an expression for $\dot{F}_p$ by noting that by problem 7.1

$$dF = \mu\, dn - p\, dv$$

for constant temperature. Suppose the pumping action requires the transfer at constant volume of $j$ particles per unit time from a region of chemical potential $\mu_1$ to one of chemical potential $\mu_2$, both regions being in the pump. Then if the regions are labelled 1 and 2, the first region gains free energy $\delta F_1 = \mu_1$ per electron, the second region gains $\delta F_2 = \mu_2$ per electron, and the free energy gained by the pumping circuit by virtue of the transfer of one electron is $\delta F_1 + \delta F_2 = \mu_2 - \mu_1$. Since $j$ electrons are transferred per unit time, free energy is given up by the pump at a rate

$$\dot{F}_p = (\mu_1 - \mu_2)j \quad . \tag{13.36}$$

As a *first* example consider an optically pumped laser - a laser powered by incident radiation. In this case $\dot{W} = 0$ (no work is done), $\dot{F} = 0$ (the laser is in a steady state), and $\dot{F}_p = \dot{F}_s \sim 0$. This last result comes from an extrapolation of $G = \mu N = 0$ for equilibrium radiation to non-equilibrium. The Gibbs free energy $G = F + pv$ vanishes therefore and $\dot{G} = \dot{F} = 0$ if $pv$ is constant in time. Thus (13.31) yields

$$T\dot{S}_g = \left(1 - \frac{T}{T_p}\right)\dot{E}_p - \left(1 - \frac{T}{T_s}\right)\dot{E}_s \quad (\dot{F}_p = \dot{F}_s = 0). \tag{13.37}$$

Also from (13.32)

$$\eta_s \leq \frac{1-T/T_p}{1-T/T_s} \tag{13.38}$$

which can exceed unity only if radiation is 'degraded' from a higher to a lower temperature $T_p > T_s > T$. In fact ideal laser radiation carries energy but no entropy (section 13.10), whence $\dot{F}_s = \dot{E}_s - T_s\dot{S}_s \sim 0$ shows that $T_s = \infty$ in this approximation. Equation (13.35) then yields

$$\dot{Q} = \left(\frac{\dot{E}_p}{T_p} + \dot{S}_g\right)T . \tag{13.39}$$

Thus as a result of being heated up the laser transfers heat and the maximum efficiency is of the Carnot type.

As our *second* example consider an electroluminescent diode which is pumped by injection of current carriers. This implies, in the present context, that the external circuit supplies the pumping action and (13.27) holds. In (13.36)

$$\mu_1 - \mu_2 = eV \tag{13.40}$$

where $e$ is the numerical charge on the electron and $V$ is the applied voltage. It follows that $\dot{F}_p = jeV$. The rate at which the external circuit supplies energy to the box is the product of the current $ej$ and the applied voltage, so that

$$\dot{E}_p = eVj = \dot{F}_p \quad \text{so that} \quad T_p\dot{S}_p = \dot{E}_p - \dot{F}_p = 0 . \tag{13.41}$$

*Whereas* $\dot{F}_p$ *was assumed to vanish in the first example,* $\dot{S}_p$ *vanishes in the present case.* If $h\nu_0$ is the mean energy of an emitted photon, we have $jh\nu_0$ as the rate at which energy is carried away by the light. However, this assumes that each electron transition in the pump creates a photon in the box. In fact this so-called quantum efficiency $\theta$ is usually less than unity, so that it is better to put

$$\dot{E}_s = \theta j h\nu_0 \ . \tag{13.42}$$

We shall assume that $T = T_D$, that no mechanical work is done, and that the emitted light carries no free energy. Inserting equations (13.27), (13.41), and (13.42) in equations (13.32) to (13.35)

$$T\dot{S}_g = eVj - \theta j h\nu_0 \tag{13.43}$$

$$\eta_s \equiv \frac{\theta h\nu_0}{eV} < \frac{1}{1-T/T_s} \tag{13.44}$$

$$-\dot{Q} = j(\theta h\nu_0 - eV) < \frac{jeV}{(T_s/T)-1} \ . \tag{13.45}$$

Diodes have been found for which $\dot{Q} < 0$, and which can therefore, by absorbing heat, act as refrigerating devices. This leads to the experimentally confirmed possibility of 'optical refrigeration'. Note that, as already anticipated, $\dot{Q} < 0$ implies $\eta_s > 1$.

The following numerical example is based on a single crystal of ZnS. With an applied voltage of 1·6V the work done per electron transition in the pump (an injection process was used) was 1·6 eV. This led to light emission with a flux of 100 photons $\mathrm{cm^{-2}s^{-1}}$ and a frequency spectrum which had a maximum at $\nu_0 = 7 \times 10^{14}\ \mathrm{s^{-1}}$ ($h\nu_0 \sim 2{\cdot}8$ eV) and a half-width of $\Delta\nu = 10^{14}\ \mathrm{s^{-1}}$. Using (13.26) with $d\nu$ replaced by $\Delta\nu$ and $\nu$ replaced by $\nu_0$, one now has a formula for an effective incident radiation temperature $T$ which can be taken to give an estimate for $T_s$. One finds $T_s \sim 640$ K, and the upper limit for the efficiency is

$$\frac{1}{1-T/T_s} = \frac{1}{1-300/640} \sim 1{\cdot}9$$

The light emitted per electron transition in the pump can therefore have a theoretical upper limit in energy given by

$$\eta \times (\text{pump energy in}) \sim 1{\cdot}9 \times 1{\cdot}6 = 3{\cdot}0 \text{ eV}.$$

The actual photon energy emitted was 2·8 eV, and so was well within the thermodynamic constraint.

As our *third* example, let $W$ represent chemical work in the production of photosynthetically produced material. Neglecting the reradiated energy $E_s$ and using $\dot{F}_p \sim 0$ for the incoming (pumping) radiation, (13.33) yields in the steady state the Carnot efficiency

$$\eta_W \le 1 - T/T_p \ .$$

This could in principle be applied for example to a leaf. In fact (13.33) lends itself to a deduction of the Carnot efficiency by specifying (i) a definite pumping temperature $T_p = \dot{E}_p/\dot{S}_p$, and (ii) the absence of a sink $\dot{E}_s = \dot{S}_s = 0$.

In the first and third examples $\dot{F}_p$ applied to radiation has been taken to be zero. This is an assumption which is not always valid. In particular, if one is dealing with black-body radiation at temperature $T$, then by (13.11)

$$\dot{F} = \dot{E} - T\dot{S} = \dot{E} - \frac{4}{3}\dot{E} = -\frac{1}{3}\dot{E} \neq 0 \ .$$

In that case the expression (13.33) is

$$\eta_W = 1 - \frac{4}{3}\frac{T}{T_p} - \left(1 - \frac{4}{3}\frac{T}{T_s}\right)\frac{\dot{E}_s}{\dot{E}_p} - \frac{\dot{F}+T\dot{S}_g}{\dot{E}_p} \ . \tag{13.46}$$

Treating the pumping and the emitted radiation as black-body, $\dot{E}_s/\dot{E}_p \sim (T_s/T_p)^4$, so that for $T \sim T_s$ we have in the steady state

$$\eta_W \le 1 - \frac{4}{3}\frac{T}{T_p} + \frac{1}{3}\left(\frac{T}{T_p}\right)^4 \equiv \eta_W^* \ . \tag{13.47}$$

This upper limit has the sensible property that (Fig.13.2)

$$\eta_W^* = 1 \quad (\text{box at } T = 0)$$

and

$$\eta_W^* = 0 \quad (T = T_p) \ .$$

It is a smaller efficiency than the Carnot efficiency, but it has to be reduced still further for photosynthesis since the ineffectiveness of low-energy solar photons, for example, has here been neglected.

## *13.9. THE EFFECTIVE TEMPERATURE CONCEPT

In this subsection we elucidate the effective temperature $T_p$ of (13.46) by first introducing the *specific intensity of radiation* $K$.

In an isotropic medium filled with radiation consider a small collection of rays (a 'ray bundle' or a 'beam of radiation'). Its

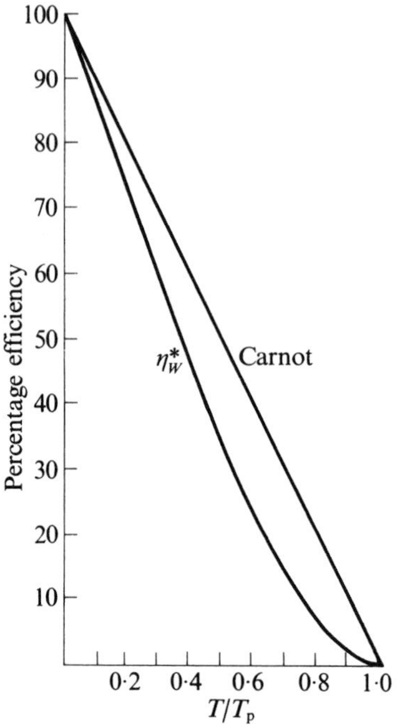


Fig.13.2. The Carnot efficiency compared with the 'black-body radiation in - black-body radiation out' efficiency $\eta_W^*$ of (13.47).

linear dimensions must be much larger than the wavelengths which dominate its properties, since this enables one to neglect diffraction and stay within ray optics. Let P and Q be two points in the beam at a distance $R$ apart (Fig.13.3). A plane through P intersects the beam in a section of area $\delta A_1$ whose normal makes an angle $\theta_1$ with the beam. The corresponding quantities at Q be $\delta A_2$ and $\theta_2$. The energy passing through $\delta A_1$ in a time interval $\delta t$, in a frequency interval $(\nu, \nu+\delta\nu)$ and in a solid angle $\delta\omega_1$ (see Appendix III) is

$$K_1(\nu;\theta_1,\phi_1;\mathrm{P},t)\cos\theta_1\ \delta\omega_1\delta A_1\delta t\delta\nu \equiv K_1\delta T_1\delta t\delta\nu \qquad (13.48)$$

where

$$\delta T_1 \equiv \cos\theta_1\ \delta\omega_1\delta A_1$$

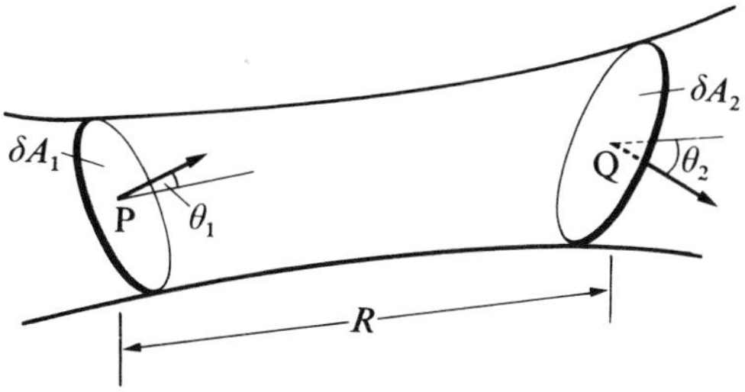


Fig.13.3. A ray bundle or beam of radiation.

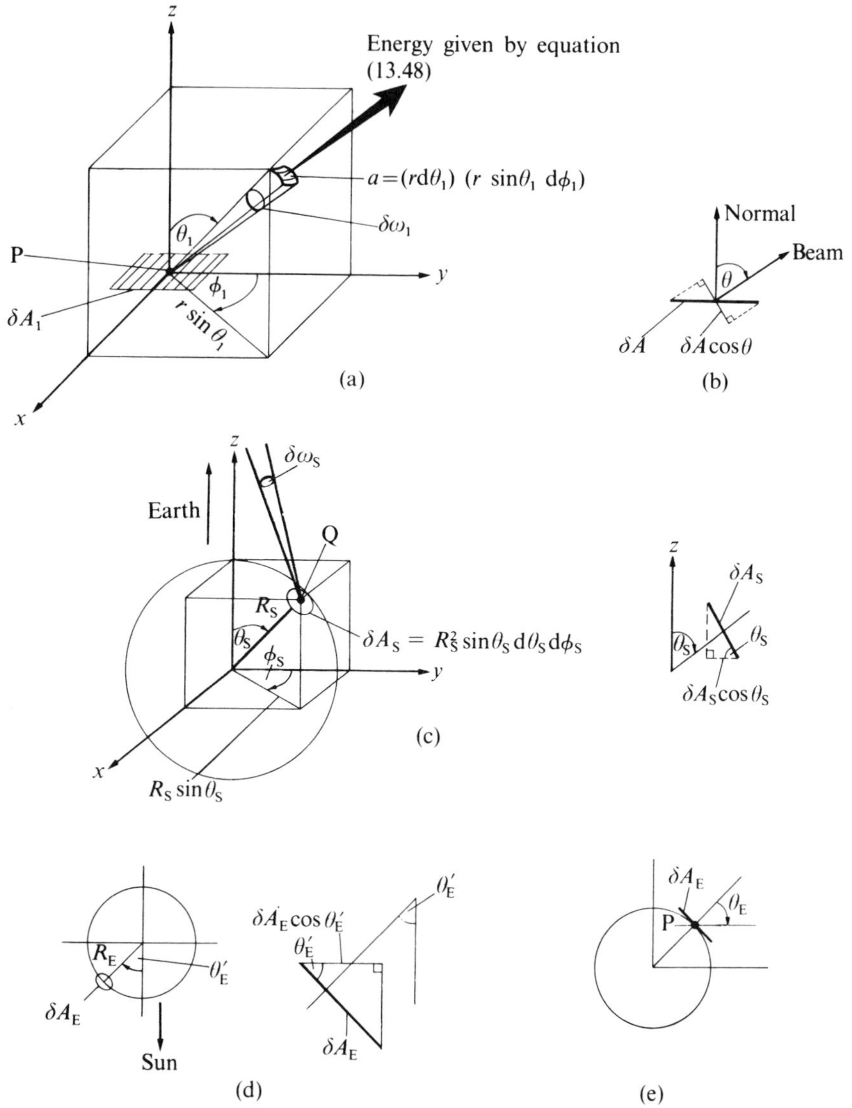


Fig. 13.4. Aids for section 13.9.

(Fig.13.4a). The quantity $K_1$ gives the energy per unit time per unit solid angle per unit frequency range per unit area when projected to be normal to the direction $(\theta,\phi)$ specifying the infinitesimal solid angle $\delta\omega_1$. As $K_1$ refers to an area normal to the beam it is multiplied by $\delta A_1 \cos\theta_1$ and not just by $\delta A_1$. The quantity (13.48) for Q is

$$K_2(\nu;\theta_2,\phi_2;Q,t)\cos\theta_2\ \delta\omega_2\delta A_2\delta t\delta\nu = K_2\delta T_2\delta t\delta\nu \quad . \tag{13.49}$$

For each interval $\delta\nu$ and $\delta t$ the energy passing through $\delta A_1$ arrives at, and passes through, $\delta A_2$. In order to calculate the energy arriving at Q one has to interpret the area $a$ of Fig. 13.4a as the projection of $\delta A_2$ to be normal to the incident radiation, so that (Fig. 13.4b)

$$\delta\omega_1 = \delta A_2 \cos\theta_2/R^2 \ .$$

Similarly to calculate the energy coming from P one has to use for $\delta\omega_2$ the solid angle subtended by $\delta A_1$ at Q:

$$\delta\omega_2 = \delta A_1 \cos\theta_1/R^2 \ .$$

With this understanding energy conservation yields

$$K_1\delta T_1 = K_2\delta T_2 \ . \tag{13.50}$$

Also

$$\delta T_1 = \delta A_1\delta A_2 \cos\theta_1 \cos\theta_2/R^2 = \delta T_2 \ , \tag{13.51}$$

so that $K_1 = K_2$. Thus *$K$ is invariant as one moves along a small beam in an isotropic medium.* For a uniform beam one can drop the parameters $P$ and $Q$ from (13.48) and (13.49). In a steady state the time parameter can be suppressed. The case of a variable refraction index is considered in problem 13.8.

$K$ is also called the *spectral brightness*. When integrated over frequencies it is called the *radiance* of a radiation field in the subject of radiometry. Illuminating engineers also require this quantity in photometry. It is then called the *luminance* of a surface (see problem 13.9).

If the radiation is isotropic, $K$ is independent of the angles $\theta$ and $\phi$, and the radiation has a certain energy per unit volume

$$\frac{1}{v} U \equiv \int u(\nu)\mathrm{d}\nu$$

which is the same throughout the medium. The energy passing through $\delta A$ in time $\delta t$ due to radiation in a frequency range $\delta\nu$ is, after integration over a solid angle of $2\pi$ corresponding to a hemisphere,

$$2\pi\ K(\nu;\theta,\phi)\ \delta A\ \delta t\ \delta\nu = \tfrac{1}{2}u(\nu)\ \delta A\ (c_0\delta t)\delta\nu \quad .$$

The right-hand side arises from the radiation passing through the surface in one direction (hence the factor $\frac{1}{2}$) in time $\delta t$. This radiation lies in a cylinder whose axis has length $c_0\ \delta t$, where $c_0$ is the velocity of the radiation in a vacuum. It follows that

$$K(\nu;\theta,\phi) = \frac{c_0}{4\pi} u(\nu) \to \frac{2h\nu^3\ c_0^{-2}}{\exp(h\nu/kT)-1} \ . \tag{13.52}$$

When one multiplies this quantity by $\cos\theta\,\mathrm{d}\omega\,\mathrm{d}\nu/h\nu$ one obtains again (13.26$a$). The parameters $\theta,\phi$ have been kept on the left-hand side even though $K$ is here independent of angles. This is done to avoid any possible confusion. On the right-hand side the spectral intensity for black-body radiation has been deduced using (13.15). This connection enables a temperature to be associated with the spectral intensity of *any* pencil of radiation by means of

$$K(\nu;\theta,\phi) = \frac{2h\nu^3 c_0^{-2}}{\exp\{h\nu/kT(\nu;\theta,\phi)\}-1} \tag{13.53}$$

so that

$$T(\nu;\theta,\phi) = \frac{h\nu}{k}\ \frac{1}{\ln\{1+2h\nu^3 c_0^{-2}/K(\nu;\theta,\phi)\}} \ . \tag{13.54}$$

We thus have a means for associating a temperature with any form of radiation. In principle, this temperature could be time-, frequency-, and angle-dependent for a given point in a medium.

If the temperature is angle-dependent, the equivalent angle-independent temperature $T(\nu)$ is obtainable by using (13.53), integrating over a solid angle of $2\pi$, and equating the result to $2h\nu^3 c_0^{-2}/[\exp h\nu/kT(\nu)\}-1]$

$$\frac{1}{2\pi}\int \frac{\mathrm{d}\omega}{\exp\{h\nu/kT(\nu;\theta,\phi)\}-1} = \frac{1}{\exp\{h\nu/kT(\nu)\}-1} \ . \tag{13.55}$$

We now estimate the effective temperature $T_E(\nu)$ for a small surface of area $\delta A_E$ on the earth when it is in a steady state with solar radiation of frequency $\nu$; all other frequencies are assumed to have no influence. The effect of the atmosphere is also neglected. The energy emitted into a solid angle $\delta\omega_S$ by an element of area $\delta A_S$ on a fictitious solar surface, in frequency range $(\nu,\nu+\delta\nu)$ and in time interval $\delta t$, is

$$\delta\nu\delta t\ \delta\omega_S K_S(\nu)\cos\theta_S\ \delta A_S\ .$$

Here $K_S$ is the specific spectral intensity of the sun. Denote by $R_S$ the radius of the sun and by $R$ the sun-earth distance. We sum over all elements $\delta A_S = R_S^2 \sin\theta_S\, d\theta_S\, d\phi_S$ of the hemisphere facing the earth, each element being projected by the factor $\cos\theta_S$ to be normal to the sun-earth axis, as required by the definition of $K_S$ (Fig.13.4c). At each element only the radiation emitted in the direction of the earth is to be included, and this can be done by taking $\delta\omega_S = \delta A_E \cos\theta_E'/R^2$. Here $\delta A_E$ is the element of area on the earth's surface which is being considered, and the factor $\cos\theta_E'$ ensures that normal incidence is used (Fig. 13.4d). The integral is

$$\int \cos\theta_S\ \delta A_S = R_S^2 \int_0^{\pi/2} \cos\theta_S \sin\theta_S\, d\theta_S \int_0^{2\pi} d\phi_S = \pi R_S^2\ .$$

The energy received at $\delta A_E$ is therefore

$$\delta\nu\delta t K_S(\nu)\ \delta A_E \cos\theta_E'(\pi R_S^2/R^2)\ .$$

For a steady state this quantity is equal to the energy emitted by $\delta A_E$, using the earth's spectral intensity $K_E$, that is to

$$\delta\nu\delta t K_E(\nu)\ \delta A_E \int \cos\theta_E\ \delta\omega_E\ .$$

The integration is carried out at the area $\delta A_E$ over all solid angles and yields (Fig.13.4e)

$$\int_0^{\pi/2} \cos\theta_E \sin\theta_E\, d\theta_E \int_0^{2\pi} d\phi_E = \pi\ ,$$

if the surface radiates from one side only. Equating the two energies for a steady state of the surface,

$$\frac{K_E(\nu)}{K_S(\nu)} = \frac{R_S^2 \cos\theta'_E}{R^2} \quad \text{or} \quad \frac{R_S^2 \cos\theta'_E}{2R^2} . \tag{13.56}$$

The second expression holds if $\delta A_E$ radiates from *both* sides, when the upper limit for $\theta_E$ is $\pi$ instead of $\pi/2$.

In order to eliminate $\cos\theta'_E$, one may (as an approximation) average it over the surface of the illuminated hemisphere to find

$$\langle \cos\theta'_E \rangle = \frac{1}{2\pi R_E^2} \int_0^{\pi/2} (\cos\theta'_E)(2\pi R_E \sin\theta'_E)(R_E d\theta'_E) = \frac{1}{2} .$$

One finds the approximate values ($R_S \sim 6{\cdot}95 \times 10^5$ km, $R \sim 1{\cdot}50 \times 10^8$ km)

$$\frac{K_E(\nu)}{K_S(\nu)} \doteqdot \frac{R_S^2}{4R^2} \quad (\sim 5{\cdot}4 \times 10^{-6})$$

for radiation from both sides. The conversion of directionally received radiation by the surface on earth into isotropic radiation from it, has led to a reduction in the spectral intensity. Using (13.53) for the effective terrestrial temperature $T_E$, this also implies that the value of $T_E$ is much less than $T_S$, and is given by

$$\frac{K_E(\nu)}{K_S(\nu)} = \frac{\exp(h\nu/kT_S)-1}{\exp(h\nu/kT_E)-1} \doteqdot \left(\frac{R_S}{2R}\right)^2 \sim 5{\cdot}4 \times 10^{-6} . \tag{13.57}$$

$T_E$ is the temperature of a terrestrial black surface maintaining a steady state with unconcentrated solar radiation via a narrow absorption band at frequency $\nu$. This is of interest for biological systems when $T_E$ can perhaps be regarded as a fictitious 'pumping temperature' for the deposition of biomass.

Pursuing this speculation for red light (6800 Å, $\nu = 4{\cdot}4 \times 10^{14}\,s^{-1}$, $h\nu = 1{\cdot}81$ ev) and $T_S \sim 5760$ K, one finds

$$\frac{h\nu}{kT_E} \sim 15.8 , \quad T_E \sim 1350 \text{ K}. \tag{13.58}$$

The Carnot efficiency $\eta_C = 1\text{-}300/5760 \sim 95\%$ is thus reduced. One finds

$$\eta_C \sim 1 - \frac{300}{1350} \sim 78\%,$$

but this is still considerably above what is achieved in biological systems (where the concept of efficiency is of course quite a complicated one). The actual temperature of the earth can be estimated by the same method, but one must integrate over all frequencies (problem 13.10). The effect of concentration of radiation by lenses or mirrors is to raise the temperature and hence the efficiency by effectively increasing the purely geometrical factor (13.57).

Note that for $h\nu/kT_E \gtrsim 3{\cdot}7$ (13.57) yields the useful approximate result

$$\frac{1}{T_E} = \frac{1}{T_S} + \frac{2k}{h\nu} \ln \frac{2R}{R_S} ,$$

provided $h\nu/kT_S \ll 1$.

Assuming the directions to be matched, any nearly monochromatic pencil of radiation can be in equilibrium with the radiation of a black body if the frequencies outside the range of interest are filtered out. Hence one can use (13.54) to define an effective temperature of a general near-monochromatic pencil of radiation. If one pieces together a whole frequency spectrum from various pencils with a set of temperatures $\{T_{eff}(\nu)\}$, then a necessary condition for all pencils to be in thermal equilibrium with each other is, by analogy with (7.7), that

$$T_{eff}(\nu) = T \qquad (\text{for all } \nu) . \tag{13.59}$$

It is *this* type of radiation, specified by only *one* temperature for the whole of its spectrum, which is called black-body radiation.

Suppose a general spectrum of incident radiation is available in the form $f(\nu)d\nu$ or $g(\lambda)d\lambda$ energy units per unit area per unit time. One can associate with it an equivalent *solar* temperature $T^*(\nu)$ as follows. The solid angle subtended by an area $A_E$ of the earth's surface at the sun is $A_E/R^2$. The power emitted by a hypothetical sun with spectral brightness $K(\nu;\theta,\phi)$ in the required frequency range and received on the area $A_E$ on earth is

$$\pi r_S^2\, K(\nu;\theta,\phi)\, d\nu \times \frac{A_E}{R^2} .$$

By equating this quantity to the function $f(\nu)d\nu$, multiplied by $A_E$, one obtains the equivalent spectral brightness of the source assumed to radiate like a black body in the position of the sun. The temperature $T^*(\nu)$ of this equivalent black body is then obtained from (13.54) to be

$$T^*(\nu) = \frac{h}{k}\left\{\ln\left\{1 + \frac{2h\nu^3}{c^2}\frac{\pi r_S^2}{R^2}\frac{1}{f(\nu)}\right\}\right\}^{-1} .$$

Bearing in mind that $f(\nu)d\nu = g(\lambda)d\lambda$, one has alternatively

$$T^*(\lambda) = \frac{hc}{k\lambda}\left\{\ln\left\{1 + \frac{2hc^2}{\lambda^5}\frac{\pi r_S^2}{R^2}\frac{1}{g(\lambda)}\right\}\right\}^{-1} .$$

If $T^*$ is expressed in degrees absolute, $\lambda$ in μm, and $g(\lambda)$ in mW cm$^{-2}$ μm$^{-1}$ one finds

$$T^*(\lambda) = \frac{14387}{\lambda}\left\{\ln\left\{1 + \frac{803\cdot 3}{\lambda^5 g(\lambda)}\right\}\right\}^{-1} .$$

This general result can also be written as

$$\frac{803\cdot 3}{\lambda^5 g(\lambda)} = \exp\left\{\frac{14\ 387}{\lambda T^*(\lambda)}\right\} -1 .$$

For $\lambda^5 g(\lambda) \gg 800$ it yields $T^* \sim 17\cdot 91\lambda^4 g(\lambda)$.

Fig. 13.5 gives some typical spectra $g(\lambda)$ of solar radiation. In a generalization of (13.58) one can find $T_E$, the effective absorber temperature for each wavelength range, as shown in Fig.13.6. The same spectra can also be interpreted as effective solar temperatures $T^*(\lambda)$ as shown in Fig:13.7.

The relation between $g(\lambda)$, $T_E(\lambda)$ and $T^*(\lambda)$ can also be seen from the relation

$$g(\lambda) = \frac{8\pi hc_0^2\lambda^{-5}}{\exp(hc/k\lambda T_S^*)-1}\left(\frac{r_S}{2R}\right)^2 = \frac{8\pi hc_0^2\lambda^{-5}}{\exp(hc/k\lambda T_E)-1} .$$

If multiplied by $d\lambda$ it gives energy per unit area per unit time in the wavelength range considered. The first expression is the incident spectral energy; the middle term utilizes an effective solar temperature noting that the relevant solid angle is $\omega/4\pi = (\pi r_S^2/R^2)(1/4\pi)$; the last expression treats the earth as a black sphere radiating uniformly in all directions in equilibrium with $g(\lambda)$ at temperature $T_E$. The factor

$$8\pi hc_0^2\lambda^{-5}d\lambda = 8\pi hc_0^{-2}\nu^3 d\nu$$

can be read off equation (13.15) if one notes that the rate of energy

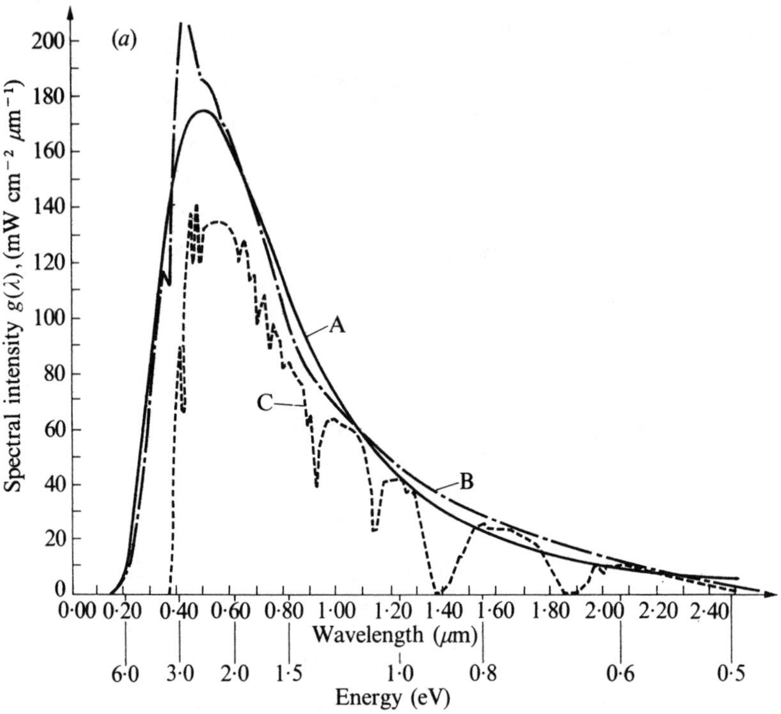


Fig.13.5. Spectral energy distributions $g(\nu)$ for various energy sources : curve A, black-body radiation at $T$ = 5760 K, normalized to 134·9 mW cm$^{-2}$; curve B, direct solar radiation above the atmosphere, normalized to 135·3 mW cm$^{-2}$; curve C, direct solar radiation on earth's surface with the sun vertically overhead on a clear day, normalized to 87·2mW cm$^{-2}$; curve D, diffuse solar radiation on a cloudy day, normalized to 30·0 mW cm$^{-2}$.

(Fig.13.5 continued on next page)

radiation per unit area of a surface can be taken to be $c_0 u(\nu,T)d\nu$.

## *13.10. THE ENTROPY OF PENCILS OF RADIATION

From (5.23), $(\partial S/\partial U)_v = 1/T$, we have

$$\left\{\frac{\partial s(\nu)}{\partial u(\nu)}\right\}_v = \frac{1}{T_{\text{eff}}(\nu)} \tag{13.60}$$

where $s(\nu)$ and $u(\nu)$ are entropy and energy in a pencil of radiation, both expressed per unit volume per unit frequency range. The entropy per unit volume is therefore given by $\int s(\nu)d\nu$. Using now (13.52) or (13.54) and the substitution

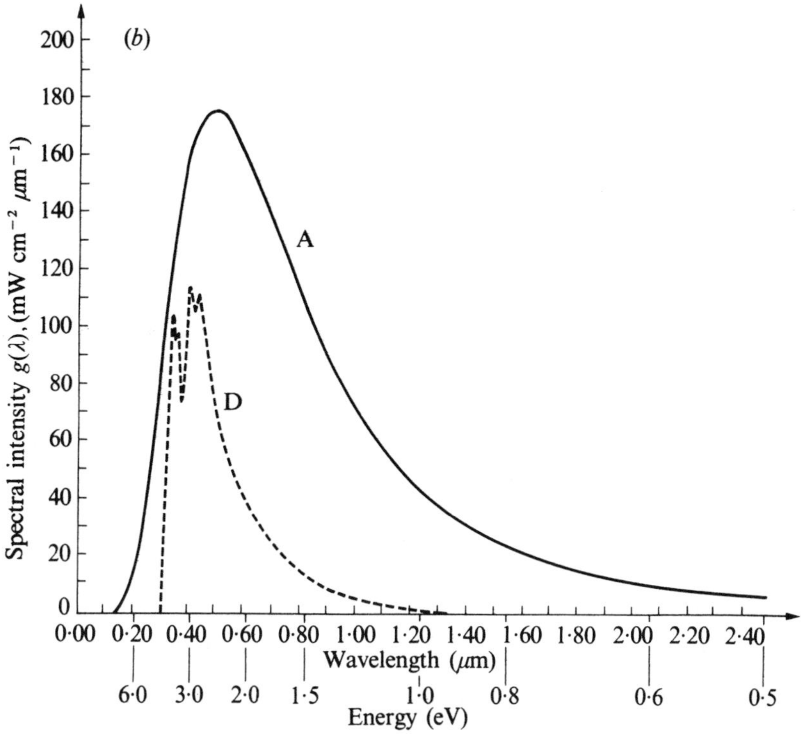


Fig.13.5(*b*).

$$x_\nu \equiv \frac{c_0^2 K(\nu;\theta,\phi)}{2h\nu^3 \partial\nu} = \frac{c_0^3 u(\nu)}{8\pi h\nu^3 \partial\nu} ,$$

$$\left\{\frac{\partial s(\nu)}{\partial x_\nu}\right\}_v = \frac{8\pi h\nu^3}{c_0^2} \left\{\frac{\partial s(\nu)}{\partial u(\nu)}\right\}_v$$

$$= \frac{8\pi k\nu^2}{c_0^3} \ln(1 + x_\nu^{-1}) \ .$$

If $f(v)$ be some function of volume, integration for given $\nu$ yields

$$s(\nu) = 8\pi k\nu^2 c_0^{-3} \left\{(1+x_\nu)\ln(1+x_\nu) - x_\nu \ln x_\nu\right\} + f(v) \ .$$

Rewriting this, and assuming that zero energy density $u(\nu)$ implies zero entropy density $s(\nu)$, one finds $f(v) = 0$ and

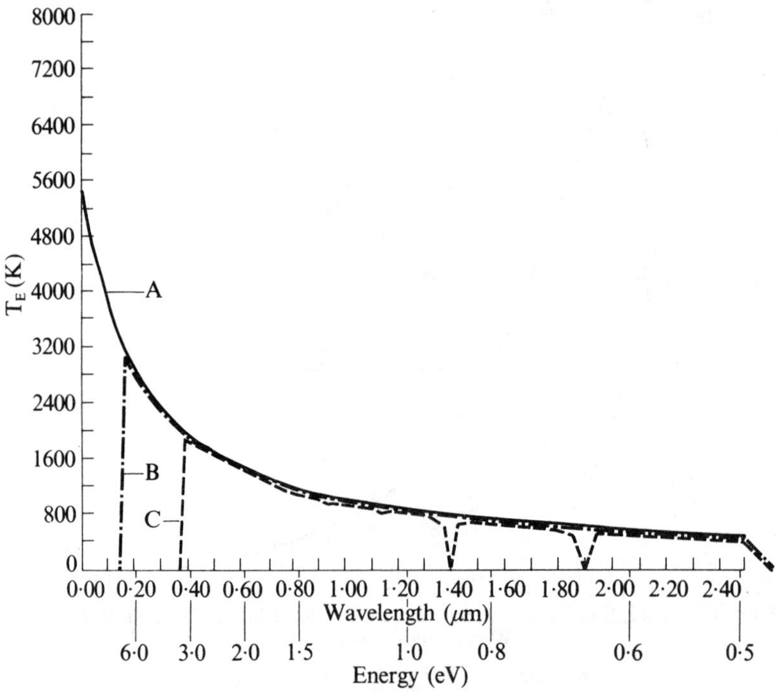


Fig.13.6. Effective black-body temperature $T_E$ on earth produced by a narrow-wave-length range of the radiation of Fig.13.5(*a*) using equation (13.57) with $T_S$ = 5760 K.

$$s(\nu) = \frac{ku(\nu)}{h\nu}\left\{(1+x_\nu^{-1})\ln(1+x_\nu^{-1}) + x_\nu^{-1}\ln x_\nu\right\} .$$

To find the unpolarized entropy $\delta S_\nu$ passing through the surface element of area $\delta A$ in time $\delta t$ due to radiation in the frequency range $\delta\nu$, we must multiply $s(\nu)$ by

$$c_0 \cos\theta \ \delta\omega \ \delta\nu \ \delta A \ \delta t$$

as suggested by (13.48). One finds

$$\delta S_\nu = \frac{k\delta E_\nu}{h\nu}\left\{(1+x_\nu^{-1})\ln(1+x_\nu^{-1}) + x_\nu^{-1}\ln x_\nu\right\} \tag{13.61}$$

where $\delta E_\nu = (\delta A\cos\theta)(c_0\ \delta t)\{u(\nu;\theta,\phi)\delta\nu\}\delta\omega$ is the corresponding energy passing through $\delta A$. If $u(\nu,\theta,\phi) \equiv u(\nu)/4\pi$, $\delta\tau \equiv 2\nu^2 c_0^{-2}\cos\theta\ \delta\nu\ \delta\omega\ \delta A\ \delta t$,

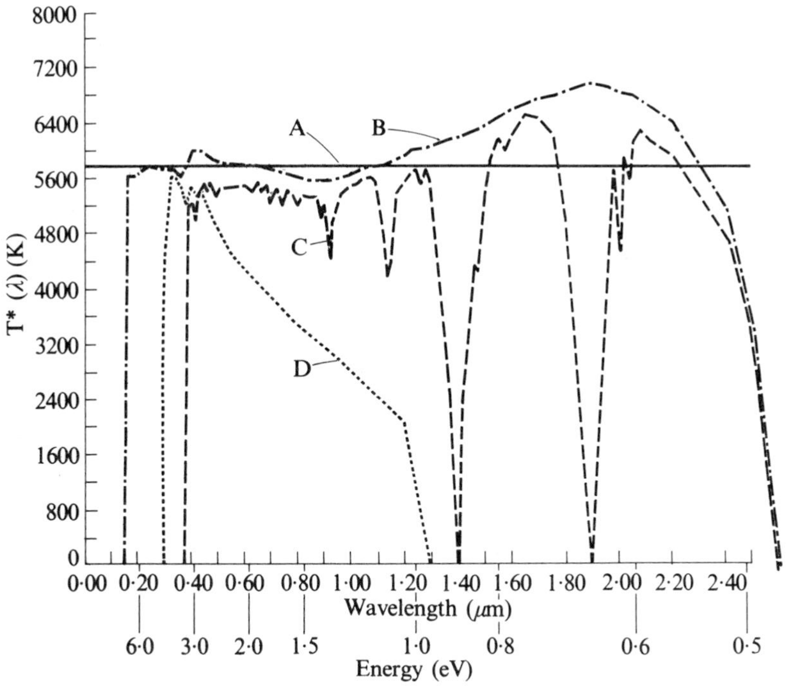


Fig.13.7. Effective solar temperatures $T^*(\lambda)$ for narrow wavelength ranges received on earth. The spectra of Fig.13.5($a$,$b$) have been used.

then, $\delta E_\nu = x_\nu \, h\nu \, \delta\tau$.

To interpret (13.61), suppose non-zero energy passes through the area but that $\delta\tau \to 0$. This can happen if the pencil is either monochromatic or unidirectional or both. Either condition can be approximated very closely by laser radiation. (They cannot be fulfilled exactly. If they *were* fulfilled, a closed system consisting of the emitter and the radiation could in the course of attaining equilibrium lose entropy, for the emitter loses energy and hence entropy, while the radiation entropy would be zero. This is clearly impossible.) Then $x_\nu \to \infty$ as $\delta\tau \to 0$ and hence $\delta S_\nu \to 0$. The entropy of such a pencil of radiation vanishes.

One can arrive at (13.59) by considering arbitrary unpolarized isotropic radiation in a reflecting enclosure of fixed volume $v$. The energy density $u(\nu)$ per unit frequency range is changed by an arbitrary small amount for various frequency ranges subject to the energy remaining constant. What is the condition for an extremal (maximum) entropy with respect to these variations? Apart from keeping the volume constant, the constraint

$$\delta U = \delta[v \int_0^\infty u(\nu)\,d\nu] = v \int_0^\infty \delta u(\nu)\,d\nu = 0$$

has to be imposed. Using a Lagrangian multiplier $\lambda$, one requires

$$v \int_0^\infty \left\{\left[\frac{\partial s(\nu)}{\partial u(\nu)}\right]_v - \lambda\right\}\delta u(\nu)\,d\nu = v \int_0^\infty \left\{T_{\mathrm{eff}}(\nu) - \lambda\right\} \delta u(\nu)\,d\nu = 0 \; .$$

This holds for arbitrary $\delta u(\nu)$ and so leads to constant $T_{\mathrm{eff}}(\nu)$. Thus the largest entropy of radiation of given volume and energy is found if the temperature is the same for all frequency ranges, and this brings us back to black-body radiation.

Equation (13.60) also suggests that if the energy and entropy of radiation in a frequency interval are increased, then the resulting curve has had an infinite slope at the origin and then increases in a concave manner towards the energy axis, since the temperature will also increase (Fig.13.8).

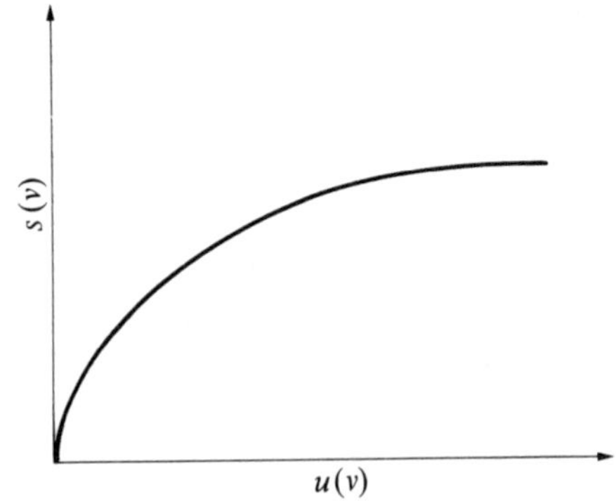


Fig.13.8. The connection between the densities of entropy and energy in radiation (schematic).

PROBLEMS

13.1. Check numerically the value of $a$ given in section 13.2.

13.2. Prove that

$$I(0,s,+) = (1-2^{-s})\zeta(s+1)$$

and hence show from tables of the Riemann zeta function $\zeta(s)$ that

$$\int_0^\infty \frac{x^s\,dx}{\exp x+1} = \begin{cases} 1\cdot 803 & \text{for } s = 2 \\ 5\cdot 681 & \text{for } s = 3 \end{cases}$$

13.3. A thermodynamic property depends on $I(0,s,\pm)$ depending on whether fermions or bosons are considered, the value of $s > -1$ being the same in both cases. Prove that the fermion property is always smaller than the corresponding property for bosons by the factor $1 - 2^{-s}$.

13.4 (*a*) Electrons ($e^-$) and positrons ($e^+$) are in thermal equilibrium with the photons ($\nu$) of black-body radiation. The temperature is so high that the rest energy of the particles is much smaller than their total energy. If electrons and positrons occur in equal numbers $N$, obtain an expression for $N$ and also for the internal energy $U$ of the electron and the positron system.
(*b*) Use the reaction $e^- + e^+ = \nu$ to show that the chemical potentials satisfy $\mu^{(+)} = \mu^{(-)} = 0$.
(*c*) Establish that $N$ and $U$ are $\frac{3}{4}$ and $\frac{7}{8}$ respectively of the corresponding values for black-body radiation in the same volume and at the same temperature.

13.5. Photons of frequencies $\nu_1,\nu_2$ and mean occupation numbers $n_1,n_2$ are assumed to create electrons and positrons with rest masses $m$, mean occupation numbers $n_+,n_-$ and kinetic energies $\varepsilon_+,\varepsilon_-$ respectively. The reverse process also occurs and pairs of processes of this type keep up thermal equilibrium in *detailed balance* between black-body radiation and electron-positron pairs.
(*a*) Show from problem 10.4 that

$$n_1 n_2 (1-n_+)(1-n_-) = n_+ n_- (1+n_1)(1+n_2)$$

and

$$h\nu_1 + h\nu_2 = 2mc^2 + \varepsilon_+ + \varepsilon_- \ .$$

Describe in words the forward and the reverse process.

(*b*) Show from part (*a*) that the chemical potential of the fermions vanishes.

(*c*) Assuming a low density $n$ of pairs show that for $T \ll mc^2/k$

$$n = 2\left(\frac{2\pi mkT}{h^2}\right)^{3/2} \exp\left(-\frac{mc^2}{kT}\right)$$

is the concentration in equilibrium with black-body radiation at temperature $T$. Use problem 12.7 for the treatment of the chemical potential in this case.

(*d*) Estimate numerically $n$ as a function of $T$.

(Problems 13.4b and 13.5b represent alternative derivations of the same result.)

13.6. In the preceding problem *two* photons were used to create an electron-positron pair. The reason is that *one* photon cannot do so. Attempt a proof as follows. Prove from equation (12.16) and conservation laws that a single particle of zero rest mass cannot have a particle of non-zero rest mass among its decay products.

13.7. A system at constant volume has energy and entropy respectively given by $bT^r$ and $cT^s$ where $b$, $c$, $r$, $s$ are positive constants. Prove that $br = cs$.

13.8. If $\mu$ is the refractive index in a medium, show that $\mu^{-2}$ multiplied by the spectral intensity $K$ is invariant along a ray. (Consider an infinitesimal cross-sectional area $\delta A$ of an elementary beam such that the refractive index is $\mu_1$ on one side and $\mu_2$ on the other side of the surface. Then apply Snell's law of refraction $\mu_1 \sin\theta_1 = \mu_2 \sin\theta_2$.)

13.9. A small surface $S$ of area $A'$ is centred on the normal to a luminous disc of radius $r$. $S$ is parallel to the disc and at a distance

$R$ from it. If $K$ is the luminance of the disc, obtain an expression for the illuminance $E$ on $S$. If the disc subtends a *small* solid angle $\omega'$ at $S$ show that

$$E = A'K\omega' .$$

13.10. It is seen in relation (13.58) that, in order to remain in a steady state with the sun at frequencies near 6800 Å, the earth must radiate in that frequency like a black body at ('colour') temperature 1350 K. Show by considering *all* frequencies that the ('bolometric') temperature of the earth, obtained from an analogous argument, is 278 K.

13.11. In this problem the total black-body radiation energy $U$ in a cavity of volume $v$ at temperature $T$ is obtained when the cavity is small enough for the first-order correction to the mode number of Table 12.1 to become relevant. For a cubic cavity whose side is $L$ the number of modes per frequency range $d\nu$ is then found to be

$$g(\nu) = g_0(\nu)\left\{1 - \left(\frac{ac}{L\nu}\right)^2\right\}, \quad a^2 = \frac{3}{8\pi} .$$

Here the asymptotic mode number per unit frequency range

$$g_0(\nu) = \frac{8\pi\nu^2 v}{c^3}$$

leads to the result (13.11)

$$U_0 = \frac{8}{15}\pi^5 k^4 h^{-3} c^{-3} T^4 v.$$

Show that for a small cavity $U/v$ is smaller than $U_0/v$ and given by

$$\frac{U}{U_0} = 1 - \frac{5}{2}\left(\frac{ach}{\pi kTL}\right)^2 .$$

(The result $\int_0^\infty x\,dx/\exp(x-1) = \pi^2/6$ has to be used. The spectral densities for other statistical mechanical theories are amended analogously if the assumption noted in connection with equation (12.7) fails.)

# 14. Heat capacities of oscillators and solids

## 14.1. THE STATISTICAL THERMODYNAMICS OF OSCILLATOR SYSTEMS

Consider a solid of $N$ molecules, each consisting of $r$ atoms, bound to fixed positions in a crystal lattice. Such a lattice consists of a basic pattern, called a *unit cell*, which is periodically repeated in three dimensions so as to generate the lattice. The $Nr$ atoms vibrate with their least possible, or *zero-point*, energy about their equilibrium positions near the absolute zero of temperature. Although this energy rises with temperature, the harmonic oscillator remains a reasonable first approximation for this motion until the solid becomes a liquid or a gas. Now *classically* the mean energy of one oscillator is $U_1 = kT$ (section 11.7), and it becomes $3kT$ for a three-dimensional oscillator which has six degrees of freedom represented by the three components of momentum and the three components of the position coordinate. It is in fact convenient to regard the $rN$ three-dimensional oscillators as $3rN$ one-dimensional ones (see problem 14.2 . One finds

$$U_{rN} = 3rN\ kT. \tag{14.1}$$

This energy goes to zero with the temperature, and no zero-point energy is present. We now consider a more general situation which does exhibit the zero-point energy.

It is shown in books on quantum mechanics that the quantum states of a three-dimensional oscillator which has frequency $\nu_1$ for the projection of its motion on the $x$-direction, $\nu_2$ for the projection on the $y$-direction, and $\nu_3$ for the projection on the $z$-direction can be specified by three non-negative integers or quantum numbers $n_1$, $n_2$, and $n_3$. The energy of such a state is

$$E(n_1,n_2,n_3) = \varepsilon(0) + \sum_{i=1}^{3} n_i\ h\nu_i \tag{14.2}$$

where $\varepsilon(0)$ is the energy when $n_1 = n_2 = n_3 = 0$. One can use (9.22) to obtain the partition function of one oscillator at temperature $T$ as

$$Z_1 = \exp\left(-\frac{\varepsilon_0}{kT}\right)\left\{\sum_{n_1=0}^{\infty}\exp\left(\frac{-n_1 h\nu_1}{kT}\right)\right\}\left\{\sum_{n_2=0}^{\infty}\exp\left(\frac{-n_2 h\nu_2}{kT}\right)\right\}\left\{\sum_{n_3=0}^{\infty}\exp\left(\frac{-n_3 h\nu_3}{kT}\right)\right\}$$

$$= \exp\left(-\frac{\varepsilon_0}{kT}\right)\prod_{i=1}^{3}\frac{1}{1-\exp(-h\nu_i/kT)} \tag{14.3}$$

where we have summed a geometrical progression. In the classical limit one treats Planck's constant as small and then finds for a three-dimensional oscillator

$$Z_1 \sim \exp\left(-\frac{\varepsilon_0}{kT}\right)\left(\frac{kT}{h}\right)^3 \frac{1}{\nu_1\nu_2\nu_3} .$$

For a one-dimensional oscillator one has similarly

$$Z_1 \sim \exp\left(-\frac{\varepsilon_0}{kT}\right)\frac{kT}{h\nu} .$$

Comparison with (11.30) shows that in classical statistical mechanics one must put

$$\varepsilon_0 \to 0 \qquad z \to h^{-1} \tag{14.4}$$

in order to have the correct limiting case of the quantum mechanical formulae.

Reverting to (14.3), one can obtain the mean energy of the oscillator by using the result $x_0 = -(\partial \ln Z/\partial\beta)$ of problem 9.1 where $x_0$ is the mean internal energy $U$, $x_i$ is an oscillator energy $E(n_1,n_2,n_3)$, and $\beta = 1/kT$. Hence

$$U_1 = kT^2\frac{\partial \ln Z}{\partial T} = kT^2\frac{\partial}{\partial T}\left\{-\frac{\varepsilon_0}{kT} - \sum_{i=1}^{3}\ln\left\{1 - \exp\left(-\frac{h\nu_i}{kT}\right)\right\}\right\}$$

$$= \varepsilon_0 + \sum_{i=1}^{3}\frac{\exp(-h\nu_i/kT)}{1-\exp(-h\nu_i/kT)}h\nu_i$$

$$\boxed{U_1 = \varepsilon_0 + \sum_{i=1}^{3}\frac{h\nu_i}{\exp(h\nu_i/kT)-1}} . \tag{14.5}$$

This energy decreases to $\varepsilon_0$ as the temperature is lowered to

$T = 0$, showing that $\varepsilon_0$ is the zero-point energy of an oscillator:

$$\varepsilon_0 \to \begin{cases} 0 & \text{(classical statistical mechanics)} \\ \frac{1}{2}h(\nu_1+\nu_2+\nu_3) & \text{(quantum statistical mechanics).} \end{cases} \tag{14.6}$$

The second entry reflects what is known about $\varepsilon_0$ from quantum mechanics.

It is instructive to rewrite the one-dimensional form of (14.5) in the following way:

$$U_1 = \varepsilon_0 + \frac{2\varepsilon_0}{\exp(2\varepsilon_0/kT)-1} \quad (\varepsilon_0 = \tfrac{1}{2}h\nu) \ . \tag{14.7}$$

If one uses $\varepsilon_0 \to 0$ one obtains (14.1) (with $3rN \to 1$) as the correct limiting case. The analogy between photons of energy $h\nu$ and the excitation states in a solid is quite close, as can be seen by comparing equations (13.15) and (14.7). The excitation in a solid is called a *phonon* of frequency $\nu$ in order to emphasize this analogy.

For a solid the $3rN$ oscillators are distinguishable by virtue of their position in the lattice and we can use from (11.35)

$$Z_{rN} = (Z_1)^{rN} \ . \tag{14.8}$$

Since the internal energy depends on $\ln Z$, one obtains for the internal energy of the solid

$$U_{rN} = rN\, U_1 \ . \tag{14.9}$$

The heat capacity of the solid due to the lattice vibrations is

$$\boxed{C_v = \frac{\partial U_{rN}}{\partial T} = rNk \sum_{i=1}^{3} E(x_i) \ , \quad x_i \equiv \frac{h\nu_i}{kT}} \tag{14.10}$$

where

$$E(x) \equiv \frac{x^2 e^x}{(e^x-1)^2} \ . \tag{14.11}$$

The function $E(x)$ is called the Einstein function. Einstein proposed in 1911, in the early days of the quantum theory, a theory of the specific heat of monatomic solids. He treated all atoms in the solid as independent oscillators on the basis of what was then called 'the new mechanics'. As $T \to 0$, $E(x) \to 0$ so that $C_v \to 0$ in agreement with the third law of thermodynamics considered in problem 6.7. The classical theory yields $C_v = 3rNk$ and so violates this law. This agrees with the general rule that low-temperature phenomena are liable to show quantum effects, whereas high temperatures are not. In fact, for high temperatures (14.11) is unity and (14.10) goes over into the classical law of Dulong and Petit which states that $C_v = 3rNk$ for solids. This law was discovered by them experimentally (1819).

The free energy is, by (9.22) and (14.3),

$$\begin{aligned} F_{rN} &= -kT \ln Z_{rN} = -kTrN \ln Z_1 \\ &= -kTrN\left[-\frac{\varepsilon_0}{kT} - \sum_{i=1}^{3} \ln\left\{1 - \exp\left(-\frac{h\nu_i}{kT}\right)\right\}\right] \\ &= rN\varepsilon_0 + rnkT \sum_{i=1}^{3} \ln\left\{1 - \exp\left(-\frac{h\nu_i}{kT}\right)\right\} \quad . \end{aligned} \tag{14.12}$$

## 14.2. DEBYE'S THEORY OF THE SPECIFIC HEATS OF SOLIDS

A real monatomic solid does not possess a spectrum which consists of a single frequency for each dimension and so becomes for an oscillator the triplet $(\nu_1,\nu_2,\nu_3)$ suggested by (14.5). One has instead to consider the equation of motion of each atom under the influence of all the other atoms and analyse the resulting differential equations in that way. A whole frequency spectrum is then found. Without going into this detail, however, one can still allow for these possibilities by replacing sums of three terms by sums of $3rN$ terms

$$rN \sum_{i=1}^{3} f(\nu_i) \to \sum_{j=1}^{3rN} f(\nu_j) \tag{14.13}$$

over all frequencies. For example (14.10) becomes

$$C_v = k \sum_{j=1}^{3rN} E\left(\frac{h\nu_j}{kT}\right) \quad . \tag{14.14}$$

In order to avoid determining the frequency spectrum, one can resort to a hybrid theory in which the atomistic viewpoint, which deals with states of excitation of frequencies $\nu_j$, is combined with a continuum model of the solid as a whole. The latter enables one to assign to the material the density of states function of the third row of Table 12.1, which is appropriate for an elastic solid. This means that (14.14) can be replaced by

$$C_v = k\,\frac{4\pi v}{h^3}\left(\frac{2}{c_t^3}+\frac{1}{c_l^3}\right)\int (h\nu)^2\,E\left(\frac{h\nu}{kT}\right)\,d(h\nu)$$

that is

$$\frac{C_v}{k} = \frac{4\pi v}{h^3}\left(\frac{2}{c_t^3}+\frac{1}{c_l^3}\right)(kT)^3\int_0^{x_D}\frac{x^4 e^x}{(e^x-1)^2}\,dx \tag{14.15}$$

where $v$ is the volume of the material. The lattice is here supposed to consist of $N$ unit cells and the quantity $r$ becomes the number of atoms per unit cell. This approach was initiated by P. Debye (1912).

The integral in (14.15) extends from $x = 0$ to an upper limit $x_D \equiv h\nu_D/kT$. This means that the total number of modes of vibration is

$$4\pi v\left(\frac{2}{c_t^3}+\frac{1}{c_l^3}\right)\int_0^{\nu_D}\nu^2 d\nu = \frac{4\pi v}{3}\left(\frac{2}{c_t^3}+\frac{1}{c_l^3}\right)\nu_D^3 \ . \tag{14.16}$$

One sees that an infinite upper limit would lead to a divergence, and (14.16) must be equated to $3Nr$ in order that the system should have no more than $3Nr$ modes of vibration. Hence the Debye frequency is given by

$$\nu_D^3 = \frac{9Nr}{4\pi v}\left(\frac{2}{c_t^3}+\frac{1}{c_l^3}\right)^{-1} \ . \tag{14.17}$$

The Debye temperature is such that

$$x_D = T_D/T \qquad T_D \equiv h\nu_D/k \ . \tag{14.18}$$

At low temperatures the integral in (14.15) extends effectively to infinity and one infers from (14.15) the Debye $T^3$-law for the low-temperature heat capacity of solids. The Einstein theory lies below this value, since it yields by (14.10)

$$\frac{C_v}{k} = rN \sum_{i=1}^{3} \left(\frac{h\nu_i}{kT}\right)^2 \exp\left(-\frac{h\nu_i}{kT}\right) . \tag{14.19}$$

The Debye theory is in good agreement with experiments on the simpler solids, and is similar to the heat capacity law (13.8) (with $s = 2$) for photons. In fact the low-temperature thermodynamic functions of the Debye theory are similar in form to the corresponding functions for photons, although the numerical coefficients are different. Combining (14.15) and (14.17),

$$\frac{C_v}{k} = \frac{9rN}{x_D^3} \int_0^{x_D} \frac{x^4 e^x}{(e^x - 1)^2} \, dx . \tag{14.20}$$

This shows that at high temperatures the classical Dulong-Petit result $C_v = 3rNk$ is recovered, as it must be.

A small inaccuracy in (14.17) must be noted. The $3rN$ modes of vibration are subject to six constraints which assert the constancy of the linear momentum (three constraints) and the angular momentum (three further constraints) of the system as a whole. The remaining $3rN$-6 equations describe the oscillatory motion inside the crystal. However, since $N$ may be of order $10^{21}$ for each cubic centimetre of crystal, the replacement of $3rN$-6 by $3rN$ is of course reasonable.

The low-temperature heat capacity of an Einstein solid and of a Debye solid vanish as $T$ goes to zero, in agreement with the third law of thermodynamics (see problem (6.7)).

Note also that the zero-point energy $\varepsilon_0$ of (14.7) or (14.12) does not contribute to the heat capacity and this term can be omitted when heat capacity problems are considered.

## 14.3. THE PHONON GAS

One can regard the energy $\varepsilon_0$ of (14.7) as belonging to a particle-like object whose rest mass is zero and which has a Bose occupation number. In the case of radiation such objects are called photons and, as already noted, in

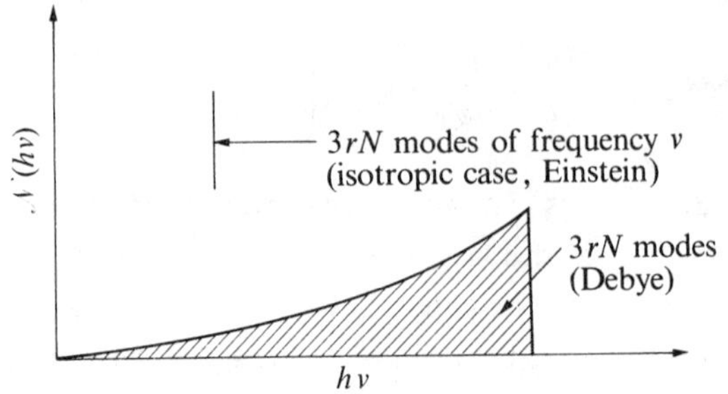


Fig.14.1. Number of modes of vibration per unit energy range for Einstein and Debye solids.

the case of a lattice, whether or not it is approximated by an elastic continuum, the objects are called *phonons*. The Debye theory becomes a theory of a *phonon gas* in which an elastic continuum has been assumed for the phonon frequency spectrum, i.e. for the normal modes. A phonon has a direction of propagation unit vector $\hat{\mathbf{k}}$ and wavelength $\lambda$. One can therefore define for a phonon a *wavevector* which is by convention $2\pi$ times the wavevector defined on p. 196

$$\mathbf{k} = \frac{2\pi}{\lambda}\,\hat{\mathbf{k}} \tag{14.21}$$

and an energy

$$\varepsilon = \hbar\omega = h\nu = \frac{hc}{\lambda}\,, \tag{14.22}$$

where $c$ is the velocity of sound in the medium. The same relations hold for a photon when $c$ is the velocity of light in the medium, i.e. $c = c_0/\mu(\nu)$ where $c_0$ is the velocity of light *in vacuo* and $\mu(\nu)$ is the refractive index of the material. The momentum of a phonon is, using the de Broglie relation of standard wave mechanics,

$$\mathbf{p} = \frac{h}{\lambda}\,\hat{\mathbf{k}} = \hbar\,\mathbf{k} \tag{14.23}$$

where (14.21) has been used. The lattice structure does, however, impose a crucial distinction between photons and phonons which is discussed in books on solid state theory and is connected with the possibility of adding reciprocal lattice vectors to $\mathbf{k}$ without yielding a physically distinct momentum. The *group velocity* of the classical wave corresponding to a phonon is, using (14.22) and (14.23),

$$\mathbf{v} = \nabla_{\mathbf{p}}\, \varepsilon(\mathbf{p}) = \nabla_{\mathbf{k}}\omega(\mathbf{k}) \ . \tag{14.24}$$

Our main concern attaches to the function $\omega(\mathbf{k})$, which is by (14.21)

$$\omega = 2\pi\nu = 2\pi\,\frac{c}{\lambda} = ck \qquad (k \equiv |\mathbf{k}|) \ . \tag{14.25}$$

For an electron or a molecule in a region where its potential energy is $V$ and its momentum is $\hbar\mathbf{k}$, as in (14.23), its energy is

$$\varepsilon = V + \frac{\hbar^2}{2m}\,k^2$$

so that

$$\omega = \frac{V}{\hbar} + \frac{\hbar}{2m}\,k^2 \ . \tag{14.26}$$

In general one may assume that, for each direction of polarization (cf. Chapter 12) denoted by $t$,

$$\omega_t(\mathbf{k}) = a_t(\hat{\mathbf{k}})k^s \qquad (t = 1,2,3) \tag{14.27}$$

and one normally finds $s = 1$ or $2$. The case $s = 2$ also applies to spin waves in a ferromagnetic solid (magnons). Relations of the type (14.27) are called *dispersion relations*. For each $\mathbf{k}$ there are in general $3r$ values of $\omega$ of which three have the property of going to zero with $k$, while the other $3r-3$ attain non-zero values. One finds from separate calculations that in the former case the limit $\mathbf{k} \to 0$ corresponds to a translation of the whole lattice (acoustic phonons), while in the latter case the limit corresponds to atoms in the same cell vibrating broadly speaking 'in opposition' (optical phonons).

The expression for the total internal energy of such oscillators may be based on equation (14.5) or on the equation given in problem 14.6. We shall adopt $V/8\pi^3$ as the number of $\mathbf{k}$-vectors per unit volume of $\mathbf{k}$-space to find

$$U = \sum_{\mathbf{k},t} \left\{\tfrac{1}{2} + \frac{1}{\exp\{\beta\hbar\omega_t(\mathbf{k})\}-1}\right\}\hbar\omega_t(\mathbf{k}) \qquad \left(\beta \equiv \frac{1}{kT}\right)$$

$$= \sum_t \frac{v}{8\pi^3} \int_0 4\pi k^2 \left\{\tfrac{1}{2} + \frac{1}{\exp\{\beta\hbar\omega_t(\mathbf{k})\}-1}\right\}\hbar\omega_t(\mathbf{k})\, dk \qquad (14.28)$$

where the upper limit is the maximum permitted value of $k$. Now let

$$x_t = \beta\hbar a_t k^s \qquad dx_t = s\beta\hbar a_t k^{s-1}\, dk \ .$$

Then

$$k^2 dk = \left(\frac{x_t}{\beta\hbar a_t}\right)^{2/s} \frac{dx_t}{s\beta\hbar a_t}\left(\frac{x_t}{\beta\hbar a_t}\right)^{(1-s)/s}$$

$$= \left(\frac{kT}{\hbar a_t}\right)^{3/s} \frac{1}{s}\, x_t^{\,3/s-1}\, dx_t \ .$$

It follows that on writing $U(0)$ for the zero-point energy (cf. section 14.2.)

$$U(T) - U(0) = \frac{vkT}{2\pi^2 s}\left(\frac{kT}{\hbar}\right)^{3/s} \sum_t \frac{1}{a_t^{3/s}} \int_0^{X_t} \frac{x_t^{3/s}}{e^{x_t}-1}\, dx_t \qquad (14.29)$$

where $X_t$ is the maximum permitted value of $\beta\hbar\omega_t$. The sum over $t$ has in the elastic continuum approximation two terms for transverse waves and one for longitudinal waves, and one then arrives again at the entry in the third row of Table 12.1. At low temperatures $X_t \to \infty$ and $U(T)-U(0) \propto T^{1+3/s}$, whence the heat capacity is seen to satisfy

$$C_v \propto T^{3/s} \ . \qquad (14.30)$$

This yields the $T^3$- law for the Debye spectrum ($s$=1), and a $T^{3/2}$ law for $s$=2 (magnons).

## 14.4. THE STATISTICAL THERMODYNAMICS OF A DEBYE SOLID

In section 14.2 the Debye $T^3$-law was obtained by using the density of states function

$$a\nu^2 d\nu\ , \quad a \equiv 4\pi v\left(2c_t^{-3} - c_l^{-3}\right)\ ,$$

with

$$3rN = \int_0^{\nu_D} a\nu^2 d\nu = \frac{1}{3}\, a\nu_D^3$$

in the general $C_v$-expression (14.14). In this way the heat capacity of a Debye solid can be obtained.

One can, however, proceed more systematically and first obtain the partition function of the system from (14.3). For a one-dimensional oscillator

$$\ln Z_1 = -\frac{h\nu}{2kT} - \ln\left\{1 - \exp\left(-\frac{h\nu}{kT}\right)\right\}\ .$$

For $rN$ three-dimensional oscillators

$$\begin{aligned}
\ln Z &= -\sum_{i=1}^{3rN}\left[\frac{h\nu_i}{2kT} + \ln\left\{1 - \exp\left(-\frac{h\nu_i}{kT}\right)\right\}\right] \\
&= -\int_0^{\nu_D} \frac{h\nu}{2kT}\, a\nu^2 d\nu - a\int_0^{\nu_D} \nu^2 \ln\left\{1 - \exp\left(-\frac{h\nu}{kT}\right)\right\} d\nu \\
&= -\frac{ah\nu_D^4}{8kT} - \frac{9rN}{x_D^3}\int_0^{x_D} x^2 \ln(1-e^{-x})\,dx \qquad \left(x \equiv \frac{h\nu}{kT}\right) \\
&= -\frac{9}{8}\, rNx_D - 3rN\ln\left(1-e^{-x_D}\right) + \frac{3rN}{x_D^3}\int_0^{x_D}\frac{x^3 dx}{e^x - 1}
\end{aligned}$$

by a partial integration. The Helmholtz free energy is $-kT\ln Z$, that is

$$F = F(0) - \frac{3rN}{x_D^3}\, kT\int_0^{x_D}\frac{x^3 dx}{e^x - 1} + 3rNkT\ln\left\{1 - \exp(-x_D)\right\} \qquad (14.31)$$

where the zero-point term is

$$F(0) \equiv \frac{9}{8}\, rNh\nu_D\ .$$

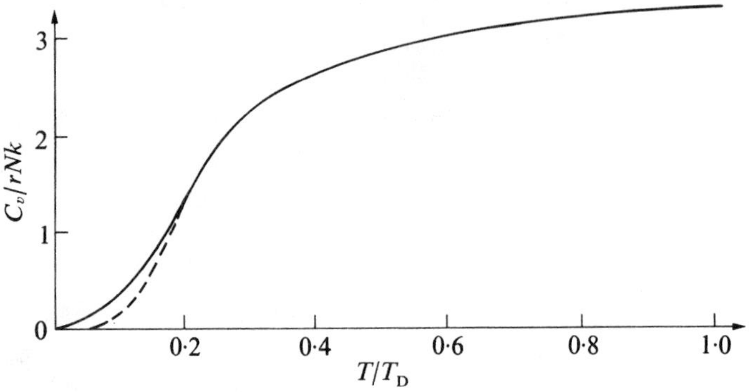


Fig.14.2. The heat capacity of a solid according to the Debye theory (solid curve) and the Einstein theory (broken curve). The latter curve has been shifted to agree with the Debye curve at high temperatures.

All thermodynamic functions can be obtained from this expression (see problem 14.8 ). If the zero-point energy is neglected and low temperatures are considered so that $x_D \to \infty$, then there are many formal similarities between the functions for the Debye theory and the corresponding functions for black-body radiation. In particular, as already observed in connection with (13.12),

$$\int_0^\infty \frac{x^3 \mathrm{d}x}{\mathrm{e}^x - 1} = \Gamma(4)\ \zeta(4) = \frac{6\pi^4}{90} = \frac{\pi^4}{15} .$$

This leads to an average energy for $s = 1$ (using, for example, equation (14.29))

$$U(T) - U(0) = \frac{v}{2\pi^2}\frac{(kT)^4}{\hbar^3}\sum_{t=1}^{3}\frac{\pi^4}{15}\frac{1}{a_t^3}$$

$$= \frac{\pi^2 v(kT)^4}{30\hbar^3}\sum_t a_t^{-3} .$$

This is of course in agreement with (14.30) if $s = 1$. In a liquid only longitudinal modes can arise and at low temperatures a Debye-type theory yields

$$U(T) = \frac{\pi^2 v(kT)^4}{30(\hbar c)^3} \tag{14.32}$$

where $c$ is the velocity of the waves.

## 14.5. THE HEAT CAPACITIES OF SOME OTHER SYSTEMS

For some *polyatomic molecules* it is a good approximation to suppose that the solutions of the Schrödinger equation may be specified by quantum numbers giving the states of translation ($t$), rotation ($r$), vibration ($v$), and electronic excitation ($e$). These quantum numbers indicate various eigenfunctions of the appropriate Hamiltonian, which could be, for example, a harmonic oscillator Hamiltonian for the vibrational states. Assuming that these quantum numbers can vary independently, the energy of the system has the form

$$E(t,r,v,e) = \varepsilon_t + \varepsilon_r + \varepsilon_v + \varepsilon_e \,.$$

The partition function is then, for a temperature $T$,

$$Z = \sum_{t,r,v,e} \exp\left(-\frac{\varepsilon_t+\varepsilon_r+\varepsilon_v+\varepsilon_e}{kT}\right)$$

$$= \left\{\sum_t \exp\left(-\frac{\varepsilon_t}{kT}\right)\right\}\left\{\sum_r \exp\left(-\frac{\varepsilon_r}{kT}\right)\right\}\left\{\sum_v \exp\left(-\frac{\varepsilon_v}{kT}\right)\right\}\left\{\sum_e \exp\left(-\frac{\varepsilon_e}{kT}\right)\right\}$$

$$= Z^{(t)}\, Z^{(r)}\, Z^{(v)}\, Z^{(e)} \,. \tag{14.33}$$

It thus decomposes into a product of partition functions, one for each of the constituent motions. Since the internal energy $U$ and hence the heat capacity $C_v$ depend on the logarithm of $Z$, these quantities are then additive:

$$C_v = C_v^{(t)} + C_v^{(r)} + C_v^{(v)} + C_v^{(e)} \,. \tag{14.34}$$

By this means, one obtains a first insight into the theory of the heat capacity of gases of polyatomic molecules. In fact, the assumption of the independence of the quantum numbers is of course not always obeyed.

For metals at low temperatures, the lattice contribution $C_v^{(\ell)} \propto T^3$ to the heat capacity also represents only one term. The free electrons contribute $C_v^{(e)} \propto T$ as discussed in Chapter

15. It follows that if $A$ and $B$ are temperature independent,

$$C_v = AT^3 + BT \ .$$

It is therefore usual to plot experimental results by showing $C_v/T$ as a function of $T^2$. This sometimes yields extraordinarily good straight lines:

$$C_v/T = AT^2 + B \ . \tag{14.35}$$

The slope $A$ of this line provides a check on the Debye (or other) theory of the lattice heat capacity. The intercept $B$ of the line provides a similar check on the theory of the electronic heat capacity. This is a very important class of experiments.

In liquid helium, simulated by a Bose gas at low temperatures, one finds that (14.29) holds for $t$=1, since only longitudinal modes of vibration can be supported in a liquid. Such a system of particles has a density-of-states law which behaves as the square root of the energy, so that the mean energy is proportional to

$$\int \frac{E^{3/2}\,\mathrm{d}E}{\exp\{(\mu-E)/kT\}-1} \ . \tag{14.36}$$

However, at a sufficiently low energy there is a 'condensed phase' for which $\mu = 0$. Then (14.29) applies with $s$=2 to this phase, and one finds at once from (14.30) that

$$C_v \propto T^{3/2} \, . \tag{14.37}$$

Above the 'condensation temperature' $\mu$ is no longer zero and a more general formula such as (11.22) has to be used (for more details see section 16.6).

As an example of a dispersion relation which does not fit into the pattern of (14.27) we consider again the condensed phase of liquid helium, but now above a temperature of about 0·6 K. The terms (14.36) in $U$ or (14.37) in $C_v$ are then beginning to be overshadowed by certain other terms which are associated with a new type of excitation. This excitation is modelled by the function

$$\varepsilon(p) = \frac{(p-p_0)^2}{2m} + \varepsilon(p_0)$$

or

$$\omega(k) = \frac{\hbar(k-k_0)^2}{2m} + \omega(k_0)$$

and is said to describe a 'roton' spectrum. The energy has a minimum at $k = k_0$ ($\sim 19\ \mathrm{nm}^{-1} \sim 1{\cdot}9\ \text{Å}^{-1}$), $m$ is $0{\cdot}16$ times the mass of an atom of $^4$He, and $\hbar\omega(k_0)$ is an energy which corresponds to about $8{\cdot}5$ K.†

In order to discuss the mean energy of a roton gas, it is instructive to start from (12.7) and use the relations

$$p = h/\lambda = \hbar\omega \quad .$$

One then finds

$$\mathrm{d}N(\lambda) = \frac{4\pi v}{\lambda^4}\,\mathrm{d}\lambda = \frac{4\pi v}{h^3}\,p^2\mathrm{d}p = \frac{v}{2\pi^2}\,\omega^2\mathrm{d}\omega \ .$$

The internal energy is (in the $p$-representation)

$$U = \frac{4\pi v}{h^3}\int\frac{\varepsilon(p)p^2\mathrm{d}p}{\exp\{\varepsilon(p)/kT\}-1} \tag{14.38}$$

where the integral can be extended at low temperatures from $p = -\infty$ to $p = \infty$. Similarly, the number of rotons is

$$N = \frac{4\pi v}{h^3}\int_{-\infty}^{\infty}\frac{p^2\mathrm{d}p}{\exp\{\varepsilon(p)/kT\}-1} \ . \tag{14.39}$$

To carry out these integrals, one can neglect the -1 in the integrand since $\varepsilon(p_0)/kT$ is always in the exponent and is reasonably large in the temperature range of interest near 2K. The greatest contribution to the integral comes therefore from $p \sim p_0$, so that $p^2$ can be taken out of the

† The wavevector $k$ will not be confused with Boltzmann's constant $k$, which is always multiplied by $T$. After equation (14.39) the wavevector disappears again.

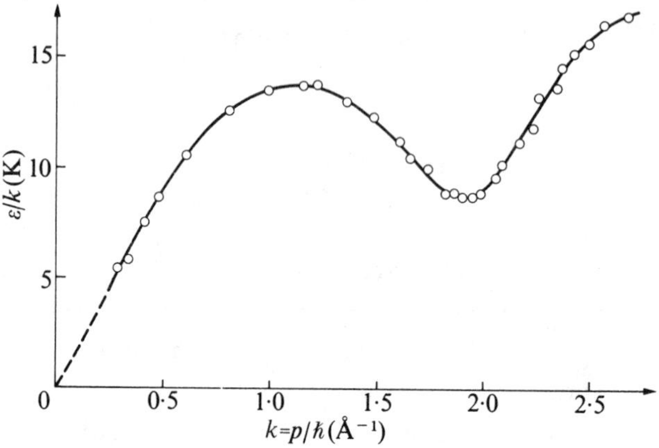


Fig.14.3. The empirical relationship between energy (in K) and wavevector (in $\text{Å}^{-1}$) for $^4$He as obtained by neutron diffraction.

integral as $p_0^2$ in reasonable approximation. Setting

$$x^2 \equiv \frac{(p-p_0)^2}{2mkT}$$

one finds

$$2x\mathrm{d}x = \frac{p-p_0}{mkT}\,\mathrm{d}p = \frac{\surd(2mkT)}{mkT}\,x\,\mathrm{d}p\ .$$

It follows that

$$\mathrm{d}p = \surd(2mkT)\mathrm{d}x\ .$$

Hence, to a sufficient approximation and noting that the remaining integral is even in $x$,

$$N = \frac{8\pi v}{h^3}\,p_0^2\,\surd(2mkT)\,\exp\left\{-\frac{\varepsilon(p_0)}{kT}\right\}\int_0^\infty \exp(-x^2)\,\mathrm{d}x$$

$$U = \frac{8\pi v}{h^3}\,p_0^2\,\surd(2mkT)\,\exp\left\{-\frac{\varepsilon(p_0)}{kT}\right\}\int_0^\infty \{\varepsilon(p_0) + kTx^2\}\exp(-x^2)\,\mathrm{d}x\ .$$

The first integral is $\sqrt{\pi}/2$; the second integral is

$$\frac{\sqrt{\pi}}{2}\, kT \left\{\frac{\varepsilon(p_0)}{kT} + \frac{1}{2}\right\}$$

so that

$$N = \frac{4\pi v}{h^3}\, p_0^2 \sqrt{(2\pi mkT)}\, \exp\left\{-\frac{\varepsilon(p_0)}{kT}\right\}$$

$$U = \left\{\frac{\varepsilon(p_0)}{kT} + \frac{1}{2}\right\} kT\, N\ .$$

The heat capacity is now obtained by a straightforward calculation, in which we write $\varepsilon_0$ for $\varepsilon(p_0)$, as follows:

$$U = (\varepsilon_0 + \tfrac{1}{2}kT)N$$

$$C_v = (\varepsilon_0 N + \tfrac{1}{2}kT\, N)\left(\frac{\partial \ln N}{\partial T}\right)_v + \tfrac{1}{2}\, kN\ .$$

If $A$ denotes a constant

$$\ln N = A + \tfrac{1}{2}\ln T - \varepsilon_0/kT$$

so that

$$\left(\frac{\partial \ln N}{\partial T}\right)_v = \left\{\frac{1}{2kT} + \frac{\varepsilon_0}{(kT)^2}\right\} k.$$

It follows that

$$\frac{C_v}{kN} = \tfrac{1}{2} + (\varepsilon_0 + \tfrac{1}{2}kT)\left\{\frac{1}{2kT} + \frac{\varepsilon_0}{(kT)^2}\right\}$$

$$= \tfrac{1}{2} + \frac{\varepsilon_0}{2kT} + \left(\frac{\varepsilon_0}{kT}\right)^2 + \tfrac{1}{4} + \tfrac{1}{2}\frac{\varepsilon_0}{kT}$$

$$= \tfrac{3}{4} + \frac{\varepsilon_0}{kT} + \left(\frac{\varepsilon_0}{kT}\right)^2\ .$$

The temperature dependence of the roton heat capacity is therefore dominated by the exponential temperature factor in the roton number. As the temperature is raised above about 0·6 K the rotons start rapidly to contribute to the heat capacity but between 2 K and 3 K the chemical potential becomes non-zero and then a more usual quantum gas theory for bosons, preferably taking interactions into account, becomes more applicable.

PROBLEMS

14.1. Show that the partition function of $N$ non-interacting distinguishable two-dimensional isotropic harmonic oscillators of frequency $\nu$ at temperature $T$ is

$$Z_N = \left\{\frac{1}{2\sinh(h\nu/2kT)}\right\}^{2N} .$$

Show that there are $j + 1$ quantum states of energy $(j+1)h\nu$ ($j = 0,1,2,\ldots$), i.e. that this energy is $(j+1)$-fold degenerate.

14.2. For $N$ $w$-dimensional isotropic non-interacting distinguishable oscillators problem 14.1 leads to

$$Z_N = \left\{\frac{1}{2\sinh(h\nu/2kT)}\right\}^{wN} .$$

Show that the energies of the oscillator may be expressed in the form $E_j = (\tfrac{1}{2}w + j)h\nu$ ($j = 0,1,2,\ldots$), and have degeneracies

$$\frac{(w+j-1)!}{j!\,(w-1)!} .$$

14.3. Find the main thermodynamic functions for the system of problem 14.2.

14.4. A classical oscillator at temperature $T$ has Hamiltonian $p^2/2m + V(x)$, where $V(x)$ is a potential energy function. Show that the partition function is

$$Z = h^{-1}\,(2\pi mkT)^{\frac{1}{2}} \int_{-\infty}^{\infty} \exp\left(-\frac{V}{kT}\right) dx .$$

Hence obtain the form

$$Z = A\,kT + B(kT)^2$$

for the partition function of an anharmonic oscillator with

$$V(x) = \tfrac{1}{2}ax^2 + bx^3 + cx^4$$

where $a, b$, and $c$ are constants and $A$ and $B$ are to be expressed in terms of them. Here $bx^3 + cx^4$ is assumed to be much smaller than $ax^2$. Establish that the heat capacity has the form $\alpha + \gamma T$ where $\alpha$ and $\gamma$ are constants.

14.5. Show that the Debye heat capacity (14.15) of a solid at temperature $T$ can be written in the form

$$\frac{C_v}{k} = 3rN\,\Delta(x_D)$$

where

$$\Delta(a) \equiv \frac{12}{a^3}\int_0^a \frac{y^3\,dy}{e^y-1} - \frac{3a}{e^a-1} .$$

14.6. By applying the argument which was used to obtain the Debye heat capacity (14.15) to the internal energy

$$U = \sum_{i=1}^{3rN}\left\{\frac{1}{2} + \frac{1}{\exp(h\nu_i/kT)-1}\right\}h\nu_i$$

show that for a Debye solid at temperature $T$

$$U = 3NkTrD(x_D)$$

where

$$D(a) \equiv \frac{3}{a^3}\int_0^a \left(\frac{1}{2} + \frac{1}{e^y-1}\right) y^3\,dy .$$

14.7. Referring to the last two problems, show that

$$\Delta(a) = D(a) - a\,D'(a) .$$

Hence verify that the relation

$$C_v = (\partial U/\partial T)_v$$

is satisfied.

14.8. Use (14.31) to show that for a Debye solid at temperature $T$, the pressure, internal energy, entropy, and chemical potential are respectively

$$p = -9rNh\left\{\tfrac{1}{8} + x_D^{-4}\int_0^{x_D}\frac{x^3\,dx}{e^x - 1}\right\}\left(\frac{\partial \nu_D}{\partial v}\right)_T$$

$$U = \frac{9}{8}rN\,h\nu_D + \frac{9rNkT}{x_D^3}\int_0^{x_D}\frac{x^3\,dx}{e^x - 1}$$

$$S = \frac{12rNk}{x_D^3}\int_0^{x_D}\frac{x^3\,dx}{e^x - 1} - 3rNk\ln(1 - e^{-x_D})$$

$$\mu = \frac{3}{2}rh\nu_D + 3kT\ln(1 - e^{-x_D})$$

14.9. ($a$) Obtain the free energy of an Einstein solid of $3N$ one-dimensional oscillators, all of frequency $\nu$, if the interaction between them lowers the free energy by the amount $AN$, where $A$ is a constant. Hence show that the chemical potential of the solid is

$$\mu_s = -kT\ln\left\{Z_1^3\exp\left(\frac{A}{kT_s}\right)\right\}$$

where $Z_1$ is the partition function of one oscillator and the suffix s is a reminder that a solid is envisaged.

($b$) Show that the chemical potential of an ideal quantum gas of $N_g$ particles in a volume $v_g$ is in the classical approximation (using (12.12))

$$\mu_g = -kT_g\ln\left(\frac{2\pi mkT_g}{h^2}\right)^{3/2}\frac{v_g g}{N_g}$$

where $T_g$ is the temperature of the gas, $m$ is the mass of one of its molecules, and $g$ is the spin degeneracy.

($c$) Assuming that the vapour of an Einstein solid satisfies $pv_g = N_g kT$ and that it is in equilibrium with the solid, show that the vapour pressure of an Einstein solid is with $T_g \equiv T$

$$p = \left(\frac{2\pi mgkT}{h^2}\right)^{3/2} gkT\left\{2\sinh\left(\frac{h\nu}{2kT}\right)\right\}^3\exp\left(-\frac{A}{kT}\right).$$

# 15. Fermi systems

## 15.1. THE STATISTICAL THERMODYNAMICS OF AN IDEAL QUANTUM GAS WHOSE PARTICLES HAVE NON-ZERO REST MASS

In its simplest form a metal is regarded as $N$ weakly interacting electrons in a field-free potential box. The atoms of the metal are regarded as having contributed their outer electrons (one electron per atom in the simplest cases) to this sea of electrons. The positively charged atoms which are left behind provide a periodic potential. However, if one approximates this situation by supposing that this positive charge is smeared out uniformly, one has in effect a field-free box. This is the picture of a metal developed initially by A. Sommerfeld (1928). The formal theory of Chapter 11 applies with the density of states function of the fourth row in Table 12.1. It is convenient to allow also for bosons and for a spin degeneracy $g$ of each single-particle energy level. Then $g = 2$ for electrons since the spin of the electron can have only two values.

Equations (11.6) and (11.7) yield for the total number of particles and the free energy of a system of $N$ particles of mass $m$ at chemical potential $\mu$ and at temperature $T$ in a volume $v$:

$$N = gv\left(\frac{2\pi mkT}{h^2}\right)^{3/2} I\left(\frac{\mu}{kT}, \tfrac{1}{2}, \pm\right), \tag{15.1}$$

$$F = \mu N - gkTv\left(\frac{2\pi mkT}{h^2}\right)^{3/2} I\left(\frac{\mu}{kT}, \frac{3}{2}, \pm\right), \tag{15.2}$$

$$= NkT\left(\frac{\mu}{kT} - z\right) \tag{15.3}$$

where $N$ is the grand canonical average value and

$$z \equiv \frac{I(\mu/kT, 3/2, \pm)}{I(\mu/kT, \frac{1}{2}, \pm)} \quad . \tag{15.4}$$

Also equations (11.19) and (11.20) yield for the internal energy and for the pressure

$$U = \frac{3}{2} NkTz = \frac{3}{2} pv \ . \tag{15.5}$$

Lastly, the entropy of the system is from (11.18)

$$S = Nk \left( \frac{5}{2} z - \frac{\mu}{kT} \right) \ . \tag{15.6}$$

It is an advantage of the approach adopted here that these results can be written down at once as deductions from a general theory and without new calculations. Note, as a check, that $U-TS+pv$ turns out to be $\mu N$, as required by thermodynamics.

## 15.2. A FERMI GAS AT ABSOLUTE ZERO (ELECTRONS IN METALS)

Reverting to electrons in metals, it is found that the chemical potential $\mu \gg kT$ at room temperature, so that the integrals $I$ have to be evaluated under this condition. One finds

$$\Gamma(s+1)\ I\left(\frac{\mu}{kT}, s, +\right) = \int_0^\infty \frac{x^s \, dx}{\exp[(e-\mu)/kT]+1} \doteqdot \int_0^{\mu/kT} x^s \, dx$$

The Fermi-Dirac function has here been replaced by unity for $-\infty<e\leq\mu$ and by zero for $\mu<e$. This is quite justified for large values of $\mu$, since the function drops from unity to zero in an energy range of order $4kT$ which is still much smaller than $\mu$. Hence for $s > -1$

$$\Gamma(s+1)\ I\left(\frac{\mu}{kT}, s, +\right) \doteqdot \frac{1}{s+1}\left(\frac{\mu}{kT}\right)^{s+1} \ . \tag{15.7}$$

It follows that

$$I\left(\frac{\mu}{kT}, \tfrac{1}{2}, +\right) \doteqdot \frac{4}{3\sqrt{\pi}}\left(\frac{\mu}{kT}\right)^{3/2} \tag{15.8}$$

since from the properties of the gamma function (see p.436)

$$\frac{3}{2}\,\Gamma\left(\frac{3}{2}\right) = \frac{3}{2} \times \tfrac{1}{2}\,\Gamma(\tfrac{1}{2}) = \tfrac{3}{4}\sqrt{\pi} \ . \tag{15.9}$$

Another useful result is that by (15.4) and (15.7)

$$z \equiv \frac{I(\mu/kT,3/2,+)}{I(\mu/kT,\frac{1}{2},+)} \doteqdot \frac{(\mu/kT)^{5/2}\big/(5/2)\Gamma(5/2)}{(\mu/kT)^{3/2}\big/(3/2)\Gamma(3/2)}$$

$$= \frac{\mu}{kT}\,\frac{3}{5}\,\frac{\Gamma(3/2)}{\Gamma(5/2)} \;.$$

By another property of the gamma function,

$$\Gamma(s+1) = s\Gamma(s) \tag{15.10}$$

so that

$$z \doteqdot \frac{2}{5}\,\frac{\mu}{kT} \;. \tag{15.11}$$

One can now use the approximations (15.8) and (15.11) in our main results (15.1) to (15.6) to find

$$\left.\begin{aligned} N &= \frac{4}{3\sqrt{\pi}}\,gv\left(\frac{2\pi m\mu}{h^2}\right)^{3/2}, \qquad S = 0\,, \\ F &= \frac{3}{5}\,\mu N = U = \frac{3}{2}\,pv\,. \end{aligned}\right\} \tag{15.12}$$

These equations describe a metal at the lowest temperatures. The vanishing of the entropy is in agreement with the third law of thermodynamics; p is called the Fermi pressure.

## 15.3. A FERMI GAS AT LOW TEMPERATURES (ELECTRONS IN METALS)

One can proceed to the next order of accuracy in (15.7) and hence obtain a theory of electrons in metals which is also valid well above the absolute zero of temperature. For this purpose one needs the result

$$I\left(\frac{\mu}{kT},s,+\right) \doteqdot \left(\frac{\mu}{kT}\right)^{s+1}\frac{1}{\Gamma(s+2)}\left\{1 + \frac{s(s+1)}{6}\left(\frac{\pi kT}{\mu}\right)^2\right\} \tag{15.13}$$

for $\mu \gg kT$ and $s > -1$. This will now be proved.

Consider therefore $J \equiv \Gamma(s+1)I(a,s,+)$ and apply a partial integration:

$$J = \int_0^\infty \frac{x^s \, dx}{e^{x-a}+1} = \left( \frac{x^{s+1}}{s+1} \frac{1}{e^{x-a}+1} \right)_0^\infty + \frac{1}{s+1} \int_0^\infty \frac{x^{s+1} e^{x-a}}{(e^{x-a}+1)^2} \, dx$$

$$= \frac{1}{s+1} \int_{-\infty}^{\infty} (y+a)^{s+1} \frac{e^y}{(e^y+1)^2} \, dy \quad .$$

In the last step we have put $y = x-a$. The lower limit of the integral then becomes $y = -a$. However, since $a \gg 1$ and the large values of $y$ do not contribute much to the integral, the lower limit has been kept at $-\infty$ as an approximation. Expanding for $s > -1$,

$$J = \frac{1}{s+1} \int_{-\infty}^{\infty} a^{s+1} \left\{ 1+(s+1) \frac{y}{a} + \frac{s(s+1)}{2} \left( \frac{y}{a} \right)^2 + \ldots \right\} \frac{e^y}{(e^y+1)^2} \, dy \, . \tag{15.14}$$

Now multiplying

$$g(y) \equiv \frac{e^y}{(e^y+1)^2}$$

by $e^{-2y}/e^{-2y}$ yields $g(-y)$, showing that $g(y)$ is an even function of $y$. The second term in (15.14) therefore vanishes as an integral of an odd function. The first term involves

$$\int_{-\infty}^{\infty} \frac{e^y}{(e^y+1)^2} \, dy = -\left( \frac{1}{e^y+1} \right)_{-\infty}^{\infty} = 1 \; . \tag{15.15}$$

The third term involves an even integral and yields

$$2 \int_0^\infty \frac{y^2 e^y}{(e^y+1)^2} \, dy = 2 \int_0^\infty y^2 e^{-y} (1-2e^{-y}+3e^{-2y} - \ldots) \, dy \tag{15.16}$$

by the binomial expansion ($z \equiv e^{-y}$)

$$(1+z)^{-2} = 1-2z + 3z^2 - \ldots \; .$$

Term-by-term integration, using the definition (11.8) of the gamma function, yields for (15.16)

$$2\int_0^\infty y^2 e^{-y}\,dy - 4\int_0^\infty y^2 e^{-2y}\,dy + 6\int_0^\infty y^2 e^{-3y}\,dy - \ldots$$

$$= 2\times 2! - \tfrac{1}{2}\int_0^\infty u^2 e^{-u}\,du + \frac{2}{9}\int_0^\infty v^2 e^{-v}\,dv - \ldots$$

$$= 4\left(1 - \frac{1}{4} + \frac{1}{9} - \ldots\right) . \tag{15.17}$$

This series can be summed by means of the theory of the Riemann zeta function

$$\zeta(r) \equiv \sum_{j=1}^{\infty} j^{-r} \qquad (r \geq 1) \tag{15.17'}$$

which for $r=2$ has the value $\pi^2/6$. It follows that

$$\frac{\zeta(2)}{2} = 2\sum_{j=1}^{\infty} \frac{1}{(2j)^2}$$

$$= 2\left(\frac{1}{4} + \frac{1}{16} + \frac{1}{36} + \frac{1}{64} + \ldots\right)$$

$$= \left(1 + \frac{1}{4} + \frac{1}{9} + \frac{1}{16} + \frac{1}{25} + \ldots\right) - \left(1 - \frac{1}{4} + \frac{1}{9} - \frac{1}{16} + \ldots\right) .$$

Hence the quantity (15.17), i.e. (15.16), is

$$4\,\{\zeta(2) - \tfrac{1}{2}\zeta(2)\} = 2\,\zeta(2) = \frac{\pi^2}{3} . \tag{15.17''}$$

Substituting this result and (15.15) into (15.14),

$$J \doteqdot \frac{a^{s+1}}{s+1} + \frac{s a^{s-1}}{2}\,\frac{\pi^2}{3} = \frac{a^{s+1}}{s+1}\left\{1 + \frac{s(s+1)}{6}\left(\frac{\pi}{a}\right)^2\right\} .$$

This establishes (15.13).

It follows that, to this approximation,

$$\frac{I(\mu/kT, 3/2, +)}{I(\mu/kT, \tfrac{1}{2}, +)} \doteqdot \frac{\mu}{kT}\,\frac{\Gamma(5/2)}{\Gamma(7/2)}\,\frac{1+\left(\frac{15}{24}\right)\left(\frac{\pi kT}{\mu}\right)^2}{1+\left(\frac{3}{24}\right)\left(\frac{\pi kT}{\mu}\right)^2} .$$

Using again $\Gamma(s+1) = s\Gamma(s)$,

$$z \doteqdot \frac{2}{5}\frac{\mu}{kT}\left[1 + \frac{1}{2}\left(\frac{\pi kT}{\mu}\right)^2\right] \qquad \left(\frac{\mu}{kT} \gg 1\right), \tag{15.18}$$

in a generalization of (15.11).

One now finds by substituting (15.13) into (15.1), and by substituting (15.18) into (15.3), (15.5) and (15.6) that

$$N \doteqdot \frac{4}{3\sqrt{\pi}}\, gv \left(\frac{2\pi m\mu}{h^2}\right)^{3/2} \left\{1 + \frac{1}{8}\left(\frac{\mu kT}{\mu}\right)^2\right\} \tag{15.19}$$

$$U \doteqdot \frac{3}{5}\,\mu N \left\{1 + \frac{1}{2}\left(\frac{\pi kT}{\mu}\right)^2\right\} = \frac{3}{2}\, pv \tag{15.20}$$

$$F \doteqdot \frac{3}{5}\,\mu N \left\{1 - \frac{1}{3}\left(\frac{\pi kT}{\mu}\right)^2\right\} \tag{15.21}$$

$$S \doteqdot \frac{1}{2}\,\frac{\pi^2 k^2}{\mu}\, NT \,. \tag{15.22}$$

The entropy goes to zero as the temperature approaches the absolute zero, as required by the third law of thermodynamics.

The variation of the chemical potential with temperature can be obtained from (15.19) to be given by

$$\mu(0)^{3/2} \doteqdot \mu(T)^{3/2}\left[1 + \frac{1}{8}\left\{\frac{\pi kT}{\mu(T)}\right\}^2\right] .$$

It follows, using $\mu(0)$ instead of $\mu(T)$ in the correction term, that

$$\mu(T)^{3/2} \doteqdot \frac{\mu(0)^{3/2}}{1+\frac{1}{8}\left(\frac{\pi kT}{\mu(0)}\right)^2} \doteqdot \mu(0)^{3/2}\left\{1 - \frac{1}{8}\left\{\frac{\pi kT}{\mu(0)}\right\}^2\right\}$$

so that

$$\mu(T) \doteqdot \mu(0)\left[1 - \frac{1}{12}\left\{\frac{\pi kT}{\mu(0)}\right\}^2\right] . \tag{15.23}$$

Thus the chemical potential of a low-temperature Fermi gas rises as the temperature falls.

One can substitute (15.23) into (15.20) to find

$$U \doteqdot \frac{3}{5}\,\mu(0)\, N \left[1 - \frac{1}{12}\left\{\frac{\pi kT}{\mu(0)}\right\}^2\right]\left[1 + \frac{1}{2}\left\{\frac{\pi kT}{\mu(0)}\right\}^2\right]$$

$$U \doteq \frac{3}{5}\mu(0)N\left[1 + \frac{5}{12}\left\{\frac{\pi kT}{\mu(0)}\right\}^2\right] .$$

The heat capacity of a low-temperature Fermi gas is accordingly a linear function of temperature:

$$C_v = \frac{\pi^2 k^2}{2\mu(0)} NT . \qquad (15.24)$$

This formula has been verified experimentally for a number of metals by first making allowance for the lattice heat capacity. By (14.15) this rises as $T^3$ at low temperatures, so that at low enough temperatures ($T \sim$ 1-10 K) the electronic specific heat dominates in metals. The value of the coefficient is changed if the Coulomb interactions between electrons are taken into account.

Classical statistical mechanics would give rise to electronic heat capacity which is much larger, namely (see section 11.7)

$$C_{v,\mathrm{cl}} = \frac{3}{2}NK$$

so that

$$\frac{C_v}{C_{v,\mathrm{cl}}} = \frac{\pi^2}{3}\frac{kT}{\mu(0)} .$$

For $\mu(0) \sim 1$ eV and room temperature ($kT \sim 1/40$ eV), this is a small quantity. Its small size arises from the fact that Fermi statistics allows only particles near the energy $\mu(0)$ to gain thermal energy since most quantum states at lower energies are already occupied. This means that particles in these states cannot be excited by small amounts of energy, there being no vacant states for them to occupy. Such vacancies do exist for particles near the energy $\mu(0)$. The effect is that only a fraction $\sim kT/\mu(0)$ of particles can be expected to contribute to the heat capacity. In classical statistical mechanics all particles in the gas contribute equally, and the electronic heat capacity is overestimated.

## 15.4. ELECTRONS AND HOLES IN SEMICONDUCTORS

A single-fermion quantum state of energy $E$ has been seen to have a mean occupation number

$$f(E) = \frac{1}{\exp(\eta-\gamma)+1} \qquad \left(\eta \equiv \frac{E}{kT}, \quad \gamma \equiv \frac{\mu}{kT}\right) \tag{15.25}$$

at temperature $T$. The density of states rises (in the simplest cases) as $\sqrt{E}$, and this has been used in (15.1). However, the zero of the density of states function is often more conveniently placed at some general energy, $E_c$ say. Then with $\eta_c \equiv E_c/kT$ one has a density of states function

$$\frac{4\pi mgv}{h^3}\sqrt{(2m)}\ \sqrt{(E-E_c)} \tag{15.26}$$

which vanishes at $E = E_c$, as required (see Fig.15.1). This yields for the number of electrons per unit volume

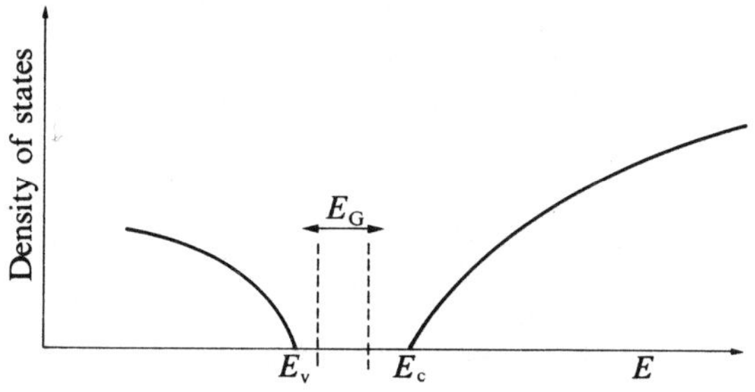


Fig.15.1. Density of states in valence and conduction bands of a simple semiconductor. For non-degeneracy $E_V + 2kT < \mu < E_C - 2kT$ is usually sufficient, and two such possible positions for $\mu$ are shown schematically.

$$n = \frac{N}{v} = \frac{4\pi mg}{h^3}\sqrt{(2m)}\ (kT)^{3/2}\int_{\eta_c}^{\infty}\frac{(\eta-\eta_c)^{\frac{1}{2}}d\eta}{\exp(\eta-\gamma)+1} \tag{15.27}$$

Dividing and multiplying by $\Gamma(3/2) = \frac{1}{2}\sqrt{\pi}$ ,

$$n = g\left(\frac{2\pi mkT}{h^2}\right)^{3/2}\frac{1}{\Gamma(3/2)}\int_0^{\infty}\frac{x^{\frac{1}{2}}\,dx}{\exp(x+\eta_c-\gamma)+1}$$

where $x \equiv \eta-\eta_c$. Taking the spin degeneracy $g = 2$,

$$n = N_c\, I(\gamma-\eta_c,\tfrac{1}{2},+) \qquad N_c \equiv 2\left(\frac{2\pi mkT}{h^2}\right)^{3/2} . \tag{15.28}$$

This is a formula frequently used for the electron concentration in the conduction band of a semiconductor at temperature $T$. The band is assumed to be confined to energies $E > E_c$. For $E < E_c$ one supposes that there is a 'forbidden gap', of energy width $E_G$, say. Broadly speaking, imperfections or interaction effects alone can introduce electronic states into this gap. One often calls $\mu = \gamma kT$ the Fermi level.

The energy gap is, say,

$$E_G = E_c - E_v \tag{15.29}$$

where $E_v$ denotes the energy at the top of the next lower band. This is called the valence band. The number density of electrons in it is given by a formula like (15.27) with the integral replaced by

$$\int_{-\infty}^{\eta_v} \frac{(\eta_v-\eta)^{\frac{1}{2}}\, d\eta}{\exp(\eta-\gamma)+1} .$$

Thus one has a square-root law for the density of states when measured towards more negative energies. The number density of electron vacancies, or 'holes', in the valence band requires an integral

$$\int_{-\infty}^{\eta_v} \frac{(\eta_v-\eta)^{\frac{1}{2}}\, d\eta}{\exp(\gamma-\eta)+1}$$

since $1-f(E)$ is the probability of an electron state being empty. The quantity $1-f(E)$ is

$$\frac{1+\exp(\eta-\gamma)-1}{\exp(\eta-\gamma)+1} = \frac{1}{\exp(\gamma-\eta)+1} .$$

The hole density in the valence band is therefore

$$p = N_v F_{\frac{1}{2}}(\eta_v-\gamma,\tfrac{1}{2},+) \qquad N_v \equiv 2\left(\frac{2\pi mkT}{h^2}\right)^{3/2} . \tag{15.30}$$

The mass in (15.28) is usually replaced by an effective mass $m_e$. This takes partial account of the fact that the electrons move in a periodic lattice. Similarly $m_h$ is used instead of $m$ in (15.30). One can think of the energy gap as also being due to the periodic lattice, and we shall not investigate this matter here.

The model of a two-band semiconductor is also useful because it broadens one's understanding of the classical approximation. In section 10.2 it was specified by $\mu \to -\infty$. However, it is sufficient to ensure that the occupation probabilities of states by electrons are small in the conduction band, for then the Maxwell-Boltzmann distribution results:

$$f(E) \propto \exp(\gamma-\eta) \quad (\eta>\eta_c) .$$

Similarly it is sufficient to think of the valence band as being practically full of electrons, for then the distribution law for holes is

$$1-f(E) \propto \exp(\eta-\gamma) \quad (\eta<\eta_v) .$$

Both bands can therefore be treated in the classical approximation if

$$\eta_c \gg \gamma \gg \eta_v .$$

that is if

$$\frac{E_G}{kT} \gg \frac{\mu-E_v}{kT} \gg 0 .$$

Thus these two conditions are sufficient. The bands are then said to be *non-degenerate*. In this approximation

$$\left.\begin{aligned} &n = N_c \exp(\gamma-\eta_c) \qquad p = N_v \exp(\eta_v-\gamma) \\ &\text{so that} \\ &np = N_c N_v \exp(-\eta_G) . \end{aligned}\right\} \qquad (15.31)$$

Even if $m_e$, $m_h$, $T$, $E_c$, $E_v$ and the condition (15.31) are given there is still not enough information available to locate the chemical potential and hence to estimate $n$ and $p$. This can however be done as follows.

Suppose that at absolute zero $n = p = 0$ and that at temperature $T$ all the electrons have come from the valence band. Then $n = p$ in (15.31) and this completes the information supplied. One then finds

$$\exp\gamma = \left(\frac{N_v}{N_c}\right)^{\frac{1}{2}} \exp\left(\frac{\eta_c + \eta_v}{2}\right) = \left(\frac{m_h}{m_e}\right)^{\frac{3}{4}} \exp\left(\frac{E_c + E_v}{2kT}\right) . \tag{15.32}$$

Substituting (15.31) into (15.32), the electron and hole densities are found to be

$$n = (N_c N_v)^{\frac{1}{2}} \exp\left(-\frac{\eta_G}{2} - \eta_c\right) = p. \tag{15.33}$$

As the temperature is raised $n$ and $p$ rise and eventually the condition of non-degeneracy is violated. The bands are then said to be *degenerate*, and (15.28) and (15.30) must be used.

## 15.5. DEFECTS IN SEMICONDUCTORS

Electronic states can be introduced in the energy gap by defects of various kinds. The rigorous periodicity of the lattice, which has led to this gap, is then broken, and defect or impurity levels can occur. Such defects can capture 0, 1, 2, ..., $M$ electrons. For a general state of capture of, say, $m$ electrons, all the electrons on the defects may be able to assume a whole spectrum of centre (i.e. many-electron) energy levels. These will be denoted by $E(m,j)$. The corresponding canonical partition function (9.22) is

$$Z_m = \sum_j \exp\left\{-\frac{E(m,j)}{kT}\right\} \quad (m = 0,1,\dots,M) . \tag{15.34}$$

With

$$\lambda \equiv \exp\gamma \equiv \exp\left(\frac{\mu}{kT}\right) \tag{15.34'}$$

the grand partition function (9.20) for one centre is

$$\Xi_1 = \sum_m \lambda^m Z_m \ . \tag{15.35}$$

For $N$ independent centres which are distinguishable by virtue of their position the general state of $N$ centres is specified by the numbers $m_1, m_2, \ldots, m_N$ of electrons captured by various centres. Given this set of numbers, each centre has an energy value distinguished by a further index. Hence (9.20) becomes with $n = m_1 + m_2 + \ldots + m_N$

$$\Xi_N = \sum_{\substack{m_1, m_2, \ldots, m_N \\ j_1, j_2, \ldots, j_N}} \lambda^{m_1+m_2+\ldots+m_N} \exp\left[-\frac{1}{kT}\left\{E(m_1,j_1)+E(m_2,j_2)+\ldots+E(m_N j_N)\right\}\right]$$

$$= \sum_{m_1, m_2, \ldots, m_N} \lambda^{m_1+m_2+\ldots+m_N} Z^{(1)}_{m_1} Z^{(2)}_{m_2} \ldots Z^{(N)}_{m_N}$$

where the superscript refers to the centre. However, as the $N$ centres are assumed identical one finds simply

$$\Xi_N = \left(\sum_m \lambda^m Z_m\right)^N \ . \tag{15.36}$$

This simple result is derived rather analogously to that obtained in (11.35).

The probability that a defect has captured $m$ electrons in a thermal equilibrium situation at temperature $T$ and chemical potential $\mu$ is $\lambda^m Z_m/\Xi$. This is by (15.35) correctly normalized. For $N$ centres, the mean number of centres to have captured $m$ electrons is simply

$$\nu_m = \left(\frac{\lambda^m Z_m}{\Xi_1}\right) N \ . \tag{15.37}$$

The mean number of trapped electrons $n_t$ held in these $N$ defects is given by

$$\frac{n_t}{N} = \frac{\sum_{m=1}^{M} m\lambda^m Z_m}{\Xi_1} \ . \tag{15.38}$$

This result may also be derived from (9.20), which applies to a general system:

$$\Xi_1 = \sum_{n,i} \exp(\gamma_n - \eta_i) = \sum_{n,i} \lambda^n \exp(-\eta_i) .$$

It shows that the mean number of particles $\bar{n}$ is

$$\lambda\left(\frac{\partial \ln \Xi_1^N}{\partial \lambda}\right)_{v,T} = N \frac{\sum_{n,i} n\lambda^n \exp(-\eta_i)}{\Xi_1} = \bar{n} . \qquad (15.39)$$

Application of this formula to (15.36) yields (15.38) provided that $\bar{n}$ is replaced by $n_t$.

If $M = 1$, (15.38) yields

$$\frac{n_t}{N} = \frac{\lambda Z_1}{Z_0 + \lambda Z_1} = \frac{1}{1+(Z_0/Z_1)\exp(-\gamma)} . \qquad (15.40)$$

This is similar in form to the *free-electron* Fermi function (10.9). The changes from (10.9) are due to the interactions among the electrons on the defect, which have been taken into account by assigning different energy-level spectra to a defect with one trapped electron as against a defect with no trapped electrons. Thus $Z_0/Z_1$ will in general not be unity. If the centre without the electron has one $a$-fold degenerate state of energy $E_0$ and the centre with the electron has one $b$-fold degenerate state of energy $E_0+E$, defined in problem 14.1, then

$$\frac{Z_0}{Z_1} = \frac{a \exp(-\eta_0)}{b \exp(-\eta_0 - \eta)} = \frac{a}{b} \exp \eta . \qquad (15.41)$$

One finds

$$\frac{n_t}{N} = \frac{1}{1+(a/b)\exp(\eta-\gamma)} . \qquad (15.42)$$

This is a frequently used result in semiconductor physics, and $a$ and $b$ are usually small integers.

## 15.6. MASS ACTION LAWS FROM STATISTICAL MECHANICS

If a semiconductor is steadily irradiated with photons whose energies exceed the band-gap energy, then electrons from the valence band can be promoted to the conduction band by the absorption of energy. This is of importance in photodectors, solar cells, etc. The radiation creates new electrons and holes. These come rapidly into equilibrium with the other current carriers (electrons or holes) in their respective bands. Excess electrons will, however, tend to drop back into the valence band in what is called a *recombination process*. The energy liberated may be given to the lattice or it may be emitted as radiation. In this way a steady state can be set up in which carriers are generated by the external source, but they also recombine, leaving the number of electrons and holes constant in time.

One would expect the rate of recombination to be proportional to the concentrations of electrons ($n$) and holes ($p$) since these govern the probability of an electron 'meeting' a hole with the possibility of a recombination act. This leads us to consider a product of (15.28) and (15.30):

$$np = N_c N_v I(\gamma_e - \eta_c, \tfrac{1}{2}, +) I(\eta_v - \gamma_h, \tfrac{1}{2}, +) \ .$$

We have here replaced the chemical potential by two distinct ones: $\mu_e$ and $\mu_h$ for electrons and holes respectively. This takes account of the fact that the carriers in each band are in equilibrium with each other, while the carriers in different bands (electrons and holes) are not in equilibrium with each other. Complete equilibrium would occur only if $\mu_e = \mu_h$ ($\equiv \mu$). The quantities $\mu_e$ and $\mu_h$ are called *quasi-Fermi levels*, and their introduction here represents an approximation.

Consider now the recombination of a hole (h) and an electron (e) to yield an electron in the valence band ($e_{vb}$) as a chemical reaction

$$e + h \rightleftharpoons e_{vb} \ . \qquad (15.43)$$

The mass action law would lead one to expect a concentration - independent quantity $np/n_{vb}$ where $n_{vb}$ is the concentration of valence-band electrons. This latter quantity is, however, so great that is in any case essentially constant when electron and hole concentrations vary, provided we assume

$$n, p \ll n_{vb} \ .$$

This enables us to confine attention to non-degenerate semiconductors. One finds

$$np = N_c N_v \exp(\gamma_e - \eta_c + \eta_v - \gamma_h) = N_c N_v \exp(\gamma_e - \gamma_h - \eta_G) \ . \quad (15.44)$$

In the limit of a near-equilibrium situation the right-hand side is indeed independent of electron and hole concentration, showing that the model of the process as a reaction is satisfactory. If the electron or hole gas have too high a concentration (so that they become degenerate), one can put

$$(f_n n)(f_p p) = N_c N_v \exp(\gamma_e - \gamma_h - \eta_G)$$

where

$$f_n \equiv \exp(\gamma_e - \eta_c)/I(\gamma_e - \gamma_c, \tfrac{1}{2}, +) = \exp(\gamma_e - \eta_c)(N_c/n)$$

$$f_p \equiv \exp(\eta_v - \gamma_h)/I(\eta_v - \gamma_h, \tfrac{1}{2}, +) = \exp(\eta_v - \gamma_h)(N_v/p) \ .$$

The concentrations are here generalized to effective concentrations or activities $f_n n$, $f_p p$, where the $f_n, f_p$ are the activity coefficients of Chapter 8. Note that $f_n$ plotted against $n/N_c$ gives a curve which rises rapidly from unity for $n \ll N_c$. With the aid of the $f$s a form of mass action law thus survives even for degenerate materials.

The recombination process may of course take place between the conduction band and impurities. Suppose the impurities are initially neutral ($D^{\times}$) and of concentration $N^{\times}$. If they can 'donate' electrons to the conduction band, they become positively charged ($D^{+}$). Let the concentration of such impurities, also called *donors*, be $N^{+}$. The recombination process is then the forward reaction, and the carrier generation process is the reverse reaction in

$$e + D^{+} \rightleftharpoons D^{\times} \ .$$

If $N$ is now the total *concentration* of donors, one needs to consider the concentrations of occupied and empty donors, which are, using (15.40),

$$N^{\times} = \frac{\lambda_D Z_1}{Z_0+\lambda_D Z_1} N \qquad (\lambda_D \equiv \exp\frac{\mu_D}{kT} \equiv \exp\gamma_D)$$

$$N^{+} = N - N^{\times} = \frac{Z_0}{Z_0+\lambda_D Z_1} N \quad .$$

Here $\mu_D$ is the quasi-Fermi level for the donors. We now consider the quantity

$$\frac{nN^{+}}{N^{\times}} = N_c \exp(\gamma_e - \eta_c) \frac{Z_0 N}{Z_0+\lambda_D Z_1} \frac{Z_0+\lambda_D Z_1}{\lambda_D Z_1 N}$$

$$= \frac{Z_0}{Z_1} \quad N_c \exp(\gamma_e - \gamma_D - \eta_c) \; .$$

One sees that the mass action law is again satisfied near equilibrium, since the right-hand side is, with $\gamma_e = \gamma_D$, independent of Fermi levels and so is independent of the concentrations which occur on the left-hand side.

## 15.7. GRAVITATIONAL COLLAPSE OF FERMI SYSTEMS TO BLACK HOLES

The description of the gravitational collapse of a system depends on the fact that a uniform sphere of radius $R$ and mass $M$ has a gravitational potential energy

$$U_G = -\frac{3}{5} G M^2/R \tag{15.45}$$

where $G$ is Newton's gravitational constant. The zero of the potential energy corresponds to infinite separation of the individual particles. Contraction below the radius $R$ in response to gravitational attraction thus lowers the potential energy of the system further. This will therefore occur until and unless other (positive) energy terms are brought into play to arrest such a collapse.

In degenerate Fermi systems one such term arises from the Fermi or quantum mechanical pressure given by (15.12), which holds for non-relativistic fermions. To cover the relativistic motion of particles as well one must go back to

$$N = A(kT)^{s+1} \Gamma(s+1) I(\gamma, s, +) \tag{11.6}$$

and problem 11.1:

$$U\ [= (s+1)pv] = A(kT)^{s+2}\,\Gamma(s+2)I(\gamma,s+1,+). \tag{15.46}$$

In the degenerate approximation (15.7)

$$N = \frac{A}{s+1}\,\mu^{s+1} \tag{15.47}$$

$$U = \frac{A}{s+2}\,\mu^{s+2} = \frac{s+1}{s+2}\,N\left[\frac{(s+1)N}{A}\right]^{1/(s+1)}. \tag{15.48}$$

From Table 12.1, with c (not now $c_0$) as the velocity of light in vacuo, one has

$$\begin{aligned} nr &: s = 1/2, \quad A = 4\pi mg\,h^{-3}\,(2m)^{1/2}\,(4\pi R^3/3) \\ r &: s = 2\ , \quad A = (4\pi g/h^3c^3)\quad(4\pi R^3/3) \end{aligned} \tag{15.49}$$

in (15.48) for the non-relativistic and the extreme relativistic approximation respectively. The volume has been taken to be a sphere of radius $R$, and it will be assumed that all particles to be considered have the same spin degeneracy $g$.

To set up the total energy of the system, let $n_G$, $n_{nr}$, $n_r$ be the number of particles, of masses $m_G$, $m_{nr}$, $m_r$ respectively, which feature in our three energy terms $U_G$, $U_{nr}$, $U_r$, $U_{nr} + U_r$ being a kinetic energy. The total energy of a system for which other energies are negligible is then

$$U \equiv \frac{A_{nr}}{R^2} + \frac{A_r - A_G}{R},\quad A_{nr} \equiv \frac{3}{5}\,\frac{h^2 n_{nr}}{8m_{nr}}\left(\frac{9n_{nr}}{2\pi^2 g}\right)^{2/3},\quad A_r \equiv \frac{3}{4}\,\frac{hcn_r}{2}\left(\frac{9n_r}{2\pi^2 g}\right)^{1/3},$$

$$A_G \equiv \frac{3}{5}\,G\,n_G^2 m_G^2\,. \tag{15.50}$$

We look for the configuration of least total energy $U$, and find three cases. (i) $A_r > A_G$. Since $U$ is lowered by expansion, $R \to \infty$. In other words the system is not gravitationally bound. Two cases occur if $A_r < A_G$. (ii) $A_r < A_G$, $A_{nr} = 0$. $U$ is lowered as $R \to 0$. Thus for large enough gravitational mass $n_G m_G$ the system collapses (into a black hole). (iii) $A_r < A_G$, $A_{nr} > 0$. A true minimum of $U$ at some finite $R$-value is possible (Fig.15.2), as will now be discussed.

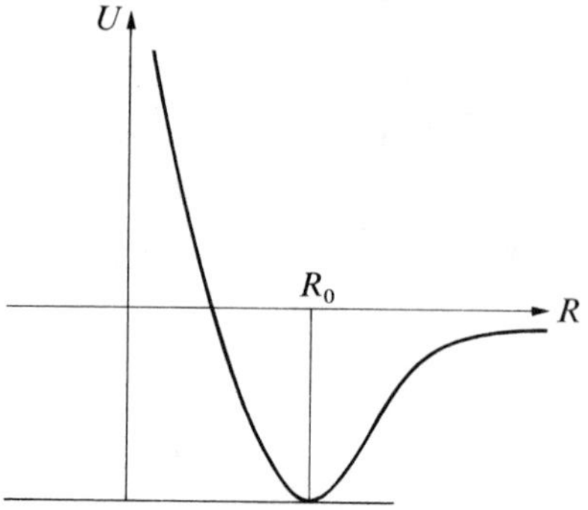


Fig.15.2. Energy of a sphere of degenerate and gravitationally interacting fermions (schematic).

The equilibrium configuration occurs for the radius $R = R_0$ which minimizes (15.50), assuming the other parameters to be kept constant. After some algebra, the result can be written in the form

$$R_0 = \frac{2A_{\mathrm{nr}}/A_{\mathrm{G}}}{1-A_{\mathrm{r}}/A_{\mathrm{G}}} = \frac{R_1}{1-(n_{\mathrm{Gcr}}/n_{\mathrm{G}})^{2/3}} \qquad (\text{if } n_{\mathrm{G}} > n_{\mathrm{Gcr}}) \tag{15.51}$$

where

$$n_{\mathrm{Gcr}} = \frac{3}{8\pi\sqrt{(2g)}}\left(\frac{5\ \hbar c}{2Gm_{\mathrm{G}}^2}\right)^{3/2}\left(\frac{n_{\mathrm{r}}}{n_{\mathrm{G}}}\right)^2 \equiv \frac{15}{8}\left(\frac{5\pi}{2g}\right)^{\frac{1}{2}} n_0 \left(\frac{n_{\mathrm{r}}}{n_{\mathrm{G}}}\right)^2 \tag{15.52}$$

and

$$R_1 \equiv \frac{\hbar^2}{4G}\,\frac{n_{\mathrm{nr}}}{n_{\mathrm{G}}^2 m_{\mathrm{G}}^2 m_{\mathrm{nr}}}\left(\frac{9n_{\mathrm{nr}}}{2\pi^2 g}\right)^{2/3} . \tag{15.53}$$

Here $n_0 \equiv (m_{\mathrm{Pl}}/m_{\mathrm{G}})^3$ where $m_{\mathrm{Pl}} \equiv (\hbar c/G)^{1/2}$ is the so-called Planck mass. At $R = R_0$ one finds

$$U_{\mathrm{nr}}(R_0) = -\frac{1}{2}\,[U_{\mathrm{G}}(R_0)+U_{\mathrm{r}}(R_0)] .$$

This is analogous to the virial theorem (cf. p. 336, below). The existence of case (iii), and hence the stability of matter, must be attributed to the Fermi pressure and hence to the operation of the Pauli exclusion principle.

The coupling constant (*fine-structure constant*) $e^2/\hbar c \sim 1/137$ for electromagnetic interaction yields as its gravitational analogue $Gm_{\mathrm{G}}^2/\hbar c \sim 6 \times 10^{-39}$. This may be called the *gravitational fine-structure constant* and it is just $n_0^{-2/3}$.

We now add a remark on case (ii). In relativity, pressure is added to energy density as being subject to gravitational interaction. As a consequence if all fermions are *(extreme) relativistic* there is no true minimum value of the energy, and, as has been seen, gravitational collapse is expected for this model. It occurs if the systems has a mass in excess of the so-called Chandrasekhar limiting mass which is roughly

$$n_0 m_G \sim 2{\cdot}2 \times 10^{57}\ m_G \sim 3{\cdot}67 \times 10^{30}\ \text{Kg (for protons)} \sim 1{\cdot}8 M_\odot. \tag{15.54}$$

Here $M_\odot$ is the solar mass and a more accurate theory gives 1.47 as coefficient of $M_\odot$. Thus degenerate and extreme relativistic systems having a mass in excess of the Chandrasekhar limit should not be found in nature, except perhaps while they are in the process of collapse. The rather rare but spectacular supernova are conjectured to have their origin in such a collapse. We now note some special cases of the above theory.

*Example 1: White dwarf stars*

About 10 per cent of stars in our galaxy are of this type. Let $N$ be the number of nucleons in such a star. Then there will be about $N/2$ of each: neutrons (suffix n), protons (suffix p) and electrons (suffix e). The occurrence of $m_{\text{nr}}$ in the denominator (15.50) means that the nucleon contribution to $U_{nr}$ can be neglected compared with that of the electrons $m_p \sim m_n \gg m_e$. On the other hand, the contribution of the electrons to $U_G$ is negligible compared with that of the nucleons for the same reason. Hence we can put, with $n_r = 0$, $m_G = m_p \sim m_n$, $n_G = n_p + n_n = \frac{N}{2} + \frac{N}{2} = N$, $m_{\text{nr}} = m_e$, $n_{\text{nr}} \sim \frac{1}{2}N$

to find

$$R_1 = \frac{h^2}{4G}\frac{(n_{\text{nr}}/N)}{Nm_p^2\ m_e}\left(\frac{9n_{\text{nr}}}{2\pi^2 g}\right)^{2/3} \sim 8000\ \text{km}. \tag{15.55}$$

From this figure the density is of order $3 \times 10^9$ kg m$^{-3}$. Such stars are therefore a million times as heavy as the earth, but of the same size.

Consider now a sequence of such stars as $n_G \sim 2n_{nr}$ increases. Then $R_1$ decreases by (15.53) as $n_G^{-1/3}$. This is a mass-radius relationship $M \propto n_G \propto R^{-3}$ showing that massive stars have smaller radii. The energy per particle increases by (15.48) as $n_G^{2/3}$. Thus the assumption $n_r = 0$ cannot remain correct since the electrons become more energetic as $n_G$ increases. Eventually, for large enough $n_G$, the electrons will be mainly relativistic and the curve of $R_0$ against $n_G$ will drop rapidly to reach the Chandrasekhar limit (see Fig.15.3). Stars above this mass are unstable and are expected to collapse into black holes, as already noted.

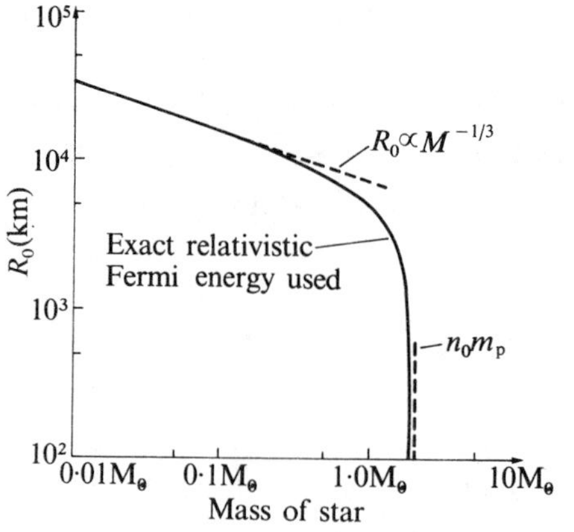


Fig.15.3. The value of $R$ which minimizes the sum of the electron Fermi energy plus the gravitational energy plotted as a function of the star's mass $M$. Uniform density is assumed. For small masses this exact curve approaches the non-relativistic equation. The limiting mass applies when the electrons become extremely relativistic.

*Example 2: Neutron stars*

In this case the total number of nucleons is $N$, made up largely of neutrons and a small number $xN$ of protons and $xN$ of electrons. The neutrons and protons dominate the gravititational interaction, the neutrons dominate in the term $U_{nr}$, and only the electrons contribute to $U_r$. Hence

$$n_G = N,\ m_G = m_p,\ n_{nr} = (1-x)N,\ m_{nr} = m_p,\ n_r = xN,\ m_r = m_e\ .$$

Thus (15.51) yields for small $x$

$$R_0 \sim R_1 \sim \frac{h^2}{4G\ Nm_p^3}\left(\frac{9N}{2\pi^2 g}\right)^{2/3} \quad .$$

This is smaller than (15.55) by the factor $m_e/m_p \sim 1/1800$, and means that a star of a solar mass ($N \sim 1{\cdot}2 \times 10^{57}$) has $R_1 \sim 12{\cdot}6$ km. This implies a density of $2{\cdot}4 \times 10^{17}$ kg m$^{-3}$ which is of the order of the density in atomic nuclei.

As in example 1, for a sequence of neutron stars of increasing $N$ one reaches the case of extreme relativistic neutrons and hence a critical mass of order (15.54). Neutron stars above this mass are expected to collapse into black holes.

One guess for the number of black holes in our galaxy is $10^9$, since this represents the estimated 10 per cent of matter which has led to stars of more than 10 solar masses and which would have had time to collapse into the black-hole state. Astronomers also discuss the existence of time-reversed objects (*white holes*) and have suggested them as responsible, for example, for the large energy emission rate from some quasars. It is to be noted that definite proof for the existence of black holes is still outstanding.

## *15.8. BLACK-HOLE THERMODYNAMICS

It appears that very massive stars must collapse into black holes unless they can lose sufficient mass before the collapse is completed. The so-called gravitational radius $r_0$ of such a hole is such that no signals, particles, or light can reach an observer from the interior. We now estimate $r_0$. If $M$ is the mass of a black hole and a particle of mass $m$ has speed $v(r)$ at a distance $r$ from it, the Newtonian energy equation for this particle is

$$E = \begin{cases} \frac{1}{2}mv(r)^2 - \frac{GMm}{r} = -\frac{GMm}{r_{\max}} & (E < 0) \\ \frac{1}{2}mv(r)^2 - \frac{GMm}{r} = \frac{1}{2}mv(\infty)^2 & (E \geq 0) \end{cases}$$

If $E < 0$ the particle can reach only a distance $r_{\max}$ from the black hole and does not escape since the gravitational energy dominates. The escape of the particle with speed $v(\infty)$ at infinity is possible only if

$v(\infty)^2 \geq 0$ so that the kinetic energy term dominates. This in turn implies that the velocity of $m$ at position $r$ satisfies

$$v(r) \geq \sqrt{\left(\frac{2GM}{r}\right)} \equiv v_e(r) \ .$$

This defines the escape velocity at position $r$. It also applies to photons of frequency $\nu$, as they have an energy $h\nu$ or an equivalent mass $h\nu/c^2$. If the escape velocity as defined above exceeds the velocity of light *in vacuo* ($c$), the lower limit is too high and nothing can escape from within this special radius $r_0$; even light is turned back by the gravitational pull. This radius is thus given by $v_c(r_0) = c$, i.e.

$$r_0 = \frac{2GM}{c^2} \sim 2{\cdot}95 \left(\frac{M}{M_\odot}\right) \text{ km} \ .$$

General relativity yields the same value of $r_0$. It is also the so-called event horizon of a Schwarzschild (i.e. uncharged and non-rotating) black hole. Particle or photon creation *near* a black hole is, however, still possible as a result of the high gravitational field gradient it produces. This effect is referred to as radiation from a black hole or as the Hawking effect after S.W. Hawking who discovered it in 1974. The above relation shows that the event horizon increases if more matter or radiation fall into the black hole.

An estimate for the gravitational field at the surface of a black hole of mass $M$ can be obtained classically to be

$$\kappa \sim \frac{GM}{r_0^2} \sim \frac{c^4}{4GM}$$

and this can be very large for black holes of small mass, thus leading to a corresponding large rate of pair creation.

An estimate for the time for stellar collapse from an initial radius $r > r_0$ can also be obtained classically by assuming free fall for the material under its own gravitational field. Hence if $x$, $\dot{x}$, and $t$ refer to an arbitrary instant during the collapse, the equation of motion for the outer layer of the star is

$$\ddot{x} = -GMx^{-2} = -\frac{r_0 c^2}{2} x^{-2} \ .$$

Hence

$$2\dot{x}\ddot{x} = -r_0c^2x^{-2}\dot{x} \quad .$$

An integration from $t = 0$ when $\dot{x} = 0$ yields $\dot{x}^2 = r_0c^2(x^{-1}-r^{-1}) \sim r_0c^2x^{-1}$. Hence

$$\dot{x} \sim -c\sqrt{r_0}\,x^{-\frac{1}{2}} \quad .$$

Integrating from $x = r$ to $x = r_0$, when $t = t_0$ say,

$$\frac{2}{3}x^{3/2}\Big|_r^{r_0} \sim -c\sqrt{r_0}(t_0-0) \quad .$$

It follows that the collapse is completed after a time

$$t_0 \sim \frac{2r}{3c}\sqrt{\frac{r}{r_0}} \sim 750 \text{ s}^{\dagger}$$

where the numerical value applies to the sun ($r \sim 7 \times 10^5$ km). At $x = r_0$, $|\dot{x}| \sim c$ and a subsequent transition of the excited black hole into a ground state takes of the order of $t_1 = r_0/c \sim 10^{-5}\ (M/M_\odot)$ s.

The time $t_1$ can be used to define a temperature $T$ of an uncharged non-rotating black hole by an argument from the uncertainty principle which asserts that $t_1 \times$ energy $\sim \hbar$. The relation obtained from quantum considerations and relativity is, for uncertainties given by $t_1$ and $kT$,

$$t_1kT = \hbar/4\pi \quad .$$

In agreement with equations used in section 13.6 it leads to

$$T = \frac{\hbar c}{4\pi r_0k} = \frac{\hbar c^3}{8\pi GMk} = \frac{\hbar\kappa}{2\pi ck} = 6\cdot 18 \times 10^{-8}\,\frac{M_\odot}{M}\ (\text{K}) \quad .$$

The relation $T = \hbar\kappa/2\pi ck$ remains valid for charged and rotating black holes. The fact that all points on the event horizon of a (stationary) black hole have the same surface gravity $\kappa$, as implicitly assumed above, implies that they all have the same temperature. Since equilibrium (on a gravitational equipotential surface, see section 18.6) implies uniform temperature by the zeroth law of thermodynamics, the uniformity of $\kappa$ on the event horizon has been called the *zeroth law of black hole thermodynamics*. We now show that a black hole is a source of radiation, subject to the key assumption that equilibrium with black body radiation is possible.

$^{\dagger}$In relativity this holds only for an observer travelling with the particle.

Suppose now that a black hole at temperature $T_1$ interacts briefly with black-body radiation at a lower temperature $T_2$. A small amount of energy $E$ will be absorbed by the black hole, so that the entropy change is

$$E\left(\frac{1}{T_1}-\frac{1}{T_2}\right) .$$

This is negative, contrary to the generalized second law, and is an indication that a black hole must be responsible for radiation of energy $R$ being emitted. The entropy change is then $(R-E)$ $(1/T_2 - 1/T_1)$ and becomes positive if $R > E$.

The radiation $R$ has to be black-body, provided that equilibrium with black-body radiation is to be possible, and it will be proportional to the area

$$A = 4\pi r_0^2$$

of a black hole. Hence from (13.11) and (13.13) the rate of energy loss due to radiation from a black hole of mass $M$ and radius $r_0$ is, if $U_r$ is the radiation energy in an enclosure of volume $v$ and temperature $T$,

$$\dot{R} = \frac{U_r}{4v}Ac = \frac{1}{4}acT^4A = \frac{2\pi^5k^4}{15h^3c^2}\left(\frac{\hbar c}{4\pi r_0 k}\right)^4 4\pi r_0^2 = \frac{\hbar}{4^5\times 15\pi}\left(\frac{c^3}{GM}\right)^2 \qquad (\equiv \alpha M^{-2}) .$$

Thus, noting that 1 erg $s^{-1} = 10^{-7}$ $Js^{-1} = 10^{-7}$W,

$$\dot{R} \sim 8{\cdot}6 \times 10^{-22}\left(\frac{M_\odot}{M}\right)^2 \text{ erg s}^{-1} = 8{\cdot}6 \times 10^{-29}\left(\frac{M_\odot}{M}\right)^2 \text{ W} .$$

The power radiated by the sun is, in contrast, $4 \times 10^{26}$ W from section 13.7. The lifetime $t_L$ of a black-hole under the condition of energy loss by radiation and negligible energy gain can be calculated from

$$\dot{M}c^2 = -\alpha M^{-2} .$$

Integration yields

$$t_L = \frac{c^2M^3}{3\alpha} \sim 6{\cdot}9 \times 10^{74}\left(\frac{M}{M_\odot}\right)^3 \text{ s.}$$

The emission of particles other than photons decreases this estimate somewhat. The long life of massive black holes derives from their slight emission rate. These non-relativistic estimates are of course only heuristic guides.

Small black holes of order $10^{-5}$g = $10^{-8}$ kg are believed to be possible - they have a rapid energy loss and a correspondingly high temperature. They expire therefore in a kind of final explosion. Assuming black-hole creation occurred during the early big bang, i.e. about $10^{17}$ s ago, the original mass of the black holes which might be expected to be exploding now is given by $t_L \sim 10^{17}$ s, i.e.

$$M/M_{\odot} \sim 3 \times 10^{-19}, \quad M \sim 10^{12} \text{ kg} .$$

Such black holes would have had an initial value of $r_0 \sim 10^{-15}$ m and a density of $10^{54}$gm cm$^{-3}$ = $10^{57}$ kg m$^{-3}$.

A caution is *not* to expect the entropy $S$ of a black hole to follow the black-body law and be proportional to $T^3$. The negative heat capacity of black holes, noted in section 13.6, is against this idea. In fact as seen there $S \propto T^{-2} \propto M^2 \propto A$. We see that the surface area of a black hole suggests the characteristics of entropy and the surface gravity suggests the characteristics of temperature. This result can be obtained from

$$dS = \frac{1}{T}\, dU = \frac{8\pi GkM}{\hbar c^3} \cdot c^2\, dM$$

whence

$$S = \frac{4\pi Gk}{\hbar c}\, M^2 = \frac{c^3 k}{4\hbar G}\, A.$$

This identification can be made more precise, allowing for the more general types of black holes suggested by general relativity. This will be shown next.

Black holes can theoretically have charge $Q$, electrostatic potential $\phi$, angular velocity $\Omega$, and angular momentum $J$. Considering the latter as scalars, the following are the four main relations which follow from general relativity theory for the seven parameters

$$M,\ \kappa,\ A,\ \Omega,\ J,\ \phi,\ Q.$$

(i) $$A = 4\pi r_0^2 = \frac{16\pi G^2 M^2}{c^4} \rightarrow \frac{4\pi G}{c^4}\{2GM^2 - Q^2 + 2(G^2M^4 - J^2c^2 - GM^2Q^2)^{1/2}\} .$$

(ii) $$\kappa = \frac{GM}{r_0^2} = \frac{4\pi GM}{A} \rightarrow \frac{4\pi}{A}\left(G^2M^2 - \frac{J^2c^2}{M^2} - GQ^2\right)^{1/2}$$

The last forms are the generalizations. They can be interpreted by remarking that the outwardly directed centrifugal force (due to the black hole angular momentum) and electrostatic repulsion of a charge distribution both tend to reduce the inwardly directed surface gravity which one would have, if $J$ and $Q$ were zero. Then there are two additional relations:

(iii) $\Omega AM = 4\pi J$

(iv) $A\phi = \frac{4\pi Q}{c^2}\left\{GM + \left(G^2M^2 - \frac{J^2c^2}{M^2} - GQ^2\right)^{1/2}\right\}$ .

In spite of these apparently complicated relations, a close analogy with thermodynamics emerges on considering differentials. This can be done as follows.

The area equation is squared to yield

$$U^2 \equiv M^2c^4 = \frac{c^8}{16\pi G^2} A + \frac{c^4}{2G} Q^2 + \frac{\pi}{A}(Q^4 + 4c^2J^2) .$$

This implies

$$dU = \left\{\frac{c^8}{16\pi G} - \frac{\pi}{A^2}(Q^4+4c^2J^2)\right\}\frac{dA}{2U} + \frac{4\pi c^2J}{AU}dJ + \left(\frac{c^4}{G} + \frac{4\pi Q^2}{A}\right)\frac{Q}{2U}dQ .$$

After some algebra, one finds

$$dU = \frac{c^2}{8\pi G}\kappa dA + \Omega dJ + \phi dQ .$$

One can identify the first term on the right-hand side as $TdS$, where

$$T \equiv \frac{\hbar\kappa}{2\pi kc}$$

$$S \equiv \frac{c^3k}{4\hbar G} A .$$

Thus the proportionality of entropy and area (but not of entropy and $M^2$) survives into the general case. The terms $\Omega dJ + \phi dQ$ are work terms of the $pdv$ variety and $dM$ is analogous to $dU$. Thus, in a curious way general relativity leads to an adiabatic expansion of the cosmological fluid (section 13.5) and to thermodynamic-type results for black holes, although thermodynamics does not seem to have been fed into the theory at the outset. But failure of the Fourth Law (pp. 79, 101) causes difficulty.

That the resulting theory is otherwise consistent is indicated by the very large entropies of black holes, suggesting that they can often feature as thermodynamically stable end-states of a physical system. We have for an uncharged non-rotating black hole

$$S = \frac{4\pi kG}{\hbar c} M^2 \sim 1{\cdot}45 \times 10^{61} \left(\frac{M}{M_\odot}\right)^2 \text{ erg K}^{-1} = 1{\cdot}45 \times 10^{54} \left(\frac{M}{M_\odot}\right)^2 \text{ JK}^{-1} .$$

For further insights the original papers should be consulted.

## *15.9. A CONNECTION BETWEEN BOSE AND FERMI STATISTICS: POPULATION INVERSION AND LASERS

In this section two strands of ideas will be pulled together. The first refers to two groups of fermion states ($i$ and $j$) specified by quasi-Fermi levels $\mu_i$ and $\mu_j$ as introduced in section 15.6. If $I$ and $J$ are states in groups $i$ and $j$ respectively, then we define a quantity

$$Z_{IJ} \equiv \frac{p_J}{1-p_J}\,\frac{1-p_I}{p_I} = \exp(F_j-F_i+\eta_I-\eta_J) . \qquad (15.56)$$

Here $p_J$ is the occupation probability of state $J$ and $1-p_J$ is the probability of a fermion vacancy at $J$, $F_j \equiv \mu_j/kT$, and $\eta_J \equiv E_J/kT$, where $E_J$ is the energy of the state $J$. It then follows from the Fermi-Dirac distribution that

$$p_J = \frac{1}{1+\exp(\eta_J-F_j)}$$

$$1-p_J = \frac{\exp(\eta_J-F_j)}{1+\exp(\eta_J-F_j)}$$

$$\frac{p_J}{1-p_J} = \exp(F_j-\eta_J)$$

and hence (15.56) follows.

The second strand of ideas applies to Bose statistics and derives from problems 10.5 to 10.7 . If $N_\nu$ is the mean number of photons in a mode of frequency $\nu$ in a material, then the stimulated emission rate, the absorption rate, and the spontaneous emission rate are respectively

$$B_{IJ}N_\nu p_I(1-p_J), \quad B_{IJ}N_\nu p_J(1-p_I), \quad B_{IJ}p_I(1-p_J) .$$

We have assumed that

$$E_I - E_J = h\nu. \tag{15.57}$$

Denoting these rates by

$$u_{IJ}^{\text{st}}, \quad u_{IJ}^{\text{abs}}, \quad u_{IJ}^{\text{sp}} ,$$

one finds that

$$\frac{1}{N_\nu} \frac{u_{IJ}^{\text{st}} - u_{IJ}^{\text{abs}}}{u_{IJ}^{\text{sp}}} = 1 - Z_{IJ} = 1 - \exp\left(F_j - F_i + \frac{h\nu}{kT}\right) \tag{15.58}$$

where we have used equations (15.56) and (15.57).

The left-hand side of (15.58) is a consequence of the Bose distribution and the right-hand side is a consequence of the Fermi distribution, and we see here that they are coupled physically in processes in which the emission or absorption of radiation depends on the transitions of electrons. The condition for stimulated emission to dominate absorption, i.e. the condition for *negative absorption*, is that (15.58) be negative. This requires the separation between the quasi-Fermi levels to *exceed* the photon energy under consideration:

$$\mu_i - \mu_j > h\nu \quad . \tag{15.59}$$

One may ask, without reference to radiative processes, for the condition that a quantum state $I$ have a *greater* occupation probability than a quantum state $J$ *lying below it* in energy by the amount (15.57). The condition is

$$\frac{1}{1+\exp(\eta_I - F_i)} > \frac{1}{1+\exp(\eta_J - F_j)} \quad .$$

This reduces exactly to (15.59), which is therefore also the condition for population inversion. Such inversion is required for the operation of semiconductor lasers above threshold.

If there are groups of $n_r$ fermions of quasi-Fermi levels $\mu_r$ ($r$=1,2,...) then the drop in what may be called the 'quasi-Gibbs free energy' as a result of a radiative emission of a photon of frequency $\nu$ is

$$\Delta G = \sum n_r \mu_r - \{n_1\mu_1 + \ldots + (n_i - 1)\mu_i + \ldots + (n_j + 1)\mu_j + \ldots\} \ .$$

It can be shown that the condition of negative absorption is in this case

$$\Delta G > h\nu \tag{15.60}$$

in generalization of (15.59). These concepts are useful in the study of lasers and masers in which population inversion plays an important part.

PROBLEMS

15.1 Prove by algebraic manipulation that

$$C_r \equiv \sum_{s=0}^{\infty} \frac{(-1)^s}{(s+1)^r} = (1-2^{1-r})\zeta(r)$$

and verify that this yields for $r = 2$ (cf. (15.17''))

$$2C_2 = 2\left(1 - \frac{1}{4} + \frac{1}{9} - \ldots\right) = \zeta(2).$$

15.2 Show that

$$J_{r-1} \equiv \int_0^{\infty} \frac{x^{r-1}\mathrm{d}x}{e^x+1} = \Gamma(r)C_r = \Gamma(r)(1-2^{1-r})\zeta(r) .$$

15.3 By using the notation of problem 15.2 show that for a function $f(x)$ such that

$$I \equiv \int_0^{\infty} \frac{f(x)\,\mathrm{d}x}{1+\exp(x-a)}$$

converges, one has for $a \gg 1$

$$I \doteqdot \int_0^a f(x)\,\mathrm{d}x + \sum_{\substack{r=1\\ r\ \text{odd}}}^{\infty} \frac{2}{r!} f^{(r)}(a) J_r$$

$$= \int_0^a f(x)\,\mathrm{d}x + \frac{\pi^2}{6} f'(a) + \frac{7\pi^4}{360} f'''(a) + \ldots .$$

15.4 Apply the result of problem 15.3 to prove equation (15.13). [This verifies that the problems 15.1 to 15.3 lead to a generalization of equation (15.13).]

15.5. Neutrinos are fermions of zero mass having the density of states which arises from the fifth line of Table 12.1. Show that

$$Ae^{s}\text{d}e = \frac{4\pi gv}{h^3c^3}\,e^2\text{d}e \ .$$

Hence verify that

$$N = 8\pi gv\left(\frac{kT}{hc}\right)^3 I\left(\frac{\mu}{kT},2,+\right)$$

$$pv = \frac{1}{3}U = 8\pi gvkT\left(\frac{kT}{hc}\right)^3 I\left(\frac{\mu}{kT},3,+\right)$$

$$F = 8\pi gvkT\left(\frac{kT}{hc}\right)^3\left\{\frac{\mu}{kT}\,I\left(\frac{\mu}{kT},2,+\right) - I\left(\frac{\mu}{kT},3,+\right)\right\} .$$

15.6. Show that for an ideal classical gas the quantity $\lambda$ of (15.34') is proportional to the pressure. Assuming that $\lambda Z_1/Z_0 = p/p_0$, where $p_0$ is independent of pressure but may depend on temperature, show that with $M = 1$ (15.40) yields

$$\frac{n_t}{N} = \frac{p}{p_0+p} \ .$$

Interpret this result in the light of the analogy: defect → adsorption site on a surface, trapped electron → adsorbed atom or molecule; degree of trapping $n_t/N$ → coverage of a surface by molecules adsorbed from an ideal gas phase.

[This is a deduction of the so-called **Langmuir** adsorption isotherm.]

15.7. Suppose that the interactions among successive particles adsorbed at the same site can be neglected in the sense that one may write

$$\frac{Z_j}{Z_0} = \left(\frac{Z_1}{Z_0}\right)^j \ .$$

Show that

$$\frac{n_t}{N} = \frac{1}{x-1} - \frac{(M+1)}{x^{M+1}-1} \qquad \left(x \equiv \frac{Z_0}{\lambda Z_1}\right)$$

so that for $M = 1$

$$\frac{n_t}{N} = \frac{1}{x+1} \ .$$

Explain why this is unlikely to be a good model for electrons trapped at defects.

15.8. In a steady-state generation - recombination situation in a semiconductor, electrons recombine with $(r-1)$-electron defects to yield $r$-electron defects . Find an expression for the reaction constant. Consider the situation if the valence band replaces the conduction band.

# 16. Fluctuations

## 16.1. VARIANCE AND STANDARD DEVIATION

In many statistical mechanical systems an extensive variable $X$ gives rise to an ensemble average $\langle X \rangle$. Although the ratio $X/\langle X \rangle$ may have wide variability, say $0 < X/\langle X \rangle < \infty$, an overwhelming majority of the systems of the ensemble have values of $X$ which cluster close to $\langle X \rangle$. One can say that the probability distribution is highly peaked near $X = \langle X \rangle$. Choosing $X$ to represent the number of particles $n$ in a system, this means that the difference between a grand canonical distribution with given $\langle n \rangle$ and a canonical distribution with $n$ fixed at $\langle n \rangle$ can often be neglected (see section 11.2). In problems 9.4 to 9.6 specific situations of this type have already been studied, and the notion of the standard deviation $\sigma$ of a normal distribution has been introduced. These ideas will now be developed further.

A more general definition of the *standard deviation* $\sigma_X$ and of the *variance* $\sigma_X^2$ of $X$ can be given as follows:

$$\sigma_X^2 \equiv \langle [X - \langle X \rangle]^2 \rangle = \langle X^2 \rangle - 2\langle (X\langle X \rangle) \rangle + \langle X \rangle^2$$

$$= \langle X^2 \rangle - (\langle X \rangle)^2 \ . \qquad (16.1)$$

Here the angular brackets denote an average over the distribution.

The *fractional deviation* $\sigma_X/\langle X \rangle$, sometimes called the coefficient of variation, is then dimensionless. It is now desirable that this quantity is in fact small under the normal conditions considered in this book. The fluctuations in statistical mechanical quantities are liable to be large for states lying near phase boundaries, and these situations furnish important exceptions (cf. section 16.6).

For a canonical ensemble with partition function $Z$

$$\langle E \rangle Z = \sum_i E_i \exp(-\eta_i)$$

$$Z = \sum_i \exp(-\eta_i) \tag{16.2}$$

where $\eta_i \equiv E_i/kT$ and the notation of section 9.3 has been used. Differentiating at constant volume so that the energy levels $E_i$ remain unchanged,

$$\left(\frac{\partial\langle E \rangle}{\partial T}\right)_{v,N} Z + \langle E \rangle \frac{1}{kT^2} \sum_i E_i \exp(-\eta_i) = \frac{1}{kT^2} \sum_i E_i^2 \exp(-\eta_i) .$$

Multiplying by $kT^2/Z$ ,

$$kT^2 \left(\frac{\partial\langle E \rangle}{\partial T}\right)_{v,N} + \langle E \rangle^2 = \langle E^2 \rangle .$$

Hence with the notation $\beta \equiv 1/kT$ ,

$$\boxed{\sigma_E^2 = kT^2 \left(\frac{\partial\langle E \rangle}{\partial T}\right)_{v,N} = kT^2 C_v = -\left(\frac{\partial\langle E \rangle}{\partial \beta}\right)_{v,N}} . \tag{16.3}$$

Now for a simple ideal gas $\langle E \rangle$ and $C_v$ are respectively of order $nkT$ and $nk$ so that

$$\frac{\sigma_E^2}{\langle E \rangle^2} \sim \frac{kT^2 nk}{(nkT)^2} = \frac{1}{n}$$

or

$$\frac{\sigma_E}{\langle E \rangle} \sim \frac{1}{\sqrt{n}} . \tag{16.4}$$

For a gas of $10^{18}$ molecules (in a cubic metre say) this number is therefore exceedingly small: $\sigma_E/\langle E \rangle \sim 10^{-7}$ per cent. When discussing the fluctuations in numbers $n$, one again finds $\sigma_n = \sqrt{n}$ for a normal distribution (cf. problem 9.6) and so $\sigma_n/\langle n \rangle \sim 1/\sqrt{n}$.

Since $\sigma_E^2 \geq 0$, (16.3) implies a proof that for a system in equilibrium with a heat reservoir at temperature $T$ the heat capacity $C_v$ is positive (see also equation (7.21)).

## 16.2. NORMAL VARIATES

A normal variate $x$ is a quantity which can vary throughout the range $(-\infty, +\infty)$, and is subject to the probability distribution

$$p(x) = a \exp\left(-\frac{x^2}{2\sigma^2}\right), \quad a \equiv (2\pi\sigma^2)^{-\frac{1}{2}} . \tag{16.5}$$

A cartesian component $V_i$ of the velocity of a molecule in a gas with a Maxwell-Boltzmann distribution has just the distribution (16.5) with

$$\sigma_{V_i}^2 = \frac{kT}{m} . \tag{16.6}$$

If one treats the three components $i = 1,2,3$ as *independent* in the probabilistic sense, one can just *multiply the three distributions* to obtain the probability of finding a molecule with velocity components in the ranges

$$(V_1, V_1 + dV_1), \quad (V_2, V_2 + dV_2), \quad (V_3, V_3 + dV_3) . \tag{16.7}$$

This leads to a probability distribution for the molecular speeds given by

$$p(V) = 4\pi V^2 \left(\frac{m}{2\pi kT}\right)^{3/2} \exp\left(-\frac{mV^2}{2kT}\right) . \tag{16.8}$$

All these results are implied by problems 9.1 and 9.7. Since $V \geq 0$, the *speed* cannot be taken as the normal variate $x$ of (16.5). Nonetheless, it is clear that normal variates are of importance in statistical mechanics, and molecular velocity components in a classical gas in equilibrium furnish just one example. We shall therefore investigate some consequences of (16.5).

The main result is that the $n$th moment of a distribution $p(x)$ is defined by

$$\langle x^n \rangle \equiv a \int_{-\infty}^{\infty} x^n \exp\left(-\frac{x^2}{2\sigma^2}\right) dx \tag{16.9}$$

and is given by

$$\left.\begin{aligned} \langle x^n \rangle &= \left(\frac{1}{2} \times \frac{3}{2} \cdots \frac{n-1}{2}\right) 2^{n/2}\, \sigma^n \qquad (n \text{ even}) \\ \langle x^n \rangle &= 0 \qquad (n \text{ odd}) \end{aligned}\right\} \tag{16.10}$$

*Proof of (16.10).* For odd $n$ the integrand in (16.9) is odd in $x$, and so the integral must vanish. If $n$ is even, the integral is even, and we may write with $y \equiv x^2/2\sigma^2$, $dx = \{\sigma/(2y)^{\frac{1}{2}}\}dy$ ,

$$\begin{aligned} \langle x^n \rangle &= 2a \int_0^\infty (2y\sigma^2)^{n/2} \exp(-y) \frac{\sigma}{\sqrt{(2y)}}\, dy \\ &= 2^{(n+1)/2}\, a\sigma^{n+1} \int_0^\infty y^{(n-1)/2} \exp(-y)\, dy \\ &= 2^{(n+1)/2}\, a\sigma^{n+1} \sqrt{\pi} \left(\frac{1}{2} \times \frac{3}{2} \cdots \times \frac{n-1}{2}\right) \end{aligned}$$

since the last integral is $\Gamma\{(n+1)/2\}$. The distribution has to be normalized so that $\langle x^0 \rangle = 1$. This yields the expression (16.5) for $a$. Substituting for $a$ one finds (16.10).

Some special cases are

$$\langle x^2 \rangle = \sigma^2 \quad \langle x^4 \rangle = 3\sigma^4 = 3(\langle x^2 \rangle)^2 . \tag{16.11}$$

For a velocity component using (16.6), one finds

$$\langle V_i^2 \rangle = \frac{kT}{m} \quad \langle V_i^4 \rangle = 3\left(\frac{kT}{m}\right)^2 . \tag{16.12}$$

The first of these results follows also from the equipartition theorem:

$$\langle \tfrac{1}{2} mV_i^2 \rangle = \tfrac{1}{2}kT, \quad \langle \tfrac{1}{2}mV^2 \rangle = \tfrac{3}{2} kT \tag{16.13}$$

where $V^2 = V_1^2 + V_2^2 + V_3^2$ . The variance of the kinetic energy $T_i$ due to a velocity component $i$ is

$$\sigma^2_{T_i} = \left\langle \left(T_i - \bar{T}_i\right)^2 \right\rangle = \langle T_i^2 \rangle - \langle T_i \rangle^2$$
$$= \left(\frac{m}{2}\right)^2 \left(\langle v_i^4 \rangle - \langle v_i^2 \rangle^2\right)$$
$$= \left(\frac{3}{4} - \frac{1}{4}\right)(kT)^2 = \frac{1}{2}(kT)^2 \; . \qquad (16.14)$$

Since the components are treated as statistically independent, the variance of the total kinetic energy T is

$$\sigma^2_T = \sum_{i=1}^{3} \sigma^2_{T_i} = \frac{3}{2}(kT)^2 \; . \qquad (16.15)$$

All these results show that the energy distributions broaden with rise in temperature. There are many experimental confirmations of this tendency.

## *16.3. OSCILLATORS AND MODES OF VIBRATION

The energy $E(p,q)$ of a one-dimensional oscillator, or equivalently of a mode of vibration, as considered in section 11.7, is

$$E(p,q) \equiv ap^2 + bq^2 \equiv T + V \qquad \left(a \equiv \frac{1}{2m}\,,\; b \equiv \frac{m\omega^2}{2}\right) \qquad (16.16)$$

We shall consider the variance of the energy of this system. One finds quite generally that if some probability distribution is appropriate to $p$ and the same or another is appropriate to $q$, then

$$\langle E \rangle^2 = \langle T \rangle^2 + \langle V \rangle^2 + 2\langle T \rangle \langle V \rangle \; .$$

Also

$$\langle E^2 \rangle = \langle T^2 \rangle + \langle V^2 \rangle + 2\langle TV \rangle \; .$$

Hence

$$\sigma^2_E = \langle E^2 \rangle - \langle E \rangle^2 = \sigma^2_T + \sigma^2_V + 2(\langle TV \rangle - \langle T \rangle \langle V \rangle) \; .$$

It is reasonable to assume that $p$ and $q$ are both normal variates such that

$$\langle ap^2 \rangle = \langle bq^2 \rangle \; . \qquad (16.17)$$

This ensures that the average kinetic and potential energies are the same, as would be required by equipartition. By virtue of (16.11) it implies furthermore that

$$\sigma_T^2 = a^2(\langle p^4\rangle - \langle p^2\rangle^2) = 2a^2\langle p^2\rangle^2 = 2b^2\langle q^2\rangle^2 = \sigma_V^2$$

and also that

$$\langle E\rangle = a\langle p^2\rangle + b\langle q^2\rangle = 2b\langle q^2\rangle .$$

It follows therefore that

$$\sigma_E^2 = \langle E\rangle^2 + 2(\langle TV\rangle - \langle T\rangle\langle V\rangle) .$$

It is also reasonable to assume that $p$ and $q$ are *independent* variates, so that

$$\langle TV\rangle = \langle T\rangle\langle V\rangle .$$

Hence one finally obtains the result

$$\sigma_E^2 = \langle E\rangle^2 \quad \text{or} \quad \langle E^2\rangle = 2\langle E\rangle^2 . \tag{16.18}$$

We can summarize our result as follows: If $p$ and $q$ are independent normal variates of zero mean, subject to (16.17), then the energy (16.16) of a one-dimensional oscillator satisfies the condition that in the relation

$$\sigma_E^2 = \langle E\rangle^2 - A^2 \tag{16.19}$$

$A$ is in fact zero.

This result can be obtained from the Maxwell distribution as in section 11.7 by showing that this distribution implies

$$\langle E^2\rangle = 2(kT)^2 , \quad \langle E\rangle = kT \tag{16.20}$$

whence, again, $A = 0$ in (16.19). However, our procedure above has been more general and has made no direct use of the Maxwell distribution, which is in any case known to be invalid in quantum statistics.

*Proof of (16.20).* It is easily shown from a distribution $A\exp(-E/kT)$ that

$$\langle E^n\rangle = A\int_0^\infty E^n \exp\left(-\frac{E}{kT}\right) dE$$

$$= A(kT)^{n+1}\int_0^\infty x^n e^{-x}\, dx = An!\,(kT)^{n+1} .$$

For normalization we need $\langle E^0\rangle = 1$ so that $A = 1/kT$. Hence one finally has

$$\langle E^2\rangle = 2(kT)^2 , \quad \langle E\rangle = kT$$

and hence $\sigma_E^2 = (kT)^2$, whence $A = 0$ in (16.19).

## *16.4. DISCRETE ENERGY LEVELS WITHOUT QUANTIZATION

In order to investigate to what extent one can go beyond classical mechanics by using (16.19), we shall now drop the classical conclusion derived in the last subsection that $A = 0$. We shall retain (16.19) and merely require $A$ to be a temperature-independent constant.

At $T = 0$ any system is in its ground state (to be denoted by gs). Since the probability of a system being in any other state is then zero, we have $\sigma_{E,\mathrm{gs}}^2 = 0$. It follows from (16.19) that

$$\langle E\rangle_{\mathrm{gs}} = A . \tag{16.21}$$

Since $A$ is temperature-independent, (16.21) furnishes a physical identification of the constant $A$. Indeed, classically $\langle E\rangle_{\mathrm{gs}} = 0$ and one recovers $A = 0$ from (16.21) in that case. We shall call $A$ the *zero-point energy*.

Using (16.3), one can now make the following interesting observation ($\beta = 1/kT$):

$$\left(\frac{\partial\langle E\rangle}{\partial\beta}\right)_{v,N} = -\sigma_E^2 = A^2 - \langle E\rangle^2 .$$

Accepting that $A$ is a constant and writing $y$ for $\langle E\rangle$, we have

$$\frac{dy}{A^2-y^2} = d\beta .$$

It follows that

$$2A\,d\beta = \left(\frac{1}{y+A} - \frac{1}{y-A}\right) dy$$

that is

$$\frac{y+A}{y-A}\exp(-2A\beta) = C$$

where $C$ is a constant independent of $\beta$. Assume that $y = \infty$ at $T = \infty$ ($\beta=0$), so that $C = 1$. Hence

$$y = \langle E \rangle = A + \frac{2A}{\exp(2A\beta)-1} = A\coth(A\beta) \ . \tag{16.22}$$

We see again from this solution that it has the property

$$\lim_{A\to 0}[\langle E \rangle] = \frac{1}{\beta} = kT$$

as required by equipartition. Thus (16.22) is a generalization of the classical result, indicating the modifications which arise from a non-zero mean ground state energy (16.21).

From problems 9.1 to 9.3 we know that

$$\langle E \rangle = -\left(\frac{\partial \ln Z}{\partial \beta}\right)_{v,N} \ .$$

Substituting (16.22), one finds

$$\ln Z = -\int A\coth(A\beta)\,d\beta = \ln\frac{D}{\sinh A\beta}$$

where $D$ is a constant of integration. Thus

$$Z = \frac{2D}{\exp(A\beta)-\exp(-A\beta)} = \frac{2D\exp(-A\beta)}{1-\exp(-2A\beta)}$$

$$= 2D\exp(-A\beta)\,\{1+\exp(-2A\beta) + \exp(-4A\beta) + \ldots\} \ .$$

It follows that

$$Z = 2D\sum_{n=0}^{\infty}\exp\left\{-\frac{2A(n+\frac{1}{2})}{kT}\right\} \ . \tag{16.23}$$

The energy levels of an oscillator specified to have zero-point energy $A$

have thus the form

$$E_n = 2A(n + \tfrac{1}{2}), \quad \text{that is } A,\ 3A,\ 5A,\ \ldots\ .$$

*We have here obtained the energy levels of the quantum mechanical harmonic oscillator from a study of fluctuations.* This demonstrates that there exists an important connection between fluctuation theory and the kind of mechanics used to describe a system.

## 16.5. PARTICLE NUMBER FLUCTUATIONS

In order to turn now to a quantitative study of particle fluctuations, already alluded to in section 16.1, we start with the partition function (9.20):

$$\Xi = \sum_{i,n} \exp\left[\frac{(\mu n - E_i)}{kT}\right] . \tag{16.24}$$

In order to keep the energy level spectrum $\{E_i\}$ unchanged we shall keep the volume $v$ fixed, and differentiate to find

$$\left(\frac{\partial \Xi}{\partial \mu}\right)_{v,T} = \frac{1}{kT} \sum_{i,n} n \exp\left(\frac{\mu n - E_i}{kT}\right)$$

$$\left(\frac{\partial^2 \Xi}{\partial \mu^2}\right)_{v,T} = \frac{1}{(kT)^2} \sum_{i,n} n^2 \exp\left(\frac{\mu n - E_i}{kT}\right) .$$

These results are equivalent to

$$\langle n \rangle = \frac{kT}{\Xi}\left(\frac{\partial \Xi}{\partial \mu}\right)_{v,T}$$

$$\langle n^2 \rangle = \frac{(kT)^2}{\Xi}\left(\frac{\partial^2 \Xi}{\partial \mu^2}\right)_{v,T} = \frac{(kT)^2}{\Xi}\left\{\frac{\partial}{\partial \mu}\left(\frac{\langle n \rangle \Xi}{kT}\right)\right\}_{v,T}$$

$$= \frac{(kT)^2}{\Xi}\,\frac{\Xi}{kT}\left(\frac{\partial \langle n \rangle}{\partial \mu}\right)_{v,T} + \frac{\langle n \rangle kT}{\Xi}\left(\frac{\partial \Xi}{\partial \mu}\right)_{v,T}$$

Since the last term is $\langle n \rangle^2$, it now follows that

$$\boxed{\sigma_n^2 = kT\left(\frac{\partial\langle n\rangle}{\partial\mu}\right)_{v,T}} \quad . \tag{16.25}$$

This result has a somewhat more general validity than suggested by its deduction. Let $\langle n_j\rangle$ be the mean occupation number of the single-particle quantum state $j$ (as for example obtained in (10.8)). Then

$$n = \sum_j n_j$$

One can replace in the above argument:

$$\mu n \text{ by } \sum \mu_j n_j \ , \qquad \frac{\partial}{\partial\mu} \text{ by } \frac{\partial}{\partial\mu_j} \ , \qquad \langle n\rangle \text{ by } \langle n_j\rangle \quad .$$

After the differentiation one can equate all the $\mu_j$s to a single $\mu$. Hence one finds the variance for the occupation number of a single quantum state $j$ to be

$$\sigma_{n_j}^2 = kT\left(\frac{\partial\langle n_j\rangle}{\partial\mu}\right)_{v,T} \quad . \tag{16.26}$$

For our main distributions one has with $t_j \equiv \exp\{(\mu-e_j)/kT\}$ and $a = 0, +1, -1,$

$$\langle n_j\rangle = \frac{1}{t_j^{-1}+a}$$

$$\sigma_{n_j}^2 = kT\left(\frac{\partial\langle n_j\rangle}{\partial\mu}\right)_{v,T} = \frac{1}{t_j^{-1}+a}\,\frac{t_j^{-1}}{t_j^{-1}+a} = \langle n_j\rangle\,(1 - a\langle n_j\rangle) \ . \tag{16.26a}$$

It follows that for Maxwell, Fermi, and Bose distributions one has respectively

$$\frac{\sigma_{n_j}}{\langle n_j\rangle} = \begin{cases} \sqrt{\left(\dfrac{1}{\langle n_j\rangle}\right)} & \text{(M-B)} \\ \sqrt{\left\{\dfrac{1}{\langle n_j\rangle} - 1\right\}} & \text{(F-D)} \\ \sqrt{\left\{\dfrac{1}{\langle n_j\rangle} + 1\right\}} & \text{(B-E)} \ . \end{cases}$$

The fractional deviations are smallest for heavily occupied states, but in the Fermi-Dirac case $\langle n_j\rangle$ always lies between

0 and 1. For given $\langle n_j \rangle$, lying between 0 and 1,

$$\sigma^2_{n_j,\text{ B-E}} > \sigma^2_{n_j,\text{ M-B}} = \langle n_j \rangle > \sigma^2_{n_j,\text{F-D}} \,. \qquad (16.27)$$

The largest standard deviations are

$$\sigma_{n_j,\text{B-E}} \sim \langle n_j \rangle \,, \qquad \sigma_{n_j,\text{M-B}} = \sqrt{\langle n_j \rangle} \,, \qquad \sigma_{n_j,\text{F-D}} = \tfrac{1}{2} \,, \qquad (16.28)$$

and occur for large occupation numbers in the first two cases and for $\langle n_j \rangle = \frac{1}{2}$ in the last case. These standard deviations are tolerable except for the case of the Bose-Einstein systems with $\langle n_j \rangle \gg 1$. The standard deviation is then of order $\langle n_j \rangle$ and the fluctuations associated with the mean occupation number are so large as to provide an exception to the rule which says that the distribution of the $n_j$ is highly peaked at $\langle n_j \rangle$, as suggested in section 16.1. These exceptions occur for large $\langle n_j \rangle$,[†] i.e. if $e_j \sim \mu$. Since, however,

$$\mu \le e_1 = e_2 = \ldots = e_{g_1} < e_{g_1+1} \le \ldots$$

it suffices to consider the lowest energy level, assumed to be $g_1$-fold degenerate. This level has the greatest equilibrium occupation, and this occupation increases as the temperature is lowered. In principle, therefore, all particles of a system will condense into this level if the temperature is low enough. This phenomenon is called *Bose-Einstein condensation*. The lowest single-particle energy level has now to be treated separately, so that the resulting theory will be somewhat more accurate than that incorporated in previous sections.

## *16.6. BOSE-EINSTEIN CONDENSATION

Let $n_1(T)$ be the number of bosons which at temperature $T$ occupy the lowest single-particle energy level of a Bose gas. Then the form of equation (11.6), which incorporates the correction for the lowest level,

[†]Note incidentally that for large $\langle n \rangle$ and $a = -1$ the $\langle n \rangle^2$-term in (16.26a) is believed to be responsible for photon bunching, i.e. photon correlation in thermal and laser radiation.

is ($A$ denotes no longer the zero-point energy in this section)

$$\frac{N}{v} = \frac{n_1(T)}{v} + \frac{A}{v}(kT)^{s+1}\,\Gamma(s+1)\,I\left(\frac{\mu}{kT},s,-\right) \quad . \tag{16.29}$$

At high temperatures $n_1(T)/v$ is negligible and (11.6) holds in its original form. As the temperature is lowered, then while $n_1/v$ is negligible the integral $I$ must increase in order that $N/v$ be kept constant. Thus the chemical potential increases. If the number of particles is finite, then $n_1(T)$ is always finite and $\mu < e_1$ in

$$n_1(T) = \frac{g_1}{\exp\{(e_1-\mu)/kT\}-1} \quad . \tag{16.30}$$

For infinite $N$, $\mu = e_1$ is just possible. To have a finite mass per unit volume, however, we then also need the volume $v$ to be infinite. Now, increase in the volume affects the energy levels, so that the choice of energy zero requires some care. We therefore make the following stipulations:

(i) $\lim_{v\to\infty} e_1(v) = 0$ (this fixes the energy zero)

(ii) The limit $v \to \infty$ shall imply $N \to \infty$ in such a way as to keep $N/v$ a finite and non-zero constant. This type of limit is called the *thermodynamic limit*.

These stipulations imply that, as the temperature of the infinite system is lowered, $\mu$ increases towards the value zero. The question is: At what temperature ($T = T_c$ say) does $\mu$ first reach the value zero? If $T_c > 0$ then $\mu$ will remain zero for $T < T_c$. We are accordingly led to the following definition of this critical temperature $T_c$ in the thermodynamic limit:

$$\mu < 0 \quad \text{for} \quad T > T_c \tag{16.31}$$

$$\mu = 0 \quad \text{for} \quad 0 \le T \le T_c \quad . \tag{16.32}$$

In order to find an equation for $T_c$ note that (16.29) and (16.32) imply

$$N - n_1(T) = A(kT)^{s+1}\,\Gamma(s+1)\,\zeta(s+1) \, ; \quad T \le T_c$$

where (13.12a) has been used. At $T = T_c$ the original theory of Chapter 11 is still valid, so that

$$n_1(T_c) \ll N \quad .$$

Hence one finds as the equation for $T_c$

$$N = A(kT_c)^{s+1}\,\Gamma(s+1)\,\zeta(s+1) \quad . \tag{16.33}$$

Adopting the density of states function in the fourth line of Table 12.1,

$$N = \frac{4\pi vg}{h^3}\,m\sqrt{2m} \times (kT_c)^{3/2} \times \frac{\sqrt{\pi}}{2} \times \zeta\left(\frac{3}{2}\right) \quad .$$

Hence

$$T_c = \frac{h^2}{2\pi mk}\left\{\frac{N}{gv\zeta(3/2)}\right\}^{2/3} = \frac{3\cdot 31\hbar^2}{m}\left(\frac{N}{gv}\right)^{2/3} \quad .$$

This yields $T_c \sim 3\cdot 14$ K for the case of $^4$He under its own vapour pressure.

Substitution for $A$ from (16.33) in (16.29) yields

$$N - n_1(T) = N\left(\frac{T}{T_c}\right)^{s+1}\frac{I(\mu/kT, s, -1)}{\zeta(s+1)} \quad ,$$

and is valid at all temperatures. It implies that for $T \le T_c$

$$n_1(T) = \left[1 - \left(\frac{T}{T_c}\right)^{s+1}\right]N \qquad (T \le T_c) \quad . \tag{16.34}$$

This shows an increase of $n_1(T)$ in the thermodynamic limit from $n_1 = 0$ at $T = T_c$ to the value $n_1 = N$ at $T = 0$. We thus have a quantitative expression for the 'momentum space condensation' of particles into the lowest energy level.

It is well known that the series

$$\zeta(1) \equiv \sum_{j=1}^{\infty} j^{-1} = 1 + \frac{1}{2} + \frac{1}{3} + \ldots$$

diverges. Thus for $s = 0$ (16.33) shows that $T_c = 0$ and the chemical potential reaches the value zero in the thermodynamic limit only at the absolute zero. Interesting Bose-Einstein condensation phenomena therefore occur for density of states functions which rise with energy so that $N(E) \propto E^s$ only if $s > 0$. In these cases the heat capacity at constant volume suffers a discontinuity at $T = T_c$ for $s > 1$. For $0 < s < 1$ there is a discontinuity only in one of the derivatives of $C_v$ with respect to temperature. In any case there is for $s > 0$ a phase transition at the temperature $T = T_c$ and this illustrates the remark in section 16.1 that the fluctuations can become large at states near phase boundaries. The temperature dependence of the specific heat of liquid $^4$He is in only rough agreement with the ideal Bose gas theory with $s = \frac{1}{2}$ as outlined here.

The superfluidity of $^4$He is often attributed to the properties of the condensed phase; this consists of the molecules which are in the lowest energy level. This is, however, dubious as the topological properties of the many-particle wavefunction are almost certainly involved but not taken into account in the above analysis. Also it does not seem to have been pointed out before that the thermodynamic limit is incompatible with the existence of limiting masses such as (15.54) above which gravitational equilibrium cannot be sustained.

Unlike $^4$He which is a boson because it has zero nuclear spin, the isotopic $^3$He is a fermion with nuclear spin $\frac{1}{2}$. It has been studied against a background of the ideal Fermi gas theory outlined in Chapter 15. These two isotopes, together with helium films and mixtures of $^3$He and $^4$He, form well-studied systems which exhibit at low temperatures the effect of quantum statistics on macroscopic properties. For a survey of this branch of modern research the specialized literature should be consulted.

## 16.7. EINSTEIN'S FORMULA FOR THE FLUCTUATION PROBABILITY

An important approximate but rather general formula for the probability $P(x)\mathrm{d}x$ that a fluctuating macroscopic variable of a system at temperature $T$ lies in the range $(x,\ x + \mathrm{d}x)$ will now be obtained. A subensemble of the canonical ensemble, consisting of systems lying within this specification, must be selected. Denoting this restricted sum over states by a prime, and the normalization constant by $C$,

$$P(x) = C \sum_i{}' \exp\left(-\frac{E_i}{kT}\right) = C \exp[\ln \sum{}' e^{-E_i/kT}]$$

$$= C \exp[-F(x)/kT] \ . \qquad (16.35)$$

Here $F(x)$ is the free energy the system would have if $x$ were the value at which the fluctuating variable is held. Thus

$$P(x)\,dx = P(x_0)\,\exp\left[-\frac{F(x)-F(x_0)}{kT}\right]\,dx \quad . \tag{16.36}$$

A Taylor expansion to second order, noting $(\partial F/\partial x)_{x=x_0}$ vanishes for an equilibrium state, yields for small fluctuations

$$P(x)\,dx \simeq P(x_0)\,\exp[-\gamma(x-x_0)^2]\,dx\ , \qquad \gamma \equiv \frac{1}{2kT}\left(\frac{\partial^2 F}{\partial x^2}\right)_{x=x_0} \ . \tag{16.37}$$

For an *isolated* system $-F(x)$ is replaced by the entropy $S(x)$, and one finds (16.37) with

$$\gamma = -\frac{1}{2k}\,(\partial^2 S/\partial x^2)_{x=x_0} \ . \tag{16.38}$$

These are formulae essentially derived by Einstein (1910) for the theory of Brownian motion.

For the mean square fluctuation one finds ($\Delta x \equiv x-x_0$)

$$\sigma_x^2 = \langle (x-x_0)^2 \rangle = \int_{-\infty}^{\infty} (\Delta x)^2\,\exp[-\gamma(\Delta x)^2]\,dx \Big/ \int_{-\infty}^{\infty} \exp[-\gamma(\Delta x)^2]\,dx$$

$$= \tfrac{1}{2}\left(\frac{\pi}{\gamma^3}\right)^{\frac{1}{2}} \Big/ \left(\frac{\pi}{\gamma}\right)^{\frac{1}{2}} = \frac{1}{2\gamma} \ . \tag{16.39}$$

The large fluctuations for which (16.37) does not hold make a negligible contribution to the integrals, and that is why approximation (16.39) is adequate for many purposes. If $x$ is interpreted as $n$, then since $(\partial F/\partial n)_{v,T} = \mu$, (16.39) goes over into (16.25).

For an isolated system with $x$ taken as the energy, (16.38) is

$$\gamma = -\frac{1}{2k}\,\frac{\partial^2 S}{\partial U^2} = -\frac{1}{2k}\left(\frac{\partial T}{\partial U}\right)^{-1} = \frac{1}{2kT^2 C_v} \ .$$

Thus (16.39) reduces to (16.3) in this case. The result

(16.39) is thus very useful for general investigations, and it remains valid in some cases when $x_0$ is a *non*-equilibrium state.

PROBLEMS

16.1. Show that $\sigma_n^2/\langle n\rangle^2 = (kT/v)K_T$, using (16.25).

16.2. Show from the grand canonical distribution that

$$kT^2\left(\frac{\partial\langle n\rangle}{\partial T}\right)_{\mu,v} = \langle nE\rangle - \langle n\rangle\langle E\rangle - \mu\sigma_n^2 .$$

16.3. Show similarly that

$$kT^2\left(\frac{\partial\langle E\rangle}{\partial T}\right)_{\mu,v} = \sigma_E^2 + \mu(\langle E\rangle\langle n\rangle - \langle En\rangle) .$$

16.4. Show from the two preceding results that

$$\frac{\sigma_E^2}{kT^2} = \left(\frac{\partial\langle E+\mu n\rangle}{\partial T}\right)_{\mu,v} + \frac{\mu^2}{T}\left(\frac{\partial\langle n\rangle}{\partial\mu}\right)_{v,T}$$

16.5. Establish the following generalization to the grand canonical ensemble of equation (16.3)

$$\frac{\sigma_E^2}{kT^2} = C_v + \frac{\{\mu(\partial\langle n\rangle/\partial\mu)_{v,T}+T(\partial\langle n\rangle/\partial T)_{\mu,v}\}^2}{T(\partial\langle n\rangle/\partial\mu)_{v,T}} .$$

16.6. (*a*) The results (16.1) and (16.26a) together suggest that for a Fermi distribution

$$\sigma_{n_j}^2 = \langle n_j^2\rangle - \langle n_j\rangle^2 = \langle n_j\rangle - \langle n_j\rangle^2$$

that is

$$\langle n_j^2\rangle = \langle n_j\rangle .$$

Verify by direct calculation that in fact $\langle n_j^b\rangle$ is independent of $b$ (an arbitrary non-zero constant) in this case.

(*b*) Verify by direct calculation the analogous result for a Bose distribution:

$$\langle n_j^2\rangle = \langle n_j\rangle + 2\langle n_j\rangle^2 .$$

# 17. Simple transport properties

## 17.1. INTRODUCTION

Spatial gradients of intensive variables can give rise to currents and these are studied in this chapter with reference to the chemical potential. The transport properties which can be understood as a result of this observation depend again on a blend of thermodynamic and statistical mechanical ideas.

## 17.2. THE ELECTROCHEMICAL POTENTIAL

For a wide class of sufficiently slow processes in a system of energy $U$, entropy $S$, volume $v$, number of particles $N$, and chemical potential $\mu$, we know that the Gibbs equation (7.5)

$$dU = TdS - pdv + \mu dN \qquad (17.1)$$

holds. This equation will now be generalized by supposing that a coordinate-dependent electrostatic potential $\phi(r)$ is also present. If the particles have a charge $q$ we have to add an additional energy term on the right-hand side of (17.1). This arises from the fact that if $dN$ particles are added to the the system the total charge of the system increases by

$$dq = qdN$$

so that the additional energy term in (17.1) is

$$\phi dq = q\ \phi\ dN\ .$$

Hence

$$dU = TdS - pdv + (\mu + q\phi)\ dN\ . \qquad (17.2)$$

This leads to generalizations of the usual thermodynamic expressions. For example the Helmholtz free energy is $F = U - TS$, so that

$$dF = dU - TdS - SdT = -pdv + \mu dN + \phi dq - SdT$$

$$= -pdv - SdT + (\mu + q\phi)\, dN \qquad (17.3)$$

The Gibbs free energy becomes similarly from $G = F + pv$

$$dG = vdp - SdT + (\mu + q\phi)\, dN\,. \qquad (17.4)$$

In other words $\mu$ has to be replaced by the *electrochemical potential*

$$\bar{\mu} \equiv \mu + q\phi \qquad (17.5)$$

and

$$\bar{\mu} = \left(\frac{\partial G}{\partial N}\right)_{p,T} = \left(\frac{\partial F}{\partial N}\right)_{v,T} = \left(\frac{\partial U}{\partial N}\right)_{v,S} \qquad (17.6)$$

Just as we showed in section 7.4 that $\mu$ is spatially constant in equilibrium, so the presence of an electrostatic potential generalizes this to the requirement that $\bar{\mu}$ be spatially constant in equilibrium.

## 17.3. THE CURRENT DENSITIES

In (15.27) and (15.30) we introduced the concentration of electrons and holes in a semiconductor and denoted these by $n$ and $p$. The general notation $\nu_j$ for the concentration of current carriers of type j will now be more convenient. The number of such current carriers crossing unit area per unit time under the influence of a concentration gradient in the $x$-direction is then $-D_j \partial\nu_j/\partial x$ where $D_j$ is the diffusion coefficient of these carriers. This is essentially Fick's law of diffusion, which will be assumed here. The negative sign shows that an increase of $\nu_j$ with $x$ implies a flow into the negative $x$-direction. The current density due to diffusion is for particles j

$$J_j^{(D)} = -q_j D_j \frac{\partial \nu_j}{\partial x} \quad \text{or} \quad \mathbf{J}_j^{(D)} = -q_j D_j \operatorname{grad} \nu_j\,. \qquad (17.7)$$

The second form (17.7) is the generalization to three dimensions of the first form since

$$\text{grad}\ \nu_j = \left(\frac{\partial \nu_j}{\partial x}, \frac{\partial \nu_j}{\partial y}, \frac{\partial \nu_j}{\partial z}\right) .$$

If an electric field $\mathbf{E}$ is present, the current carriers attain an average drift velocity $\mathbf{v}_d$ which is limited by scattering with impurities and lattice waves:

$$\mathbf{v}_{d,j} = v_j \mathbf{E} = -v_j\ \text{grad}\ \phi . \qquad (17.8)$$

Here we have used

$$\mathbf{E} = -\left(\frac{\partial \phi}{\partial x}, \frac{\partial \phi}{\partial y}, \frac{\partial \phi}{\partial z}\right)$$

and $v_j$ is called the *mobility* of carriers j. Its unit is usually m $s^{-1}/Vm^{-1}$ ($m^2V^{-1}s^{-1}$) since it is a velocity divided by an electric field. Mobility values range from $10^{-5}$ to 10 $m^2V^{-1}s^{-1}$ and can lie even outside this range. The associated conduction current density is

$$\mathbf{J}_j^{(c)} = \sigma_j \mathbf{E} = -\sigma_j\ \text{grad}\ \phi, \quad \sigma_j \equiv |q_j| v_j \nu_j . \qquad (17.9)$$

Positive carriers move with the field and negative current carriers move against the field, but both contribute a conventional current density in the direction of the field. Hence the absolute value of the charge appears here.

## 17.4. BAND EDGES, ELECTROSTATIC POTENTIAL AND ENERGY SCALES

Let $B_c$, $B_v$ be the band edges of a semiconductor in the absence of an electroststic potential $\phi$, i.e. they are the potential energies of an electron in the conduction band and of a hole in a valence band on an energy scale in which *electron energies increase vertically upwards.* The effect of the electrostatic potential is then to effect a shift by $-|q_c|\phi$ in the electron potential energy. For a positive charge such as a hole (charge $q_v$ say) there is a shift $q_v\phi = |q_v|\phi$ on the hole energy scale, i.e. a shift $-|q_v|\phi$ on the electron energy scale. Since $\phi = \phi(\mathbf{r})$ this leads to spatially-dependent energy band edges on the electron energy scale:

$$E_c(\mathbf{r}) = B_c - |q_c|\phi(\mathbf{r}), \quad E_v(\mathbf{r}) = B_v - |q_v|\phi(\mathbf{r}) \ . \qquad (17.10)$$

The energy gap is

$$E_c(\mathbf{r}) - E_v(\mathbf{r}) = B_c - B_v + (|q_v| - |q_c|)\phi(\mathbf{r}) \qquad (17.11)$$

and this is a constant for all $\mathbf{r}$ if, as is always assumed, electrons and holes have the same numerical charge.

We therefore write generally for the band edge for carriers of type j

$$E_j(\mathbf{r}) = B_j - |q_j|\phi(\mathbf{r}) \qquad (17.12)$$

$$\text{grad } E_j(\mathbf{r}) = |q_j|\mathrm{E} = \mathrm{J}_j^{(c)}/v_j\nu_j \qquad (17.13)$$

where we have used (17.8) and (17.9).

Now (17.5) is $\bar{\mu}_j = \mu_j + q_j\phi$ for particles of charge $q_j$ on their own energy scale. This is

$\bar{\mu}_j = \mu_j - |q_j|\phi$ for electrons on their energy scale

$\bar{\mu}_j = \mu_j + |q_j|\phi$ for holes on their energy scale.

From this relation one infers, using subscripts to indicate which energy scale has been used, that for holes j = h

$$-(\bar{\mu}_h)_h = -(\mu_h)_h - |q_h|\phi \ .$$

Hence

$$(\bar{\mu}_h)_e = (\mu_h)_e - |q_h|\phi$$

after inverting the energy scale so as to make it applicable to electrons. For the electron energy scale, which will now be used,

$$\boxed{\bar{\mu}_j(\mathbf{r}) = \mu_j(\mathbf{r}) - |q_j|\phi(\mathbf{r})} \qquad (17.14)$$

is a relation for positive *and* negative charge carriers.

From (17.12) and (17.14)

$$-|q_j|\phi = E_j(\mathbf{r}) - B_j = \bar{\mu}_j(\mathbf{r}) - \mu_j(\mathbf{r}) \tag{17.15}$$

so that

$$\frac{\partial(E_j-\bar{\mu}_j)}{\partial\nu_j} = -\frac{\partial\mu_j}{\partial\nu_j}\ .$$

Note for later reference that therefore

$$\left.\begin{aligned} -\frac{|q_j|}{q_j}\frac{\partial\mu_j}{\partial\nu_j} &= kT_j\left(\frac{\partial\gamma_j}{\partial\nu_j}\right) \\ \text{where} \qquad & \\ \gamma_j &\equiv \frac{q_j}{|q_j|}\frac{E_j-\bar{\mu}_j}{kT_j}\ . \end{aligned}\right\} \tag{17.16}$$

If equilibrium holds $\bar{\mu}$ is independent of $\mathbf{r}$. If in addition $\phi(\mathbf{r})$ is zero, then

$$\left.\begin{aligned} E_j(\mathbf{r}) &= B_j \\ \mu_j(\mathbf{r}) &= \bar{\mu}_j \end{aligned}\right\} \text{ independent of } \mathbf{r}$$

and we are back to the theory of section 15.4.

## 17.5. THE FERMI LEVEL IS THE ELECTROCHEMICAL POTENTIAL

The replacement of $\mu$ by $\bar{\mu}$ in the presence of a spatially varying electrostatic potential in accordance with (17.5) and (17.16) suggests that the parameter $\mu$ in equations such as (11.7) should be replaced by $\bar{\mu}$. This will now be discussed.

For a conduction band in a semiconductor at a definite position $\mathbf{r}$ and with a density of states function $A_c\{e-E_c(\mathbf{r})\}^s$, where $s$ is a constant, one finds for the total number of electrons per unit volume in a volume element at $\mathbf{r}$

$$\nu_c(\mathbf{r}) = \frac{1}{v}\int_{E_c(\mathbf{r})}^{\infty} \frac{A_c(e-E_c)^s\, de}{1+\exp[(1/kT_c)\{(e-E_c)-(\bar{\mu}_c-E_c)\}]} .$$

We have replaced the chemical potential by the electrochemical potential, as required by our observations in section 17.2. As in section 15.4, we have also taken explicit account of the fact that the conduction band of a semiconductor terminates at a certain energy $E_c$. Here the density of states is zero for $e \le E_c$. Above this energy the density of states is assumed to rise as $A_c(e-E_c)^s$ as in Chapter 11. In the denominator a term $E_c$ has been added and subtracted for convenience. Using equation (17.16) with negative $q_j$, one sees that $-\gamma_c$ enters the exponent.

Using $(e-E_c)/kT_c$ as a variable of integration one finds

$$v\nu_c(\mathbf{r}) = A_c(kT_c)^{s+1}\,\Gamma(s+1)\, I(\gamma_c(\mathbf{r}),s,+) .$$

Note that the integration is carried out over the energy *and at a given position* $\mathbf{r}$ .

An analogous argument applies to holes and in a valence band when the forbidden energies satisfy $e \ge E_v$ where $E_v$ is another energy. It lies below $E_c$ on an electron energy scale. One finds for the volume density of holes in a valence band

$$\nu(r) = \frac{1}{v}\int_{-\infty}^{E_v} \frac{A_v(E_v-e)^s de}{1+\exp\{(1/kT_v)\,[(\bar{\mu}-E_v)-(e-E_v)]\}} . \tag{17.17}$$

Here we have used the fact that quantum states of the valence band exist only for $e < E_v$. Using $(E_v-e)/kT_v$ as the variable of integration and equation (17.16) one finds

$$v\nu_v(\mathbf{r}) = A_v(kT_v)^{s+1}\,\Gamma(s+1) I(\gamma_v(\mathbf{r}),s,+) .$$

Thus one can write more generally

$$v\nu_j(\mathbf{r}) = A_j(kT_j)^{s+1}\,\Gamma(s+1)I(\gamma_j(\mathbf{r}),s,+)$$

$$\equiv N_j I(\gamma_j(\mathbf{r}),s,+) \ . \qquad (17.18)$$

Here

$$N_j = A_j(kT_j)^{s+1}\,\Gamma(s+1) \to 2\left(\frac{2\pi m_j kT_j}{h^2}\right)^{3/2}$$

where the last form holds for $s = \frac{1}{2}$ as seen in section 15.4.

In the probability distributions

$$\frac{1}{1+\exp\{(e-E_F)/kT\}} \ , \quad \frac{1}{1+\exp\{E_F-e)/kT\}} \ ,$$

for electrons and holes respectively $E_F$ is often called the *Fermi level*. One sees that $E_F$ is $\bar{\mu}$, the electrochemical potential.

Note that the parameter $\gamma_j$ in integrals such as that occurring in (17.17) involves

$$E_j(\underline{r}) - \bar{\mu}_j = B_j - \mu_j(\underline{r}) \qquad (17.19)$$

where (17.15) has been used.

## 17.6. EINSTEIN RELATIONS

Expression such as (17.18) show that

$$d\nu = \left(\frac{\partial\nu}{\partial\mu}\right)_{v,T} d\mu + \left(\frac{\partial\nu}{\partial T}\right)_{\mu,v} dT + \left(\frac{\partial\nu}{\partial v}\right)_{T,\mu} dv \ .$$

It follows that

$$\left(\frac{\partial\nu}{\partial x}\right)_{y,z} = \left(\frac{\partial\nu}{\partial\mu}\right)_{v,T}\left(\frac{\partial\mu}{\partial x}\right)_{y,z} + \left(\frac{\partial\nu}{\partial T}\right)_{\mu,v}\left(\frac{\partial T}{\partial x}\right)_{y,z} + \left(\frac{\partial\nu}{\partial v}\right)_{T,\mu}\left(\frac{\partial v}{\partial x}\right)_{y,z} \ . \qquad (17.20)$$

The diffusion current density is, by (17.7) and (17.20),

$$\mathbf{J}_j^{(D)} = -q_j D_j \operatorname{grad} \nu_j = -q_j D_j \left\{ \left(\frac{\partial \nu_j}{\partial \mu_j}\right)_{T,v} \operatorname{grad} \mu_j + \left(\frac{\partial \nu_j}{\partial T}\right)_{\mu_j, v} \operatorname{grad} T \right\}. \tag{17.21}$$

Also by (17.9) and (17.14)

$$\begin{aligned} \mathbf{J}_j^{(c)} &= |q_j| v_j \nu_j \mathbf{E} = -|q_j| v_j \nu_j \operatorname{grad} \phi \\ &= v_j \nu_j \operatorname{grad}(\bar{\mu}_j - \mu_j) \end{aligned} \tag{17.22}$$

so that the total current density is the sum of (17.21) and (17.22). In equilibrium

$$\mathbf{J}_j = \mathbf{J}_j^{(D)} + \mathbf{J}_j^{(c)} = 0,$$

and also

$$\operatorname{grad} \bar{\mu}_j = 0,$$

so that

$$-v_j \nu_j \operatorname{grad} \mu_j - q_j D_j \left(\frac{\partial \nu_j}{\partial \mu_j}\right)_{T,v} \operatorname{grad} \mu_j = 0 .$$

It follows that

$$\boxed{\begin{aligned} \lambda_j &\equiv \frac{|q_j| D_j}{v_j} = -\frac{|q_j|}{q_j} \frac{\nu_j}{(\partial \nu_j / \partial \mu_j)_{T,v}} \\ &= kT_j \nu_j \frac{\partial \gamma_j}{\partial \nu_j} . \end{aligned}} \tag{17.23}$$

where (17.16) has been used in the last step.

The right-hand side is positive for electrons since an increase in $\bar{\mu}_j$ or $\mu_j$ causes an increase in the electron concentration $\nu_j$. The hole concentration, however, decreases as $\bar{\mu}_j$ or $\mu_j$ rises. Hence the right-hand side of (17.23) is positive in both cases.

The connection between diffusion coefficient and mobility given in (17.23) is called the *Einstein mobility relation*.

It can be derived and discussed by a number of other methods, but the above is probably the simplest. Equation (17.23) is the most general known form of the Einstein relation. Note that it makes no reference to microscopic properties.

To find a more explicit expression for the right-hand side of (17.23), note that by (17.18) for $s = \frac{1}{2}$

$$v\left(\frac{\partial \nu_j}{\partial \gamma_j}\right)_{T,v} = N_j I(\gamma_j, -\tfrac{1}{2}) \ .$$

Hence

$$\lambda_j = kT_j \frac{I(\gamma_j, \frac{1}{2})}{I(\gamma_j, -\frac{1}{2})} \ . \tag{17.24}$$

In particular for non-degenerate semiconductors, using (11.12),

$$\lambda_j = kT_j \ . \tag{17.25}$$

These relations (17.23) to (17.25) hold also in non-equilibrium situations.

The electrical mobilities $v_j$ are readily obtained experimentally from measurements of conductivity, so that the diffusion coefficient of electron or hole can be inferred by virtue of the Einstein relation. The calculation of the effective temperatures $T_j$ raises separate problems if a strong electric field is present. Near equilibrium, however, these temperatures are all equal to the lattice temperature. In other applications, for example to amorphous materials, one can use random walk models to estimate diffusion coefficients. The Einstein relation is then available to calculate mobilities.

## 17.7. THE CONNECTION BETWEEN CURRENT DENSITY, ELECTROCHEMICAL POTENTIAL, AND EINSTEIN RELATION

Since the chemical potential $\mu_j$ of a group of current carriers j is spatially constant in equilibrium in the absence of an electric field, one may try to put for the current density in the presence of an electric field **E**

$$\mathbf{J}_j = \sigma_j \mathbf{E} + k_j \operatorname{grad} \mu_j = \operatorname{grad}(k_j \mu_j - \sigma_j \phi) \tag{17.26}$$

where $k_j$ is a constant. In order to identify it, note that in the presence of an electric field it is the electrochemical potential (17.14)

$$\bar{\mu}_j = \mu_j - |q_j|\phi$$

which is spatially constant in equilibrium. It then follows that

$$\text{grad}(k_j\mu_j - k_j|q_j|\phi) = 0. \qquad (17.27)$$

Comparing (17.27) and the equilibrium form of (17.26) one sees that

$$k_j = \sigma_j/|q_j| = v_j\nu_j .$$

It follows that

$$\mathbf{J}_j = v_j\nu_j\,\text{grad}\,\bar{\mu}_j . \qquad (17.28)$$

It can be shown that this result is intimately connected with the Einstein relation. The form of this connection can be exhibited as a set of logical implications (see problem 17.5.

PROBLEMS

17.1. It has been suggested that a term $\partial(D\nu)/\partial x$ should occur in (17.7) instead of the term $\partial D/\partial x$. Analyse this question for a constant-temperature system by defining two diffusion coefficients $D$ and $D'$ and a function $U$ by

$$J_d \equiv -qD\left(\frac{\partial\nu}{\partial x}\right)_{y,z} \equiv -q\left(\frac{\partial(D'\nu)}{\partial x}\right)_{y,z} + q\nu U .$$

Hence prove that if $D$ depends on $\gamma = \mu/kT$, $T$, and an unspecified function $\tau_0(x)$, then

$$D = D' + \nu\left(\frac{\partial D'}{\partial\gamma}\right)_{T,\tau_0}\left(\frac{\partial\gamma}{\partial\nu}\right)_{T,v}$$

$$U = \left(\frac{\partial D'}{\partial \tau_0}\right)_{T,\gamma} \left(\frac{\partial \tau_0}{\partial x}\right)_{y,z}$$

where $U$ is assumed to be a function of $\gamma$, $T$, $\tau_0$, and $\partial\tau_0/\partial x$.

17.2. Suppose $\nu$ is given by (17.17) with $s = \frac{1}{2}$ and that one finds

$$D'(x) = \frac{2}{3}\frac{kT}{m}\tau_0(x)\frac{\Gamma(t+5/2)}{\Gamma(3/2)}\frac{I(\gamma,t+3/2)}{I(\gamma,1/2)}$$

where $t$ is a constant which arises from a certain dependence $\tau = \tau_0(x)\,(E/kT)^t$ of the relaxation time, $E$ being the electron kinetic energy. Use problem 17.1 and the Einstein relation to show that $D$ and the mobility are respectively given by

$$\frac{2}{3}\frac{kT}{m}\tau_0(x)\frac{\Gamma(t+5/2)}{\Gamma(3/2)}\begin{cases} I(\gamma,t+\frac{1}{2})/I(\gamma,-\frac{1}{2}) \\ \{I(\gamma,t+\frac{1}{2})/I(\gamma,\frac{1}{2})\}\;e/kT \end{cases}.$$

(The expression for $D'$ or $D$ can be obtained from the Boltzmann transport equation, which is not discussed in this book.)

17.3. Assuming (17.23) and $\mathbf{J}_j = \mathbf{J}_j^{(D)} + \mathbf{J}_j^{(c)}$, as given by (17.21) and (17.22), deduce that

$$\mathbf{J}_j = v_j \upsilon_j \,\mathrm{grad}\,\bar{\mu}_j \;.$$

17.4. Generalize the result of problem 17.3 to the case when a temperature gradient exists by adding to $\mathbf{J}_j$ a term $-q_j \nu_j D_j^T \,\mathrm{grad}\, T$ where $D^T$ is a thermal diffusion coefficient. Show that this leads to

$$\mathbf{J}_j = v_j \nu_j \,\mathrm{grad}\,\bar{\mu}_j - [v_j \nu_j \left(\frac{\partial \mu_j}{\partial T}\right)_{\nu_j, v} + q_j \nu_j D_j^T]\,\mathrm{grad}\, T\,.$$

17.5. Consider the result stated in problem 17.4 and

$$\mathbf{J}_j = |q_j| v_j \nu_j \mathbf{E} - q_j D_j \,\mathrm{grad}\,\nu_j - q_j \nu_j D_j^T \,\mathrm{grad}\, T$$

and

$$\frac{q_j D_j}{v_j} = -\nu_j \left(\frac{\partial \mu_j}{\partial \nu_j}\right)_{T,v}.$$

Show that any two imply the third.

(This result makes clear the logical relation between the macroscopic expression for the current, the alternative expression in terms of the gradient of the electrochemical potential, and the general Einstein relation.)

# 18. Outlook

## 18.1. CLOSED OR OPEN?

This question is sometimes asked about the universe: Will it always expand or will there be a contraction? It applies also to our subject: Is it complete or is it still growing? The answer is of course that it is still growing. The parts which were once flourishing may only be growing slowly now, and as time progresses new emphases emerge, and each contribution affects the nature of the emphasis which is adopted. The parts which are growing strongly tend to be more difficult to explain, and tend to be omitted from an introductory book.

In this book there are also important omissions, and in this section some of these will be briefly indicated and additional applications will be noted.

## 18.2. INTERACTING SYSTEMS AND THE VIRIAL THEOREM

The above introduction to statistical mechanics has not shown the reader how to handle interacting systems. One way of dealing with such systems is via the *virial theorem*. This states that if $\mathbf{F}$ is the force which acts on a particle located at a position $\mathbf{r}_i$ $(i=1,2,\ldots,N)$ then the kinetic energy $T$ of the system satisfies

$$\langle T \rangle = \langle -\tfrac{1}{2} \sum_{i=1}^{N} \mathbf{F}_i \cdot \mathbf{r}_i \rangle \quad \begin{Bmatrix}\text{bounded motion or} \\ \text{periodic motion}\end{Bmatrix} \tag{18.1}$$

where the angular brackets denote the time average, and

$$C \equiv -\tfrac{1}{2} \sum_{i=1}^{N} \mathbf{F}_i \cdot \mathbf{r}_i \tag{18.2}$$

is the *virial* of the system of particles.

To prove (18.1) consider a quantity $G$ defined by

$$\frac{dG}{dt} \equiv \frac{d}{dt}\left(\sum_i \mathbf{p}_i \cdot \mathbf{r}_i\right)$$

$$= \sum_i (\dot{\mathbf{p}}_i \cdot \mathbf{r}_i + \mathbf{p}_i \cdot \dot{\mathbf{r}}_i) = \sum_i \mathbf{F}_i \cdot \mathbf{r}_i + 2T \ .$$

Its time average over a period $\tau$ is, using (18.2),

$$-\tfrac{1}{2}\left\langle \frac{dG}{dt}\right\rangle = -\frac{1}{2\tau}\int_0^{\tau}\frac{dG}{dt}\,dt = \langle C\rangle - \langle T\rangle \ .$$

If there is a finite upper bound to $G$, then the left-hand side vanishes for large enough $\tau$. Alternatively, if the motion is periodic so that all co-ordinates repeat after some time $T$, then one can choose $\tau = T$ and the left-hand side vanishes again. For bounded motion or periodic motion (18.1) follows.

We now compute two important contributions to the virial $C$. The first arises from the force exerted by a container of volume $v$ on a gas at pressure $p$. If $\mathbf{n}$ be the outward drawn normal, the force on an element of area $da$ is $-p\mathbf{n}\,da$. The contribution to the virial is therefore, by (18.2)

$$C_1 = \tfrac{1}{2}p\int_S \mathbf{n}\cdot\mathbf{r}\,da = \tfrac{1}{2}p\int_S \mathbf{r}\cdot d\mathbf{S}$$

where $\mathbf{r}$ is the position of the element relative to some origin and the integral is over the container. The divergence theorem theorem of Gauss will now be used. It states that the surface integral of any vector $\mathbf{r}$, taken over a closed surface $S$, is equal to the volume integral of the divergence of $\mathbf{r}$ taken over the enclosed volume. Now for any vector $\mathbf{v}$

$$\operatorname{div}\mathbf{v} = \left(\frac{\partial v_x}{\partial x}\right)_{y,z} + \left(\frac{\partial v_y}{\partial y}\right)_{z,x} + \left(\frac{\partial v_z}{\partial z}\right)_{x,y}$$

where $v_x$, $v_y$ and $v_z$ are the components of $\mathbf{v}$. The partial derivatives are all unity for $\mathbf{r} = (x,y,z)$ so that

$$C_1 = \frac{3}{2}p\int_v d\tau = \frac{3}{2}pv \ .$$

By using the pressure $p$, a time average over all the collisions with the container has here been implied.

The second contribution to the virial to be computed arises from a distance-dependent interaction force $f(|\mathbf{r}_i - \mathbf{r}_j|) \equiv f(\mathbf{r}_{ij})$ between the molecules. Taking $f$ as positive for repulsive interactions, the forces on particles $i$ and $j$ due to the other particle are respectively

$$\mathbf{F}_i \equiv \frac{\mathbf{r}_i - \mathbf{r}_j}{r_{ij}} f(r_{ij}) \quad \text{and} \quad \mathbf{F}_j = - \mathbf{F}_i \ .$$

They contribute to the virial (18.2):

$$C_2 = -\tfrac{1}{2}(\mathbf{r}_i \cdot \mathbf{F}_i + \mathbf{r}_j \cdot \mathbf{F}_j) = -\tfrac{1}{2}(\mathbf{r}_i - \mathbf{r}_j) \cdot \mathbf{F}_i = - \ r_{ij} \ f(r_{ij}) \ .$$

Denoting the sum over all pairs of particles by a prime and taking the time average

$$C_2 = -\tfrac{1}{2} \langle \ \textstyle\sum' \ r_{ij} \cdot f(r_{ij}) \rangle \ . \tag{18.3}$$

We consider now an equilibrium situation for a gas of identical molecules. Then each molecule is, surface effects apart, in identical surroundings, and each pair contributes the same amount to $C_2$. Since there are $\frac{1}{2}N(N-1)$ pairs

$$C_2 = -\tfrac{1}{4} N(N-1) \langle r_{12} \ f(r_{12}) \rangle \ . \tag{18.4}$$

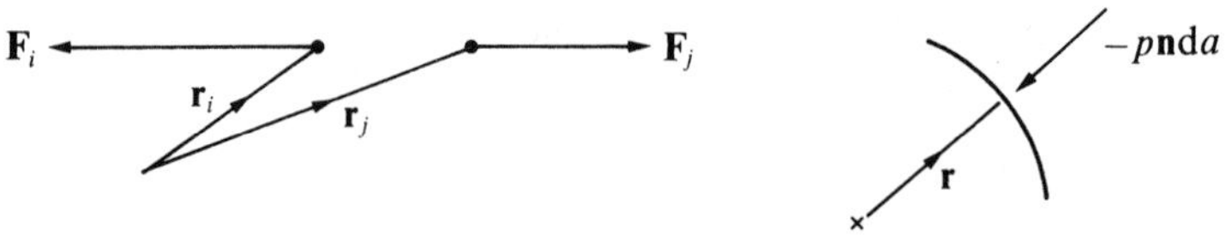


Fig.18.1. Forces for the virial theorem.

For a gas of identical interacting molecules in equilibrium we have

$$\langle T \rangle = C_1 + C_2$$

that is

$$pv = \frac{2}{3} \langle T \rangle + \frac{1}{6} \ N(N-1) \ \langle r_{12} f(r_{12}) \rangle \ . \tag{18.5}$$

If the gas can be treated by classical statistical mechanics and the molecules are treated as points, then the equipartition theorem gives us

$$\frac{2}{3}\langle \mathrm{T} \rangle = NkT \quad .$$

Hence (18.5), can be specialized to

$$pv = NkT + \frac{1}{6}N(N-1)\langle r_{12}f(r_{12})\rangle \quad . \tag{18.6}$$

Suppose that the potential energy due to the forces of interaction in the gas is

$$V = \sum{}'(a\,r_{ij}^{u})$$

where $a$ and $u$ are constants. Then by the argument leading from (18.3) to (18.4)

$$\langle V \rangle = \tfrac{1}{2}N(N-1)a\langle r_{12}^{u}\rangle \quad .$$

It follows that with $f(r_{ij}) = -\partial V/\partial r_{ij}$ ,

$$\tfrac{1}{2}N(N-1)\langle r_{12}f(r_{12})\rangle = -\tfrac{1}{2}N(N-1)\langle uar_{12}^{u}\rangle = -u\langle V\rangle \ . \tag{18.7}$$

Thus (18.5) becomes with (18.7)

$$\langle \mathrm{T} \rangle - \frac{u}{2}\langle V \rangle = \frac{3}{2}pv. \tag{18.8}$$

Since $\mathrm{T} + V$ is the total energy $E$ of the gas, (18.8) can be written as

$$2\langle \mathrm{T} \rangle - u\langle E - \mathrm{T}\rangle = 3pv \quad .$$

One finally arrives at

$$(u+2)\langle T \rangle = u\langle E \rangle + 3pv \quad . \tag{18.9}$$

The results (18.8) and (18.9) are a special case of (18.1) and even of (18.6). All these forms are referred to as the virial theorem, $p$ being the pressure at the boundary.

If equipartition holds for a system of $N$ particles each with $f$ square terms (or degrees of freedom) in its kinetic energy, and if the virial theorem holds as well, then the total average energy is classically with $pv = gU$

$$U \equiv \langle E \rangle = \frac{1+2/u}{1+3g/u} \frac{f}{2} NkT .$$

This implies a heat capacity

$$C_V = \frac{1+2/u}{1+3g/u} \frac{f}{2} Nk . \tag{18.10}$$

In the classical limit of a solid of $N$ atoms, each vibrating as a three-dimensional harmonic oscillator, this yields with $g = 0$: $u = 2$, $f = 3$, i.e. $C_V = 3nk$ in agreement with what has been found in section 14.1. For gravitational interaction, to which we may now have the confidence to apply this formula, though the relation $pv = gU$ is not correct in this case,

$$u = -1, \quad \text{i.e. } C_V < 0 \quad \text{provided } g < \frac{1}{3} .$$

For a free or unconfined system we certainly have $pv = gU$ with $g = 0$ and one has then a *negative* heat capacity. For example a star whose nuclear fuel is exhausted, radiates energy during a part of its life. It then contracts under self-gravitation, and this causes its temperature to rise. It thus behaves as if it had a negative heat capacity. Although this is an important example, the simultaneous application of the virial and the equipartition theorem to a non-equilibrium situation represents of course a strong assumption about the system.

As a second example consider again the system (15.50). If all particles are non-relativistic there is a finite and non-zero equilibrium radius $R_1$. The motion is bounded in the sense $\sum_i \mathbf{p}_i \cdot \mathbf{r}_i < \infty$, and the virial theorem holds with $U_{nr}$ interpreted as the average kinetic energy. If all particles are subject to extreme relativistic motion, however, the model system is unbounded if the system is not massive enough ($A_G < A_r$, i.e. $n_{Gcr} < n_r$) and collapses if it is massive enough ($A_G > A_r$, i.e. $n_{Gcr} > n_r$). In both cases there is no assurance that $\Sigma_i\, \mathbf{p}_i \cdot \mathbf{r}_i$ is bounded and the virial theorem does not apply.

As a third example, consider an electron gas in $d$-dimensions, with

$$V = e^2 \sum{}' \frac{1}{r_{ij}^{d-2}}$$

so that $u$ is replaced by $2-d$. One finds from (18.8)

$$\langle T \rangle + \frac{d-2}{2} \langle V \rangle = \frac{d}{2}\, pv \;,$$

i.e.

$$(d-4)\langle T \rangle = (d-2)\langle E \rangle - d\; pv$$

$$(4-d)\langle V \rangle = 2\langle E \rangle - d\; pv \;.$$

The four-dimensional situation is seen to be rather special, and is the subject of much current work on phase transitions. One finds in that case

$$\langle E \rangle = 2\; pv \;.$$

In recent work expansion with respect to $d$ about the value $d$=4 have proved to be important.

The correction term in (18.6) has been investigated experimentally and theoretically, and the results have been expressed in the form

$$pv = NkT \left\{1 + \frac{N}{v} B(T) + \left(\frac{N}{v}\right)^2 C(T) + \ldots \right\}$$

or

$$NkT\{1+pB^*(T) + p^2C^*(T) + \ldots\}$$

where $B$ and $C$ or $B^*$ and $C^*$ are second and third virial coefficients, as already discussed in problem 7.5. The ideal gas equation is recovered in the limit of zero density or zero pressure. This is the limit of *infinite dilution* when $N/v \to 0$.

## 18.3. EXACT RESULTS IN STATISTICAL MECHANICS

Approximations are always needed to give statistical mechanical descriptions of real systems. In order to estimate errors and also in order to develop more realistic models, models with interactions have been developed which can be treated exactly.

Such treatments start in classical statistical mechanics with a *Hamiltonian*. This is the total energy of the system expressed in terms of generalized moments $p_1,\ldots,p_f$ and generalized co-ordinates $q_1,\ldots,q_f$:

$$E(p,q) = T(p_1,\ldots,p_f) + V(q_1,\ldots,q_f)$$

where $T$ and $V$ are kinetic and potential energies of the system. The partition function is, by (11.29),

$$Z = z^f \int\cdots\int \exp\left(-\frac{T}{kT}\right) dp_1\ldots dp_f \int\cdots\int \exp\left(-\frac{V}{kT}\right) dq_1\ldots dq_f.$$

For an $N$-particle system $T = \Sigma_1^N p_i^2/2m$ in the simplest cases so that the $N$ integrals can be evaluated simply as $N$ products of

$$z^3 \int \exp\left(-\frac{p^2}{2mkT}\right) d\mathbf{p} = z^3 (2\pi mkT)^{3/2} .$$

We have here put $f = 3N$. The value of $z^3$ is $g/h^3$ where $h$ is Planck's constant and $g$ is the spin degeneracy of the particles, as may be seen by comparing the last formula of section 11.8 with the fourth row of Table 12.1. Hence we have arrived at

$$Z = g^N\left(\frac{2\pi mkT}{h^2}\right)^{3N/2} \int\cdots\int \exp\left(-\frac{V}{kT}\right) dq_1\ldots dq_{3N} .$$

The last term is part of the classical partition function which treats particles as distinguishable and has to be corrected as seen in section 11.8. Hence

$$Z = g^N\left(\frac{2\pi mkT}{h^2}\right)^{3N/2} \frac{Q}{N!} \tag{18.11}$$

where we can write

$$Q = \int\!\!\ldots\!\!\int \exp\left\{-\frac{V(r_1,\ldots,r_N)}{kT}\right\} d\mathbf{r}_1\ldots d\mathbf{r}_N \quad . \qquad (18.12)$$

Consider now a gas of $N$ positive and negative charges $q_i$ $(i=1,2,\ldots,2N)$, which can move in an $s$-dimensional sphere of of radius $L_0$ with potential energy

$$V(\mathbf{r}_1,\ldots,\mathbf{r}_N) = -\sum{}' aq_iq_j \ln\frac{r_{ij}}{L}\ , \ (L > L_0) \qquad (18.13)$$

The sum is over all pairs, $a$ and $L$ being positive constants. If $q_iq_j > 0$, a pair $(i,j)$ contributes positively to the force $-dV/dr_{ij}$ (if $L>L_0$) and there is therefore repulsion. If $q_iq_j < 0$ there is attraction. This is therefore a model for the motion of charges in a thin-monomolecular disc-like film $(s=2)$ or in a sphere $(s=3)$. By (18.12)

$$Q = \int\!\!\ldots\!\!\int \exp\left[\frac{1}{kT}\sum{}' aq_iq_j(\ln r_{ij} - \ln L)\right] d\mathbf{r}_1\ldots d\mathbf{r}_N \quad .$$

Let $\mathbf{R}_i \equiv \mathbf{r}_i/L_0$ be a dimensionless $s$-component vector. Then $d\mathbf{r}_i = L_0^s d\mathbf{R}_i$, the volume of the sphere has the form $v = bL_0^s$, where $b$ is an appropriate number which depends on $s$, and

$$\begin{aligned} Q &= \left(\frac{v}{b}\right)^N \int\!\!\ldots\!\!\int \exp\left\{\frac{1}{kT}\sum{}' aq_iq_j \ln\left(R_{ij}\frac{L_0}{L}\right)\right\} d\mathbf{R}_1\ldots d\mathbf{R}_N \\ &= \left(\frac{v}{b}\right)^N \int\!\!\ldots\!\!\int \prod_{i<j}\left\{\left(R_{ij}\frac{L_0}{L}\right)^{(a/kT)q_iq_j}\right\} d\mathbf{R}_1\ldots d\mathbf{R}_N \ . \end{aligned}$$

Let $d \equiv (a/kT)\sum{}' q_iq_j$; then

$$\left.\begin{aligned} Q &= \frac{v^{N+d/s}}{b^{N+d/s}L^d} A(T)\ , \\ A(T) &\equiv \int\!\!\ldots\!\!\int \prod_{i<j} R_{ij}^{aq_iq_j/kT}\, d\mathbf{R}_1 \ldots d\mathbf{R}_N \quad . \end{aligned}\right\} \qquad (18.14)$$

The multiple integral is a function of $T$. If $j$ and $l$ are a pair of oppositely charged charges, their interaction with each other contributes to $A(T)$

$$\int R_{jl}^{-a|q_jq_l|/kT}\, d\mathbf{R}_j\, d\mathbf{R}_l \, .$$

This integral is of the type

$$\int r^{-b} d\mathbf{r} \qquad (b \equiv a|q_j q_l|/kT)$$

and diverges for $b \geq s$. Hence for $Q$ to exist we must have that for all $j$ and $l$

$$b < s,$$

that is

$$T > -\frac{aq_j q_l}{sk}.$$

This is a constraint only if $q_j q_l$ is negative. For $Q$ to exist, we therefore require that, if the system contains opposite charges, $T > \tau_1 \equiv -\frac{aQ_- Q_+}{sk} = \frac{a|Q_- Q_+|}{sk}$

where $Q_-$, $Q_+$ are the most negative and most positive charges in the system.

The dimensionless integral $A(T)$ depends on the various dispositions of particles in a sphere whose radius has been normalized to unity, and so it is independent of $L_0$ and therefore of $v$.

By problem 7.1

$$p = -\left(\frac{\partial F}{\partial v}\right)_{T,N}$$

and by equation (9.22) $F = -kT \ln Z$. Hence

$$\frac{p}{kT} = \left(\frac{\partial \ln Z}{\partial v}\right)_{T,N} = \left(\frac{\partial \ln Q}{\partial v}\right)_{T,N}.$$

Applying this equation to (18.14)

$$\ln Q = \left(N + \frac{d}{s}\right) \ln v + v\text{-independent terms}$$

and one finds

$$p = k\rho\left(T + \frac{Td}{sN}\right)$$

where $\rho$ is the number of particles per unit volume. The equation of state can be written as

$$p = k\rho(T - \tau_2), \quad \tau_2 \equiv -\frac{Td}{sN} . \tag{18.15}$$

This is the ideal gas equation at a high enough temperature, but (18.15) clearly fails at $T \le \tau_2$ as one would then have $p \le 0$. Thus one sees that the model fails at the higher of the two temperatures $\tau_1, \tau_2$. If oppositely charged charges are present they presumably collapse to form neutral pairs in the neighbourhood of these temperatures.

Most models of interacting systems are more complicated than this one and cannot be solved exactly. The potential (18.13) is realistic for $s = 1$, since the potential of a charged straight line is logarithmic. It is also realistic for $s = 2$ when the potential chosen is a solution of Poisson's equation, thus corresponding to the law $V(r) \propto 1/r$ in three dimensions.

## 18.4. SOLAR CELL EFFICIENCIES

Turning to thermodynamic efficiencies, let us ask with what efficiency we could expect to convert solar energy into mechanical work. Modelling the sun as a black body at temperature 6000 K and taking the temperature at which the mechanical work is required as 300K the Carnot efficiency presents an upper bound, which is (problems 5.5 and 5.6)

$$\eta_C = 1 - \frac{300}{6000} = 95\% . \tag{18.16}$$

This is unrealistically high, and we proceed to a more specific model.

Suppose a device accepts photons of energy above an energy $E_G$ and then converts exactly the energy $E_G$ into useful work (or into electricity as in a solar cell). Let the incident photon spectrum $f(x)$ be specified by the parameter $x \equiv h\nu/kT_S$ where $T_S$ is some equivalent source temperature and $\nu$ is the photon frequency. The useful work or electricity produced is then $E_G \int_{(E_G/h)}^{\infty} f(x)\, d\nu$, and the total incident energy is

$$\int h\nu\ f(x)\,d\nu = \frac{kT_S}{h}\int_0^\infty xf(x)\,dx \quad .$$

The efficiency with $x_G \equiv E_G/kT_S$ is

$$\eta = \frac{(E_G/h)\int_{x_G}^\infty f(x)\,dx}{(kT_S/h)\int_0^\infty xf(x)\,dx} = x_G\,\frac{\int_{x_G}^\infty f(x)\,dx}{\int_0^\infty xf(x)\,dx} \quad . \tag{18.17}$$

The beauty of this model is that it involves only one parameter ($x_G$) and that it can be applied to any incident spectrum. This efficiency can therefore be maximized by matching the spectrum to the value of $E_G$ which yields maximum efficiency:

$$\frac{d\eta}{dx_G} \propto \int_{x_G}^\infty f(x)\,dx - x_G\,f(x_G) = 0 \ .$$

The optimum energy $E_G$, which can represent the main energy gap in a semiconductor used in a solar cell, must therefore satisfy

$$x_G f(x_G) = \int_{x_G}^\infty f(x)\,dx \ . \tag{18.18}$$

There clearly is a maximum for all incident spectra since (18.17) is zero as $x_G \to 0$ and as $x_G \to \infty$. Equation (18.18) has to be solved numerically and for black-body radiation it is well known to yield $x_G = 2{\cdot}17$. At $T_S = 6000$ K this gives an optimum energy gap appropriate for silicon: If $T_R$ be room temperature, for which $kT_R = (1/40)$ eV, one finds

$$E_G = 2{\cdot}17kT_S = 2{\cdot}17(kT_R)\,(T_S/T_R)$$

$$= 2{\cdot}17 \times \frac{1}{40} \times \frac{6000}{300} = 1{\cdot}1\ \text{eV}.$$

For this value of $E_G$ one finds from (18.17) that $\eta = 44\%$ (see Fig.18.2). This model thus gives a distinctly lower maximum efficiency than (18.16). More accurate modelling reduces the maximum efficiency further.

If one uses a typical spectrum for diffuse rather than

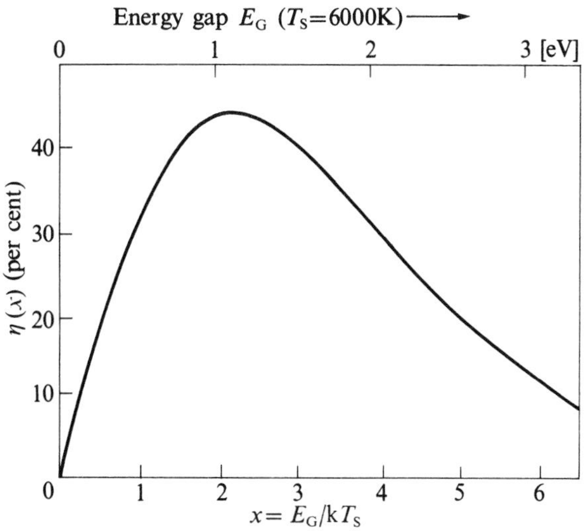


Fig.18.2. The energy efficiency (equation (18.17)) of a solar cell for black-body radiation at $T_S$ = 6000 K as a function of energy gap $E_G$, showing a maximum at $E_G/kT_S \sim 2{\cdot}17$.

black-body radiation, this spectrum has to be fed into (18.17) numerically and a different curve results.

## 18.5. SPECIAL RELATIVISTIC EFFECTS

In many branches of physics it is important to be able to transform equations from a set of co-ordinates $x,y,z,t$ to an alternative set $x'$, $y'$, $z'$, $t'$ which is in uniform motion relative to the first one. If the $y$-and $z$-axes are parallel for both systems and the relative velocity $v$ is parallel to a common $x$- and $x'$-axis, then one has in classical mechanics

$$x' = x-vt, \quad y' = y, \quad z' = z, \quad t' = t.$$

In relativistic mechanics one has, with $\gamma \equiv (1-v^2/c^2)^{-1/2}$,

$$x' = \gamma(x-vt), \quad y' = y, \quad z' = z, \quad t' = \gamma(t-vx/c^2)$$

where $c$ is the velocity of light *in vacuo*. This second

transformation brings in its train some interesting and important consequences which have modified the understanding one had earlier of the laws of mechanics, electromagnetism, and optics: this is the famous Lorentz transformation.

The effect of relative motion on thermodynamic laws has been discussed over the years in a variety of ways. We here give a brief indication of the problem. Let $\theta(\gamma)$ and $f(\gamma)$ be two undetermined functions and suppose that if heat $dQ_0$ is passed between two systems in some *inertial* frame $I_0$ (one in which Newton's laws of motion hold) in which they are both at rest, then it is judged to be $dQ$ in the relatively moving inertial frame $I$, where

$$dQ = f(\gamma)dQ_0 \; . \tag{18.19}$$

If $v \to 0$, $\gamma \to 1$, so that we require

$$f(1) = 1 \; .$$

Suppose furthermore that the second law for quasi-static processes in the form $T_0 dS_0 = dQ_0$ in $I_0$ becomes in $I$

$$TdS = \theta(\gamma)dQ \; , \quad \theta(1) = 1. \tag{18.20}$$

We shall assume *invariance* of pressure and entropy in the sense

$$p = p_0, \quad S = S_0 \; , \tag{18.21}$$

These relations are assumed by most authors. The condition $p = p_0$ can be derived from special relativity, and the condition $S = S_0$ seems reasonable since the entropy depends on invariant probabilities. Special relativity always teaches one that the volume and the energy of an unconfined system (i.e. a system which has no need of a container) transform according to

$$v = \frac{v_0}{\gamma} \; , \quad dE_0 = \gamma(dE - \mathbf{v}.d\mathbf{p})$$

where $\mathbf{p}$ is the momentum of the system in the second set of co-ordinates. It follows that one can write for quasi-static processes

$$TdS = \theta dQ = \theta f dQ_0$$

$$= \theta f(dE_0 + p dv_0) = \theta f\gamma(dE - \mathbf{v}.d\mathbf{P} + p dv) \ . \qquad (18.22)$$

Also one can write

$$TdS = \theta f \ dQ_0 = \theta f T_0 dS_0 \ . \qquad (18.23)$$

It follows from (18.22), and from (18.23) and (18.21), that

$$\left(\frac{\partial S}{\partial v}\right)_{E,P} = \theta f\gamma \frac{p}{T} \quad \text{and} \quad T = \theta f T_0 \ . \qquad (18.24)$$

This must be compared with the usual result valid for an observer at rest in the thermodynamic system under discussion

$$\left(\frac{\partial S_0}{\partial v_0}\right)_{E_0} = \frac{p_0}{T_0} \ .$$

We must therefore ask what functions are $\theta(\gamma)$ and $f(\gamma)$, for they determine how thermodynamic relations are amended if thermodynamic observations are made from a relatively moving frame of reference.

Three answers have been proposed:

(i) Thermodynamics is form-invariant. Then by (18.20) and (18.24)

$$\theta(\gamma) = 1, \quad \gamma\theta(\gamma)f(\gamma) = 1, \quad \text{i.e. } f(\gamma) = 1/\gamma.$$

It follows from (18.24) that

$$T = \frac{T_0}{\gamma} \leq T_0 \ . \qquad (18.25)$$

A moving body thus appears cooler than a stationary one. This Planck-Einstein theory of 1907 has been criticized in recent proposals .

(ii) The second law is invariant ($\theta(\gamma) = 1$) and $f(\gamma) = \gamma$. It then follows from (18.24) that

$$\left(\frac{\partial S}{\partial v}\right)_{E,P} = \gamma^2 \frac{p}{T}$$

so that some thermodynamic relations are changed for a moving frame of reference. It follows from (18.24) that

$$T = \gamma T_0 \geq T_0 \qquad (18.26)$$

A moving body thus appears hotter than a stationary one.

(iii) Temperature is invariant ($\theta f = 1$) and $f(\gamma) = 1/\gamma$. In this case

$$\theta(\gamma) = \gamma$$

and some thermodynamic relations, for example the second law, are affected by passing to a moving frame of reference.

There is no explicit consensus on these matters. Some people claim that there is no need to ask questions about these transformations, others claim that all answers stated above are permissible, and yet others take the views (i), (ii), or (iii). What is certain is that the literature is somewhat confused in places and that work is proceeding in this area. Corresponding work on relativistic statistical mechanics and kinetic theory is also in progress.

We shall proceed to supply arguments for the choice of $f(\gamma)$ and $\theta(\gamma)$, although these are possibly not universally accepted as unique. First we shall provide an argument for the choice $f(\gamma) = 1/\gamma$ which rules out view (ii). Consider an inertial frame $I_0$. An example of a non-inertial frame would be the inside of a car which is cornering, as a free body would not move in a straight line in such a car. More precise definitions of inertial frames are given in books on relatively. Let two systems $(\alpha,\beta)$, both at rest in $I_0$, be subject to a slow and small transfer of heat $Q_{\alpha 0}$ from $\beta$ to $\alpha$.

The changes in energy and momentum are, for $\alpha$ and $\beta$ in $I_0$:

$$\Delta E_{\alpha 0} = Q_{\alpha 0} = -\Delta E_{\beta 0}$$

$$\Delta \mathbf{p}_{\alpha 0} = \Delta \mathbf{p}_{\beta 0} = 0 \ .$$

This states that the energy lost by one body is gained by the other and that they are both at rest in $I_0$. For an inertial frame $I$ both systems have velocity $v$ (say) along the positive $x$-direction. In $I$ we have, by analogy with the space and time Lorentz transformation,

$$P_x = \gamma\left(P_{x0} + \frac{v}{c^2}E_0\right), \quad P_y = P_{y0}, \quad P_z = P_{z0}$$

$$E = \gamma\left(E_0 + \frac{v^2}{c}P_{x0}\right) \ .$$

Omitting the subscripts $x,y,z$, as only $P_x$ is tranformed, one finds for the changes in energy and momentum of the bodies $\alpha,\beta$ as judged in $I$

$$\Delta E_\alpha = \gamma\Delta E_{\alpha 0} = \gamma Q_{\alpha 0} = -\Delta E_\beta$$

$$\Delta P_\alpha = \frac{v}{c^2}\gamma Q_{\alpha 0} = -\Delta P_\beta \ .$$

Thus in $I$ the systems are in motion and their change in energy implies a change of relativistic mass and so a change in momentum. This in turn implies the action of a force $f = \mathrm{d}P/\mathrm{d}t$ in the $x$-direction where $f,P$, and $t$ are measured in $I$. This means that in passing up heat $\beta$ does some work on $\alpha$. The amount in $I$ is, for a movement through the distance $\Delta s$ in the $x$-direction,

$$\Delta W_\alpha = f\Delta s = \frac{\Delta P_\alpha}{\Delta t}\Delta s = \frac{\Delta s}{\Delta t}\Delta P_\alpha = v\Delta P_\alpha = \frac{v^2}{c^2}\gamma Q_{\alpha 0} \ .$$

The amount of energy gained by system $\alpha$ therefore contains an amount of mechanical work $\Delta W_\alpha$. We now *define* the amount of heat gained in frame $I$ by

$$\Delta Q_\alpha \equiv \Delta E_\alpha - \Delta W_\alpha = \gamma Q_{\alpha 0} - \frac{v^2}{c^2}\gamma Q_{\alpha 0} = \frac{Q_{\alpha 0}}{\gamma}.$$

This gives precisely $f(\gamma) = 1/\gamma$.

Next we provide an argument for $\theta(\gamma)f(\gamma) = 1$, i.e. for $T = T_0$. This then determines

$$\theta(\gamma) = 1/f(\gamma) = \gamma.$$

Suppose for simplicity that the transformation law for temperature is

$$T = \gamma^a T_0 .$$

The moving body has then the temperature $\gamma^a T_0$ in $I$ and it has the temperature $T_0$ in $I_0$. Consider two identical bodies $\mu,\nu$ in uniform relative motion, each associated with an observer sitting on it. Then each observer sees the other body as hotter if $a>0$ and would therefore expect to gain heat from it if the bodies are allowed to interact. Each would expect to lose heat to the other body if $a<0$. Such heat flows lead to effects which can be stated in a Lorentz-invariant way. If $a<0$ then in $\mu$'s rest frame $\mu$'s temperature should increase at the expense of $\nu$'s temperature. This leads to a change in rest-frame temperatures as indicated:

$$T_{\mu 0}\uparrow, \quad T_{\nu 0}\downarrow \quad (\mu\text{'s argument}).$$

By a similar argument based on $\nu$'s rest frame one would expect similarly

$$T_{\nu 0}\uparrow, \quad T_{\mu 0}\downarrow \quad (\nu\text{'s argument}).$$

This contradiction is avoided only if $a = 0$. Whereas in special relativity theory each inertial observer judges a moving clock as slow, no energy flow is expected to result.

If, however, observers judge moving bodies as of lower temperature (for example) than the same body when at rest, then this is expected to imply heat flows in some sense. This leads to the question of wider ranging relativistic definitions of heat flows etc. which are beyond the present scope. In any case, definitions leading to non-invariant temperatures are possible, but (in the present author's view) not desirable. Appendix IV deals with some additional topics

## 18.6. GENERAL RELATIVISTIC EFFECTS

We now turn to general relativity. What is thermal equilibrium? In order of increasing complexity one might answer:

(a) Thermal equilibrium is equivalent to having a system at uniform temperature throughout.

(b) Heat does not flow between any two points of the system.

(c) Modify any two regions of the system to be heat reservoirs; than a Carnot engine working between them produces no mechanical work, i.e.it has zero efficiency.

It is one of the charms of general relativity that it leads one to unexpected modifications of normal physical ideas (such as those mentioned above). We shall see that in a gravitational field (a) is wrong, (c) remains correct, and we shall not discuss (b): it involves relativistic irreversible thermodynamics and is as such beyond our scope.

In order to see the effect of a gravitational field, let its gravitational potential be $\phi_u$, $\phi_\ell$ ($\phi_u > \phi_\ell$) respectively at an 'upper' and 'lower' level. The lower level corresponds to the hot reservoir, the upper to the cold reservoir, of the normal Carnot cycle, as will be seen in equation (18.27). We need a reservoir $R$ which is always at the temperature $T_u$ of the upper level and is initially resting there (see Table 18.1 step 1). It gives some heat energy $Q_u$ to this level (step 2), and is then lowered to level $\ell$ which is at temperature $T_\ell$ (step 3). Such a process does not change the entropy of $R$ if it is conducted by a familiar friend, the quasistatic adiabatic process. We now want to return $R$ to its original thermal state by giving it heat $Q_u$ so as to replenish its entropy by that lost earlier. To achieve this a Carnot engine is arranged to work quasi-statically between the lower level and the reservoir. There is no change of entropy in this process so that the heat supplied to $R$ is given by $Q_u/T_u = Q_\ell/T_\ell$. Of course the engine performs work which is as usual $Q_\ell - Q_u$(step 4). The whole system is now returned to its original state by raising $R$, now in its original thermal state, to level u (step 5).

One can find the total work liberated, and hence one can define an efficiency for the cycle by

$$\eta \equiv \frac{W}{Q_\ell} = 1 - \frac{T_u}{T_\ell} - \frac{\phi_u - \phi_\ell}{c^2}\frac{T_u}{T_\ell} \quad . \tag{18.27}$$

One sees that the lower level corresponds to the 'hot' reservoir, for we then have the usual Carnot efficiency when there is no gravitational field ($\phi_u = \phi_\ell$). Now (a) and (c) are seen to be contradictory: If equilibrium means $T_u = T_\ell$, then the efficiency $\eta$ is meaningless (unless $\phi_u = \phi_\ell$). Negative efficiencies can arise, and this possibility must be rejected. If equilibrium means $\eta = 0$, then $T_u \neq T_\ell$, in fact

$$\frac{T_\ell}{T_u} = 1 + \frac{\phi_u - \phi_\ell}{c^2}$$

which is not unity if $\phi_u \neq \phi_\ell$. As this is not unreasonable, we adopt $\eta = 0$ as the definition of thermal equilibrium.

For $\phi/c^2 \ll 1$, one can write this as

$$T_\ell\left(1 + \frac{\phi_\ell}{c^2}\right) = T_u\left(1 + \frac{\phi_u}{c^2}\right) \quad .$$

Thus thermal equilibrium in a region implies that for small and stationary gravitational fields

$$T_0 \equiv f(\mathbf{r})\, T(\mathbf{r}) \quad \text{is independent of } \mathbf{r} \text{ in the region,} \tag{18.28}$$

where $f(\mathbf{r}) \equiv 1 + \dfrac{\phi(\mathbf{r})}{c^2}$ .

This relation is due to R.C. Tolman. Here $T_0$ is a so-called

TABLE 18.1

*A gravitational thermodynamic cycle*

$R$ : A thermal reservoir at temperature $T_u$ 　　 [C.E.] : A Carnot engine

| | | | | | |
|---|---|---|---|---|---|
| $\phi_u$, $T_u$ upper level | Ⓡ | $Q_u$ ↖ Ⓡ | ↓ | $Q_u = \frac{T_u}{T_\ell} Q_\ell$ | Ⓡ ↑ |
| $\phi_\ell$, $T_\ell$ lower level | | | Ⓡ | $Q_\ell$ → [C.E.] $\xrightarrow{Q_u}$ Ⓡ | |
| Step number (Final states are shown) | 1 | 2 | 3 | 4 | 5 |
| Reservoir — Mass: | $M$ | $M - Q_u/c^2$ | | $M$ | $M$ |
| Reservoir — Entropy: | | $S - Q_u/T_u$ | | $S$ | $S$ |
| Process liberates work | - | - | $\left(M-\frac{Q_u}{c^2}\right)(\phi_u-\phi_\ell)$ | $Q_\ell - Q_u = \left(1-\frac{T_u}{T_\ell}\right) Q_\ell$ | $-M(\phi_u-\phi_\ell)$ |

$$W = \left[\left(1 - \frac{T_u}{T_\ell}\right) - \frac{\phi_u-\phi_\ell}{c^2}\frac{T_u}{T_\ell}\right] Q_\ell$$

local temperature, namely that which has to be uniform in equilibrium. In contrast, we call $T$ the 'measured' temperature. In the absence of gravitation $T = T_0$. (18.28) states that the 'lower lying' particles in an equilibrium system are part of the hotter material, as expected from the sedimentation property of mass-energy and therefore of heat in a gravitational field .

Using a metric $(-+++)A$ ($A = \pm 1$ depending on convention) one knows from general relativity that for weak gravitational fields

$$f = 1 + \phi/c^2 \sim \sqrt{(1 + 2\phi/c^2)} \sim \sqrt{(-A\, g_{00})} \qquad (18.29)$$

where $g_{00}$ is the time component of the metric tensor $g_{ij}$ at the observer. In relativity one has to distinguish the energy $U$ of a small part of a body as measured by a stationary observer and the local (i.e. flat space-time) energy $U_0$ entering the law of energy conservation. The temperature $T$ measured by an observer is then obtained from $U$ while the temperature $T_0$, which is uniform in thermal equilibrium, is obtained from $U_0$.

The analogous entropies are assumed equal ($S = S_0$); since entropy depends on the internal state of the body it is usual to assume it independent of weak gravitational fields (by no means a convincing procedure). In any case, one finds for constant external parameters

$$\frac{1}{T_0} = \left(\frac{\partial S}{\partial U_0}\right)_{v,n} , \quad \frac{1}{T} = \frac{f}{T_0} = f\left(\frac{\partial S}{\partial U_0}\right)_{v,n} = \left(\frac{\partial S}{\partial U}\right)_{v,n} .$$

This suggests that we can put

$$U_0 = Uf . \qquad (18.30)$$

In more general theories $f$ depends on the components of the metric tensor, and in the absence of gravitational fields $U = U_0$.

We now turn to the chemical potential $\mu(\mathbf{r})$, assuming that the number of particles in the system are not affected by a weak gravitational field. Then

$$\frac{1}{\mu} = \left(\frac{\partial n}{\partial U}\right)_{S,v} = f\left(\frac{\partial n_0}{\partial U_0}\right)_{v,S} = \frac{f}{\mu_0} \tag{18.31}$$

It follows that in a region at thermal equilibrium

$$\mu_0 = f(\mathbf{r})\ \mu(\mathbf{r}) \quad \text{is independent of } \mathbf{r}\ .$$

This relation is due to O. Klein and is quite analogous to the Tolman relation.

Consider now a rigid disc rotating with constant angular velocity $\omega$ about its centre. What is the equilibrium temperature distribution? There is no gravitational field, but one can introduce one by adopting a frame of reference in which the disc is permanently at rest. In this frame of reference there then acts an outward radial field of force equal to the familiar centrifugal force. It has at distance r from the centre the magnitude

$$\frac{mv^2}{r} = m\omega^2 r\ ,$$

when acting on a mass $m$. This corresponds to a potential energy

$$-\frac{1}{2} m\omega^2 r^2 = m\phi(r) \tag{18.32}$$

which decreases as $r$ increases, thus indicating that it leads to outward motion. Thus $\phi(r)$ is the radial gravitational potential which describes the dynamics of a particle on the disc. Substituting (18.32) into (18.28) yields the equilibrium condition at a general radius $r$:

$$T(r)\left[1 - \frac{1}{2}\omega^2 \frac{r^2}{c^2}\right] = T(0)\ . \tag{18.33}$$

This result remains valid on returning to the frame of reference in which the disc is rotating, since this is a physically equivalent situation. In fact, the equivalence principle, expressed loosely, states that a uniform gravitational field with no acceleration in one frame and the corresponding uniform acceleration in a system without gravitation give equivalent results. Equation (18.33) indicates a sedimentation of heat energy near the outer radius of a rotating disc, and can be understood in terms of $E = mc^2$, i.e. the inertial property of heat energy. The faster moving particles of the disc are thus part of the hotter material.

The following important amendments which general relativity and cosmology contribute to classical thermodynamics should be noted. First, thermal equilibrium is governed by (18.28) rather than by $T$ = constant. Secondly, in cosmological models one frequently uses a gas of matter and/or radiation which is supposed to fill the whole universe. Such a thermodynamic system is *unenclosed* in the sense that there is no friction with the wall and no pressure drop at the wall because there is no enclosure. In such a model it appears possible to have non-static processes at constant entropy which were not envisaged by classical thermodynamics. Thirdly, since energy conservation as such does not hold in general relativity it appears possible to have cosmological models such that the processes which take place in them raise the entropy without any equilibrium state being approached. These matters, still subjects of current research, are, however, beyond the scope of this book.

## 18.7. A MAXIMUM TEMPERATURE?

Relatively simple density of states functions $N(E)$ are adequate to arrive at important results in statistical thermodynamics, and attention has been confined to them in this book. In both nuclear physics and solid state theory more complicated functions appear. For electrons in metals and semiconductors they are inferred either from experiments or from the band theory of solids. While these studies are outside the present scope, an exponential spectrum

$$N(E) = AE^s \exp\left(\frac{E}{kT_0}\right) \qquad (A, T_0 > 0)$$

is a natural, and still simple, generalization of the spectra used in this book. Here $A$, $s$, $T_0$ are constants. It is immediately clear that if the *canonical* partition function

$$Z = A \int^{\infty} E^{s} \exp\left[E\left(\frac{1}{kT_0} - \frac{1}{kT}\right)\right] \mathrm{d}E$$

of such a system is to converge, then $T < T_0$, i.e. $T_0$ plays the part of maximum temperature. Evidence for the existence of such a situation comes from theory and experiments on high-energy collisions of hadrons, when the relativistic energy must be used for $E$, and one finds $T_0 \sim 200$ MeV. This corresponds to temperatures of order $2 \times 10^{12}$K which are believed to have existed in the very early universe. They lie near the very limit where present physical theory can be applied. Such a spectrum means that, on increasing the acceleration for collision events, the kinetic energy of the particles produced will eventually remain constant and the additional energy available will be absorbed by the creation of new particles. It has been suggested that the existence of a maximum temperature might be the subject of a new law of thermodynamics. In our method of presenting thermodynamics the third law deals with all matters connected with the boundary points in thermodynamic phase spaces, and hence the existence of a maximum temperature may be appropriately regarded as coming within the ambit of a reformulated third law.

## 18.8. SYMMETRIES IN THERMODYNAMICS AND STATISTICAL MECHANICS

It is often possible to choose new variables $x_i'$ in the equations of motion of a system as studied in classical or quantum mechanics or in statistical mechanics. The new variables are then functions of the old ones $x_i' = x_i'(x_1, x_2, \ldots)$ . If the system has a Lagrangian density $L(x_1, x_2, \ldots)$, this is replaced by a new Lagrangian density $L'(x_1', x_2', \ldots)$ such that the *functional form* of $L'$ is different, though for corresponding values the *numerical values* are the same:

$$L'(x_1', \ldots) = L(x_1, \ldots) \quad .$$

Symmetry transformations $x_i \to x_i'$ are of such a type that the equations of motion derived from $L$ and from $L'$ are identical. Most of the relevant symmetry transformations can be generated by iteration of infinitesimal transformations. It is the invariance of the equations of motion under these infinitesimal transformations which leads to conservation laws by virtue of a theorem associated with the name of Emmy Noether. In Newtonian and special relativistic physics space-time is (i) homogeneous in time, (ii) homogeneous in space, (iii) isotropic in space, and (iv) descriptions of a system in different inertial frames are identical. These symmetries lead for closed systems to conservation of (i) energy, (ii) linear momentum, (iii) angular momentum and (iv) constancy of the centre-of mass velocity respectively. In this sense the first law of thermodynamics can in an extended form be regarded as asserting these four symmetries. It then fails in some sense in general relativity since in that theory space-time is 'curved' and hence neither homogeneous nor isotropic.

From the Lorentz transformation between energy and momentum one would expect the Boltzmann factor $\exp(-E/kT)$ so characteristic of the whole of statistical mechanics to be reinterpreted as $\exp(-E_0/kT_0)$ where the suffix 0 denotes a measurement in the rest frame. In any other inertial frame this factor becomes

$$\exp\left\{-\frac{\gamma}{kT_0}\left(E-\frac{v^2}{c}P_x\right)\right\}$$

so that relativistic statistical mechanics would look rather different from the *statistical mechanics of the rest frame*, which is what we have studied in this book.

The second law of thermodynamics corresponds to the $H$-theorem in statistical mechanics as discussed in section 9.5. There we saw in equations (9.32) and (9.34) that if $\rho$ is the fine-grained density matrix and $Y$ is the operator representing an observable, then its expectation value has an ensemble average $y = Tr(\rho Y) = Tr(U\rho U^{-1}\, UYU^{-1})$ which is invariant under rotations in Hilbert space induced by a unitary operator $U$. In the case of complete quantum-mechanical information it leads to fine-grained phase-space distributions which have constant entropy, and so they neither diffuse (increasing the entropy) nor coalesce (decreasing the entropy) with lapse of time. However, note that (9.42) holds not only for time translation but for *any* unitary transformation:

$$Tr\{US_0U^{-1}\ \ln(US_0U^{-1})\} = Tr\{S_0 \ln S_0\}.$$

Violation of this 'unitary' symmetry arises from incomplete information. This leads to coarse graining, to entropy increase, and hence to diffusion of phase-space probability distributions with lapse of time. Thus we can regard 'unitary' symmetry as leading to entropy conservation.

Thus statistical thermodynamics like many other areas of physics can be related to symmetry principles, as was already noted in Chapter 7 in connection with phase transitions, and illustrated here by reference to energy and entropy.

## 18.9. ORDER FROM FLUCTUATIONS AND GROWTH

In irreversible thermodynamics the entropy gain in time $\delta t$ of an open system is written as

$$\delta S = \delta_e S + \delta_i S \quad (\delta_i S \geq 0) \tag{18.34}$$

where $\delta_e S$ is the entropy supplied from the outside. The special cases

$$\delta_e S = 0 \quad \text{and} \quad \delta S = 0$$

represent respectively (i) a formulation of the normal second law and (ii) the possibility of maintaining any state steady. Such a state can in particular be an orderly non-equilibrium state, and it can be kept steady by pumping negative entropy

$$\delta_e S = -\delta_i S$$

into it. If this quantity is large, or if the state remains reasonably steady for a time long compared with the molecular relaxation times, the system is called *dissipative*. Indeed our use of equation (13.29) can be justified in terms of (18.34). If one drives a system to a state which is far from equilibrium, previously stable states become unstable: the *control parameter* (s) enters a new range of values which can open up new stable values for an *order parameter*. If the equations are non-linear (cf. section 7.8-7.11) quite new phenomena can be initiated by the phase transition. These new states can be reached by fluctuations about a stable state when they reach the neighbourhood of an unstable state. They are then amplified and the new regime establishes itself by virtue of its interaction with its surroundings. If these new modes are 'orderly' as in hydrodynamic phenomena (vortex formation, Bénard cells), or in oscillating chemical reactions, one has a production of order via fluctuations, a concept pioneered by I. Prigogine and his collaborators.

The word 'order' has of course to be used with care. It can, however, be given a quantitative definition which differs from that of entropy thus allowing *order and entropy* to increase together in special cases. These cases, which have not yet been discussed widely, can be considered to represent the emergence of *order from growth*. Some relevant mathematical concepts are given in Appendix I.5.

PROBLEMS

18.1. Show that for a neutral system of $2N$ particles of charges $q_i$ ($q_i = \pm q$) the characteristic temperatures introduced in section 18.3 satisfy

$$\tau_1 = 2\tau_2 = \frac{aq^2}{sk}\ .$$

18.2. Show that the equation of state for the system discussed in section 18.3 becomes identical with that for an ideal gas (above some characteristic temperature) if $L = L_0$.

(Note that the occurrence of the container dimension in the interaction is unphysical.)

18.3. Use equations (13.26), (18.17), and (18.18) to obtain the optimum values of $x_G$ and the corresponding solar cell efficiency. Assume the irradiation is black-body at temperature $T_S$ and solve (18.18) numerically to find the optimum value $x_G = 2{\cdot}17$ and hence the corresponding efficiency of 47%.

(Other factors serve to bring down the efficiency further. For example, not all electron-hole pairs created by the light are able to reach the electrodes because of recombination and inefficient collection.)

# PART C: APPENDICES

# Appendix I Information theory

## I.1. THE IDEA OF 'INFORMATION'

In this section we shall derive an expression for the 'information' of a system by statistical considerations. The equation will be identical in form to the expression for the entropy

$$S = -k \sum_{i=1}^{N} p_i \ln p_i$$

where $k$ is Boltzmann's constant, $p_i$ is the probability of the system being in state $i$, and $N$ is the total number of equilibrium states which the system can achieve. To do this we shall need to be able to quantify the notion of 'information', regarded as a technical term. This problem was first solved by an American, Claude Shannon, who in 1949 laid down the foundations of information theory.

We look for a function $I(p)$ satisfying the general conditions expressed in Fig. I.1. The amount of surprise generated

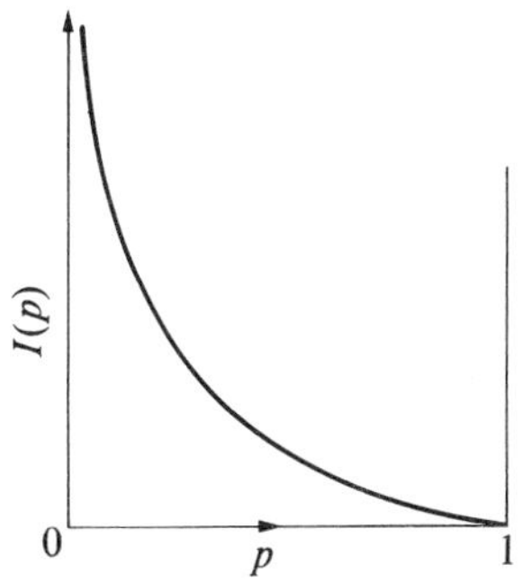


Fig.I.1. The schematic relationship between the information $I(p)$ obtained if an event having *a priori* probability $p$ is known to have occurred.

i.e. of information conveyed, when a *certain* event has occurred is zero. For example, if the sun is reported to have risen yesterday, the information conveyed is zero. Hence we put $I(1) = 0$. However, the information conveyed if less probable events are reported is greater, and $I(p)$ should therefore rise as $p$ decreases.

Try two possible functions:

$$\text{(i)}\quad I(p) = 1 - p. \quad \text{Then} \begin{cases} p = 0 \Rightarrow I = 1 \times \\ p = 1 \Rightarrow I = 0 \end{cases}$$

$$\text{(ii)}\quad I(p) = 1/p\ . \quad \text{Then} \begin{cases} p = 0 \Rightarrow I = \infty \\ p = 1 \Rightarrow I = 1 \times \end{cases}$$

Both of these formulae are unsatisfactory for the values of $p$ marked by a cross. The formula which we shall adopt is

$$\underline{I(p) = -a \ln p\ ,} \qquad a > 0\ .$$

This satisfies the conditions that $p = 0 \Rightarrow I = \infty$; $p = 1 \Rightarrow I = 0$.

## I.2. INDEPENDENT AND DEPENDENT EVENTS

If two unbiased coins each with a 'head' (H) and a 'tail' (T) are tossed then there are four possible outcomes denoted by HH, HT, TH, and TT, each of probability $\frac{1}{4}$, where the probability $p$ is defined as the ratio of the number of favourable events divided by the number of possible outcomes.

Now let the events E, F, E', and F' be defined as in Table I.1.

TABLE I.1

| | Event class | Favourable outcomes | Probability |
|---|---|---|---|
| $E$ , | no more than one head | (HT; TH; TT) | $\frac{3}{4}$ |
| $F$ , | at least one of each face | (HT; TH) | $\frac{2}{4}$ |
| $E'$, | first coin is 'head' | (HT; HH) | $\frac{2}{4}$ |
| $F'$, | second coin is 'head' | (TH; HH) | $\frac{2}{4}$ |

The probability that an outcome is found which is both $E$ and $F$ is denoted by $p(EF)$. Hence $p(EF) = 2/4$. Also $p(E) \times p(F) = 3/8$. Therefore the product of the probabilities of the separate events is not equal to the probability of both events occurring together. We say that such events are *dependent*. However, since $p(E'F') = p(E')p(F')$, the events $E'$ and $F'$ are said to be *independent*.

## I.3. A DERIVATION OF $I(p)$

Now let us demand that for two *independent* events

$$\underline{I(pq) = I(p) + I(q)}$$

where $p$ and $q$ are the probabilities of these events with $0 \leq p \leq 1$ and $0 \leq q \leq 1$. This condition states that the information gained can be added if two independent events both occur together. Then, differentiating with respect to $p$ at constant $q$, we have

$$qI'(pq) = I'(p)$$

where

$$I'(pq) = \mathrm{d}I(pq)/\mathrm{d}(pq) \ .$$

Differentiating with respect to $q$,

$$pI'(pq) = I'(q)$$

that is

$$pqI'(pq) = pI'(p) = qI'(q)$$

and this is a constant, $-c$ say, since the right-hand sides cannot depend on either $p$ or $q$. Integrating we have

$$\int \mathrm{d}I(p) = -c \int \frac{\mathrm{d}p}{p}$$

that is

$$I(p) = -c \log p + d \ .$$

Since

$$I(pq) = -c \log pq + d = -c \log p + d - c \log q + d$$

it follows that $d$ vanishes. Using logarithms to the base $b$, we have that

$$I(p) = -c \log_b p \ .$$

In information theory it is usual to choose $c = 1/\log_b 2$ so that

$$I(p) = \frac{-\log_b p}{\log_b 2} = -\log_2 p$$

since $\log_d A = \log_b A/\log_b d$. Then it follows that for two events of equal probability $\frac{1}{2}$

$$I(\tfrac{1}{2}) = -\log_2(\tfrac{1}{2}) = \log_2 2$$

$$= 1 \text{ bit of information.}$$

A bit is a *B*inary dig*IT*, and ordinary numbers are represented in binary digit form as shown in Table I.2.

As an application of this simple theory, note that it is now easy to see how one can find any integer in the range 0,1,...,8,9 by asking four questions to which the answer must be 'yes' or 'no'. We simply stipulate that the number be represented in the binary system, and then ask:

Is the first digit 0?
Is the second digit 0?
Is the third digit 0?
Is the fourth digit 0?

Each answer, yes or no, determines the digit, and four questions will always suffice. The information obtained by

TABLE I.2

| $2^3$ | $2^2$ | $2^1$ | $2^0$ | ordinary number |
|---|---|---|---|---|
| 0 | 0 | 0 | 0 | 0 |
| 0 | 0 | 0 | 1 | 1 |
| 0 | 0 | 1 | 0 | 2 |
| 0 | 0 | 1 | 1 | 3 |
| 0 | 1 | 0 | 0 | 4 |
| 0 | 1 | 0 | 1 | 5 |
| 0 | 1 | 1 | 0 | 6 |
| 0 | 1 | 1 | 1 | 7 |
| 1 | 0 | 0 | 0 | 8 |
| 1 | 0 | 0 | 1 | 9 |

knowing which of the 10 *a priori* equally likely events has occurred is

$$I\left(\frac{1}{10}\right) = -\log_2 \left(\frac{1}{10}\right) = \log_2 10 = \frac{\log_{10}10}{\log_{10}2} = \frac{1}{0\cdot3010} = 3\cdot32 \ .$$

Thus three yes-no questions are *not* enough to extract the required information with certainty, *however the questions are framed*. But, with the correct sort of questions, four questions will always suffice.

## I.4. CONNECTION BETWEEN ENTROPY AND INFORMATION

If a certain situation has $n$ possible outcomes of *a priori* probabilities $p_1, p_2, \ldots, p_n$, the information obtained if the $i$th outcome actually occurs is $I(p_i)$. The expected, or average, information obtained per trial in a large number of trials is then

$$\sum_{i=1}^{n} p_i I(p_i) = -c \sum_{i=1}^{n} p_i \log p_i$$

and this is just the expression for the entropy. A great deal has been written on the relation between entropy and information, but we shall not go beyond this point.

## I.5. CONNECTION BETWEEN ENTROPY AND 'ORDER'

A great deal has also been written on the relation betwe 'order' and entropy. In equilibrium one has the largest disorder (in some general ill-defined sense), and so people often regard disorder and entropy as equivalent. This is unsatisfactory, and we shall define a 'disorder' of a system of $n$ distinguishable states to be a *multiple* of the entropy

$$D(n) = S(n)/a(n) \tag{I.1}$$

where $a(n)$ is undetermined. On combining two identical non-interacting systems into one system, we know that $n$ is replaced by $n^2$ (since each state of the first system may be combined with each state of the second system) and $S(n)$ by $S(n^2) = 2S(n)$ so that in this case the disorder is

$$D(n^2) = \frac{S(n^2)}{a(n^2)} = \frac{2S(n)}{a(n^2)} \quad . \tag{I.2}$$

However, it is reasonable to expect 'disorder' to be an intensive variable. The combination of two identical systems, of disorder $D(n)$ each, should yield a new system of the same disorder. Thus we require

$$D(n^2) = D(n) \ . \tag{I.3}$$

Results (I.1) to (I.3) yield an equation for the unknown function:

$$a(n^2) = 2\ a(n) \qquad \text{(all positive integral } n) \ .$$

It has been studied in the form (9.2) and implies

$$a(n) = c \ln n$$

where $c$ is a constant. It follows that the constraint (I.3) yields

$$D(n) = \frac{S(n)}{c \ln n} = -\frac{\Sigma_{i=1}^{n} p_i \ln p_i}{\ln n} = -\sum_{i=1}^{n} p_i \log_n p_i \tag{I.4}$$

where $c$ has been chosen to be Boltzmann's constant to make $D(n)$ dimensionless, and $\log_n p_i \equiv \ln p_i / \ln n$ has been used.

In a biological system growth entails an increase of $n$ with time and the rate of change of disorder is, by (I.4),

$$\dot{D}(n) = \left\{\frac{\dot{S}(n)}{S(n)} - \frac{\dot{n}}{n \ln n}\right\} D(n) \; . \tag{I.5}$$

Thus one can have decreasing disorder, even though the entropy increases. The condition for this is that the second term in (I.5) is greater than the first term in the braces of (I.5) i.e. that $S$ should increase less rapidly with $n$ than $k \ln n$ .

PROBLEMS

I.1. In a certain system entropy increases and disorder decreases according to

$$\dot{S}(n) = s\, S(n) \; , \qquad \dot{D}(n) = -d\, D(n)$$

where $s$ and $d$ are positive constants. Prove that if $n = n_0$ at $t = t_0$ then $n$ increases according to

$$\ln n = (\ln n_0) \exp\{(s+d)(t-t_0)\} \; .$$

I.2. Two non-interacting systems of entropies $S_i(N_i)$ and disorders $D_i$ ($i = 1,2$) are combined to form a system of entropy $S$ and disorder $D$. Prove that

$$D = q_1 D_1 + q_2 D_2 \qquad (q_1 + q_2) = 1$$

where

$$q_i \equiv \frac{\ln N_i}{\ln N_1 N_2} \qquad (i = 1,2) \; .$$

# Appendix II The algebra of Lagrangian multipliers

Suppose $m$ variables $x_j$ $(j = 1,2,\ldots,m)$ are to satisfy the relation

$$\sum_{j=1}^{m} A_j x_j = 0 \tag{II.1}$$

for all values of the $x_j$'s, subject only to $k$ relations between the variables:

$$\sum_{j=1}^{m} B_{rj} x_j = 0 \quad (r=1,2,\ldots,k) . \tag{II.2}$$

It is desired to find the relations which the $A$'s must satisfy. Note that the $A$'s and $B$'s can be functions of the $x$'s.

Multiply the relations (II.2) by undetermined multipliers $\lambda_r (r=1,2,\ldots,k)$, and add the resulting equations (II.2) to (II.1), so that

$$\sum_{s=1}^{m} \sum_{r=1}^{k} (A_s + \lambda_r B_{rs}) x_s = 0. \tag{II.3}$$

By (II.2) $k$ of the $x_j$'s can be determined in terms of the remainder, say $x_{k+1}, x_{k+2},\ldots,x_m$, which are arbitrary. Let the $k$ arbitrary $\lambda$'s be determined by the relation

$$A_s + \sum_{r=1}^{k} \lambda_r B_{rs} = 0 \quad (s=1,2,\ldots,k) . \tag{II.4}$$

The only variables $x_j$ which survive in (II.3) are then the arbitrary ones. The condition that (II.3) be satisfied for these arbitrary $x$'s is that each surviving coefficient vanishes separately. This yields (II.4) for $s = k+1, k+2,\ldots,m$. Thus (II.4) is the relation to be satisfied by *all* $A_s$, i.e. for $s=1,2,\ldots,m$.

In this book the method described here is used for entropy maximization under constraints such as (II.2). It is therefore useful in determining equilibrium states. This procedure certainly works in normal thermodynamics whose extensive functions are homogeneous of order unity:

$$f(\lambda x) = \lambda f(x) \text{ for all } \lambda > 0.$$

For example functions such as entropy and energy are doubled when the system is doubled. However homogeneity can fail, as noted on p. 102, when the heat capacity is negative, in the case of black holes (p. 227), etc. In such cases more sophisticated arguments would appear to be required to determine equilibrium states, but these are beyond the present scope.

PROBLEM

II.1. (i) Find the maximum of the function $f = x_1^2 x_2^2 \ldots x_n^2$ on the hypersphere $g \equiv \Sigma_{i=1}^{n} x_i^2 = c^2 > 0$.

(ii) Hence show that the geometric mean $(A_1 A_2 \ldots A_n)^{1/n}$ cannot exceed the arithmetic mean for any $n$ positive numbers

$$A_1, A_2, \ldots, A_n$$

# Appendix III Solid angles

A solid angle is the area intercepted by a cone of arbitrary base area on the surface of a sphere whose radius is unity and whose centre is at the vertex of the cone.

In spherical polar co-ordinates an element of area on a sphere of radius $r$ is (Fig.III.1).

$$r\,d\theta\;r\sin\theta\;d\phi = r^2\sin\theta\;d\theta\;d\phi\;.$$

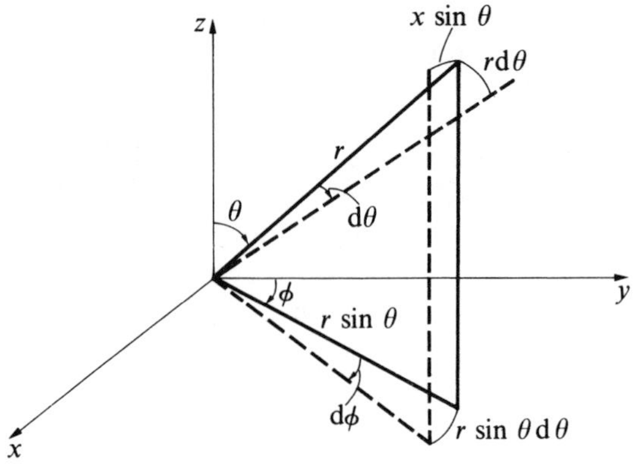


FIG. III.I. A solid angle in spherical polars.

Therefore, putting $r = 1$ for the unit sphere, an element of solid angle is

$$d\omega_1 = \sin\theta\;d\theta\;d\phi = -d(\cos\theta)d\phi\;. \qquad \text{(III.1)}$$

*Example III.1.* What is the solid angle subtended by an elemental spherical zone at the centre of the sphere?

For clarity, work with general $r$ first. The area intercepted on the sphere is the area for a right circular zone of semi-vertical zone $\theta+d\theta$ less the corresponding quantity for $\theta$. This area is

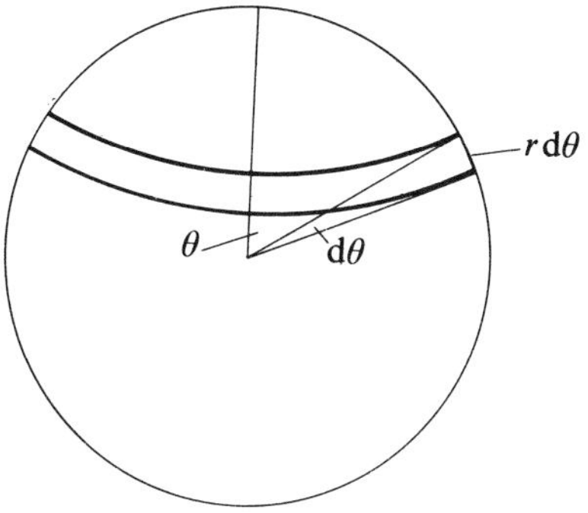


Fig.III.2. Solid angle of a spherical zone.

$$2\pi\ (r\sin\theta)r\ \mathrm{d}\theta = 2\pi r^2 \sin\theta\ \mathrm{d}\theta\ .$$

Therefore

$$\mathrm{d}\omega_2 = 2\pi \sin\theta\ \mathrm{d}\theta\ . \qquad \text{(III.2)}$$

Equation (III.2) can also be obtained from (III.1) by an integration in which $\theta$ is kept constant:

$$\mathrm{d}\omega_2 = \int_{\phi=0}^{\phi=2\pi} \mathrm{d}\omega_2 = \sin\theta\ \mathrm{d}\theta \times 2\pi\ .$$

*Example III.2.* What is the solid angle subtended by a right circular cone at the vertex? For general $r$ the area to be calculated is that of the spherical cap ABC. This is

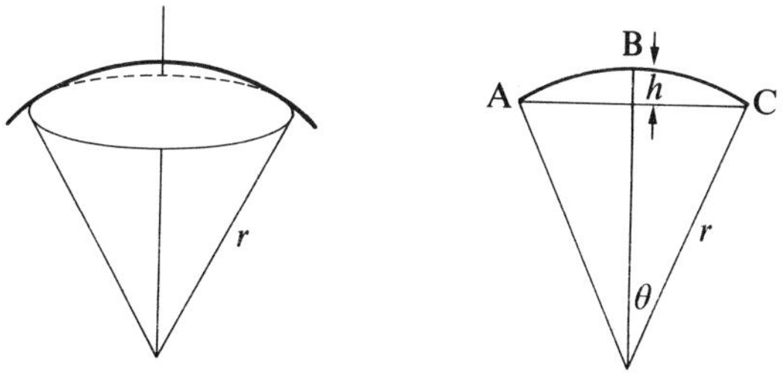


Fig.III.3. Solid angle of a spherical cap.

$$2\pi rh = 2\pi r(r - r\cos\theta) = 2\pi r^2(1-\cos\theta).$$

The solid angle is

$$\omega_3 = 2\pi(1-\cos\theta). \quad \text{(III.3)}$$

Equation (III.3) can also be obtained by integration of (III.2):

$$\omega_3 = \int_{\theta=0}^{\theta} 2\pi \sin\theta d\theta = -2\pi \Big|\cos\theta\Big|_0^\theta = -2\pi(\cos\theta - 1).$$

*Example III.3.* Consider a surface $S$ in space and an element d$S$ at $Q$. What solid angle does d$S$ subtend at a point P? Let $PQ \equiv \mathbf{r}$ and let $\hat{\mathbf{r}}$ be the corresponding unit vector. Let $\hat{\mathbf{n}}$ be the unit outward drawn normal. Let $\theta$ be the angle between $\mathbf{n}$ and $\mathbf{r}$. Then d$S$ cos $\theta$ is the area of the projection of d$S$ on a plane perpendicular to $\mathbf{r}$. The area intercepted on a unit sphere with centre P is therefore d$S \cos\theta/r^2$. Hence

$$d\omega = r^{-2}\cos\theta \, dS = \frac{\hat{\mathbf{n}}.\hat{\mathbf{r}}}{r^2} dS .$$

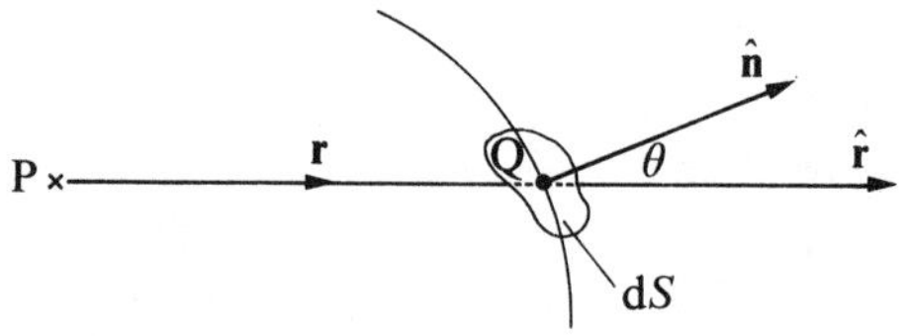


Fig.III.4. Solid angle subtended by an element of surface at a point.

# Appendix IV Special relativistic thermodynamics; some additional topics

## IV.1. THE SPECIAL RELATIVISTIC CARNOT CYCLE

The relativistic Carnot cycle can be represented by each of two diagrams (Fig.IV.1 and Fig.IV.2) which are drawn for the working fluid. Fig.IV.1 gives the pressure as a function of

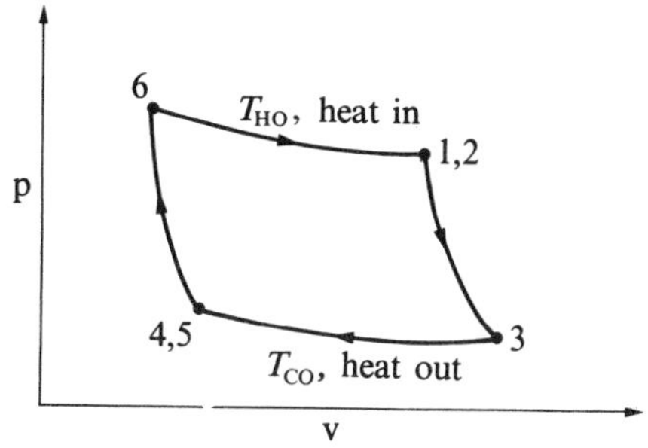


Fig.IV.1. Carnot cycle in terms of pressure and volume.

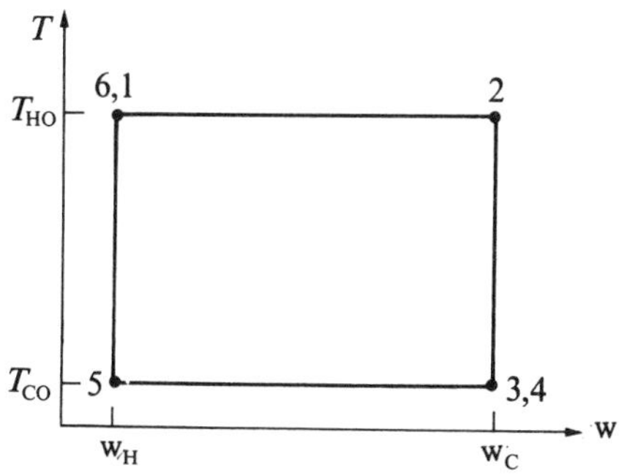


Fig.IV.2. Carnot cycle in terms of temperature and velocity.

volume, and Fig.IV.2. gives proper temperature against velocity of the fluid. Both figures are drawn for an inertial frame $I$ in which the reservoirs have constant velocities $w_H$ and $w_C$ respectively. The processes involved are as follows:

(i) Accelerations of the fluid so as to come to rest relative to the reservoir with which it is about to interact as in processes $1 \to 2$ and $4 \to 5$. The appropriate inertial frames are denoted as $I_C$ and $I_H$.

(ii) Quasi-static isothermal compression ($3 \to 4$) and expansion ($6 \to 1$) in contact with a reservoir with respect to which the fluid is at rest.

(iii) Quasi-static adiabatic expansion ($2 \to 3$) and compression ($5 \to 6$) in which the fluid temperature drops or rises.

If points 1 and 2 coincide, and points 4 and 5 coincide both reservoirs are at rest in the same inertial frame: $I_C = I_H$, and it is then usual to describe the process in that frame, denoted by

$$I = I_C = I_H \ .$$

This leads to the normal Carnot cycle and Fig.IV.2 collapses to a vertical line.

One of the simplest quantities to discuss is the entropy. If the initial and final proper entropies of the working fluids are $S_{i0}$ and $S_{f0}$ then

$$S_{f0} = S_{i0} + S_{12} + S_{C0} + S_{45} + S_{H0}$$

where $S_{i\,i+1}$ is the gain in proper entropy of the fluid in the process $i \to i+1$, and the gain in proper entropy by the fluid while in contact with the cold reservoir is

$$S_{34} = S_{C0} \ (< 0) \ .$$

This is negative since the fluid gives heat to this reservoir. Similarly,

$$S_{61} = S_{H0} \ (> 0) \ .$$

In order that the cycle be closed $S_{f0} = S_{i0}$, i.e.

$$S_{C0} + S_{H0} + S_{12} + S_{45} = 0 \tag{IV.1}$$

Of the relativistic transformations of energy $E$ and momentum $\mathbf{p}$ measured in $I$ in terms of $E'$, $\mathbf{p}'$ measured in $I'$, we need

$$E = \gamma(E' + \mathbf{w}\cdot\mathbf{P}) , \qquad \gamma \equiv \left(1 - \frac{w^2}{c^2}\right)^{-\frac{1}{2}} \tag{IV.2}$$

where $w$ is the velocity of $I'$ in $I$ along the common $x$-axis of the frames. Let $W_Q$ be the mechanical work done on the fluid as a result of heat $Q$ being transferred to it. If the reservoirs have zero momentum in inertial frames $I_{C0}$ and $I_{H0}$, we then chose $I'$ first to be one of these frames and then to be the other to find for the energy gained by the fluid as a result of the two heat transfers

$$W_Q + Q = \gamma_C T_{C0} S_{C0} + \gamma_H T_{H0} S_{H0} . \tag{IV.3}$$

The work done on the fluid due to acceleration ($W_a$) and compression and expansion ($W_c$) also occurs in the cycle. The total mechanical work done on the fluid per cycle is therefore

$$W = W_a + W_c + W_Q .$$

The first law of thermodynamics in frame $I$ now furnishes the result $Q + W = 0$, i.e.

$$-(W_a + W_c) = Q + W_Q = \gamma_C T_{C0} S_{C0} + \gamma_H T_{H0} S_{H0} . \tag{IV.4}$$

Also easy to discuss is the mechanical work ($W_R$) done by the reservoirs in frame $I$. If their rest masses be $M_{H0}$ and $M_{C0}$ initially, and their proper temperatures be $T_{H0}$ and $T_{C0}$ respectively, then their final kinetic energies in frame $I$ are

$$K_f = (\gamma_H - 1)(M_{H0}c^2 - T_{H0}S_{H0}) + (\gamma_C - 1)(M_{C0}c^2 - T_{C0}S_{C0}) .$$

Their initial kinetic energies in $I$ are

$$K_i = (\gamma_H - 1)\, M_{H0}c^2 + (\gamma_C - 1)\, M_{C0}c^2 \, .$$

Hence

$$W_R = K_i - K_f = (\gamma_H T_{H0} S_{H0} + \gamma_C T_{C0} S_{C0}) - (T_{H0} S_{H0} + T_{C0} S_{C0}) \, .$$

$$= -(W_a + W_c) + W_0 \qquad \text{(IV.5)}$$

where

$$W_0 \equiv -(T_{H0} S_{H0} + T_{C0} S_{C0}) \qquad \text{(IV.6)}$$

is the work done per cycle by the fluid in a normal cycle with stationary reservoirs. (By virtue of (IV.1) with $S_{12} = S_{45} = 0$ this becomes of course

$$W_0 = -(T_{H0} - T_{C0}) S_{H0} = -Q_0$$

the nett heat given to the fluid in the cycle. However we shall postpone this step to later.)

The efficiency of the cycle is in frame $I$

$$\eta = \frac{-W}{T_{H0} S_{H0}}$$

where $-W$ is the useful mechanical work done by the fluid in frame $I$. We divide this by the proper energy withdrawn as heat from the hot reservoir, since this is what has furnished the thermal energy for the cycle. There is, however, the question as to what should be included in $-W$. Reasonable items are as follows:

$-W_a$: This is work done by the fluid in being decelerated minus the work done on it to accelerate it. This can be absorbed by a mechanical engine coupled to the Carnot engine, for example by a compression spring. It should therefore appear in $-W$.

$-W_C$: This is the work done by the fluid in expansion minus the work done in order to compress the fluid. This is also storable energy and should appear in $-W$.

$-W_Q$: This is mechanical work done by the fluid due to heat transfer. We have not given an expression for it as yet, and in fact this mechanical work does not involve the mechanical engine. This work can therefore not be utilized and ought not to appear in $-W$. Its origin is in the change of rest mass of a fluid by heat transfer. This implies a change of momentum in $I$ and hence corresponds to work having been done, even though this work was performed without the aid of any mechanical engine.

$-W_R$: The work done on the reservoirs is clearly recoverable in frame $I$, since the reservoirs could in principle be decelerated in this frame by compressing an appropriate spring. Hence this term should also appear in $-W$.

One finds, using equations (IV.5) and (IV.6)

$$\eta = \frac{-W_a - W_C - W_R}{T_{H0}S_{H0}} = \frac{-W_0}{T_{H0}S_{H0}} = 1 + \frac{T_{C0}S_{C0}}{T_{H0}S_{H0}} \quad .$$

Assuming that proper entropy is not changed by acceleration, (IV.1) shows that $-S_{C0} = S_{H0}$ and hence

$$\eta = 1 - \frac{T_{C0}}{T_{H0}} \quad .$$

Thus one finds a sensible answer: the usual Carnot efficiency involving the proper temperatures. The argument shows that there is no need to make any decision whatsoever on two controversial questions:

(*a*) When heat energy transfer $E_0 = Q_0$ takes place between two bodies at rest and is described from another inertial frame, how is $E$ to be divided into mechanical work $W_Q$ and heat $Q$ in that frame?

(*b*) What is the relativistic transformation of temperature? Both questions have been considered in section 18.5.

## IV.2. THE CONDITION FOR THERMAL EQUILIBRIUM

We now prove the assertion that two bodies in relative motion cannot be in thermal equilibrium. For this purpose we assume a fluid in its rest frame for which the total energy is $U_0$. For an incremental change one has as in (IV.2)

$$dU_0 = T_0 dS_0 - p_0 dv_0 = \gamma(dU - \mathbf{w}.d\mathbf{P}) .$$

For two isolated fluids, A and B, which are in a general state at a certain instant, one has therefore

$$dS = dS_0^A + dS_0^B = \left\{\frac{\gamma}{T_0}\, dU - \frac{\gamma}{T_0}\, \mathbf{w}.d\mathbf{P} + \frac{p_0}{T_0}\, dv_0\right\}^A + \left\{\ \right\}^B \geq 0$$

where the Lorentz invariance of entropy has been assumed and $\{\}^B$ denotes the terms for system B. Suppose $dv_0^A = dv_0^B = 0$ and energy and momentum is conserved in an interaction between A and B which is allowed to take place for a short increment of time. Then

$$dU^A + dU^B = 0 \qquad \text{(IV.7)}$$

$$d\mathbf{P}^A + d\mathbf{P}^B = 0 . \qquad \text{(IV.8)}$$

The required condition for a spontaneous change is

$$\left\{\left(\frac{\gamma}{T_0}\right)^A - \left(\frac{\gamma}{T_0}\right)^B\right\} dU^A - \left\{\left(\frac{\gamma}{T_0}\,\mathbf{w}\right)^A - \left(\frac{\gamma}{T_0}\,\mathbf{w}\right)^B\right\} .d\mathbf{P}^A \geq 0. \qquad \text{(IV.9)}$$

As $dU^A$, $d\mathbf{P}^A$ are arbitrary, the two coefficients must vanish. This implies $\mathbf{w}^A = \mathbf{w}^B$, i.e. $\gamma^A = \gamma^B$, and therefore also $T_0^A = T_0^B$. The condition for thermal equilibrium is accordingly,' even in the general frame $I$,

$$T_0^A = T_0^B \qquad \text{(IV.10)}$$

$$\mathbf{w}^A = \mathbf{w}^B . \qquad \text{(IV.11)}$$

Hence $\mathbf{w}^A \neq \mathbf{w}^B$ excludes the possibility of thermal equilibrium.

If one considers an interaction with zero momentum change, $d\mathbf{P}^A = d\mathbf{P}^B = 0$ instead of (IV.8), one has a problem which singles out some inertial frames among all possible ones, since $dP^A = dP^B = 0$ cannot be valid in all inertial frames. The result one obtains is therefore unsatisfactory. One does not find (IV.11) and the single remaining equilibrium condition in (IV.9) states equality of $\gamma/T$ for both systems. Hence one would erroneously conclude

$$T^A = T^B \text{ is the same as } \frac{\gamma^A}{T_0^A} = \frac{\gamma^B}{T_0^B}$$

so that $T = T_0/\gamma$ . This is the Planck-Einstein transformation law, but it is obtained by a faulty argument. One has to accept (IV.10) and (IV.11) instead.

# PART D: SOLUTIONS OF PROBLEMS

# SOLUTIONS OF PROBLEMS

CHAPTER 3.

3.1 By equations (3.1) and (3.3)

$$\left(\frac{dp}{dv}\right)_a = -\frac{m_v}{m_p} = \frac{C_p l_v}{C_v l_p} \qquad \left(\frac{\partial p}{\partial v}\right)_t = \frac{l_v}{l_p}$$

$$\left(\frac{dv}{dt}\right)_a = -\frac{C_v}{l_v} \qquad \left(\frac{\partial v}{\partial t}\right)_p = \frac{C_p}{m_v} = \frac{C_p - C_v}{l_v}$$

$$\left(\frac{dp}{dt}\right)_a = -\frac{C_p}{l_p} \qquad \left(\frac{\partial p}{\partial t}\right)_v = -\frac{C_p - C_v}{l_p}.$$

Now divide each term on the left by the corresponding term on the right. The first row yields $\gamma$ at once, as required. The second row yields

$$-\frac{C_v m_v}{C_p l_v} = -\frac{C_v}{C_p}\frac{C_p}{C_p - C_v} \quad \text{(by (3.3))} = -\frac{1}{\gamma-1}$$

The third row yields

$$\frac{C_p}{C_p - C_v} = \frac{\gamma}{\gamma-1}.$$

These are the required expressions. If $pv = At$, $(\partial p/\partial v)_t = -p/v$, so that for quasi-static adiabatic processes $dp/dv = -\gamma p/v$, i.e. $pv^\gamma$ is a constant. The other two results are derived similarly.

3.2. By definition and using the first row in the solution to problem 3.1 .

$$\frac{K_t}{K_a} = \frac{-v^{-1}(\partial v/\partial p)_t}{-v^{-1}(dv/dp)_a} = \gamma.$$

3.3. For a wire under tension $F$, of length $l$ and temperature $t$, one can use the equations for a fluid with the replacement $pdv \rightarrow -Fdl$ since on expansion a fluid



performs work while a wire has work done on it when it is stretched. Hence one can chose $p \rightarrow -F$, $v \rightarrow l$, so that

$$d'Q = C_l dt + l_l dl = C_F dt - l_F dF = dU - Fdl.$$

By (3.16)

$$\frac{Y_t}{Y_a} = \frac{(\partial F/\partial L)_t}{(\partial F/\partial l)_a} = \frac{-l_l/l_F}{(dF/dt)_a(dt/dl)_a} = -\frac{l_l/l_F}{(C_F/l_F)(-l_l/C_l)} = \frac{C_l}{C_F}.$$

3.4. By (3.1)

$$\left(\frac{\partial p}{\partial t}\right)_v = -\frac{1}{(\partial v/\partial p)_t(\partial t/\partial v)_p} = -\frac{1}{-vK_t}\alpha_p v = \frac{\alpha_p}{K_t}.$$

From problem 3.1

$$\left(\frac{dp}{dt}\right)_a = \frac{\gamma}{\gamma-1}\left(\frac{\partial p}{\partial t}\right)_v = \frac{\gamma}{\gamma-1}\frac{\alpha_p}{K_t}.$$

Hence for a given $\gamma$ and $K_t$, $(dp/dt)_a$ changes sign if $\alpha_p$ changes sign. Thus adiabatic compression can yield cooling or heating. If $\gamma > 1$, and $K_t > 0$, then heating occurs if $\alpha_p > 0$, and cooling if $\alpha_p < 0$.

3.5. One needs the independent variables $t$ and $p$. Consider therefore

$$d\ln v = \frac{1}{v}dv = \frac{1}{v}\left\{\left(\frac{\partial v}{\partial t}\right)_p dt + \left(\frac{\partial v}{\partial p}\right)_t dp\right\}$$

$$= \alpha_p\, dt - K_t\, dp.$$

The result follows.

3.6. If the conditions are such that $pv^n = A$ ($n$ constant), as may happen if the temperature is kept constant, then

$$p = B\rho^n$$

where $A$, $B$ are different constants. By (3.21)

$$V^2 = \frac{dp}{d\rho} = nB\rho^{n-1} = \frac{np}{\rho} .$$

Thus we have

$$n = \frac{\rho V^2}{p} = \frac{0\cdot 00129\times(33200)^2}{13\cdot 6\times 981\times 76}\left\{\frac{(\text{g cm}^{-3})\times(\text{cm s}^{-1})^2}{(\text{g cm}^{-3})\times(\text{cm s}^{-2})\times\text{cm}}\right\} = 1\cdot 402.$$

3.7. If $\alpha_p$ is approximately constant

$$\frac{dv}{v} = \alpha_p\, dt$$

so that

$$\ln\frac{v}{v_0} = \alpha_p(t - t_0)$$

that is

$$v = v_0 \exp\{\alpha_p(t - t_0)\} .$$

If $\alpha_p$ is small enough, expanding the exponential yields

$$v \doteqdot v_0 + v_0\alpha_p(t - t_0) .$$

3.8. Along line $y = 2x$ from (0,0) to (1,2)

$$dW_1 = \int \mathbf{F}_1 . d\mathbf{r} = \int_0^1 (x^2 dx + 4x^2\, 2dx) = \int_0^1 9x^2 dx = 3x^3 \Big|_0^1 = 3$$

$$dW_2 = \int \mathbf{F}_2 . d\mathbf{r} = \int_0^1 (4x^2 dx + x^2\, 2dx) = \int_0^1 6x^2 dx = 2x^3 \Big|_0^1 = 2 .$$

Along the parabola $y = 2x^2$

$$dW_1 = \int_0^1 (x^2 dx + 4x^4 . 4x\, dx) = \int x^2 dx + 16\int x^5 dx = \frac{1}{3} + \frac{8}{3} = 3$$

$$dW_2 = \int_0^1 (4x^4 dx + x^2 . 4x\, dx) = 4\int x^4 dx + 4\int x^3 dx = \frac{4}{5} + 1 = \frac{9}{5}.$$

Thus $dW_1$ is the same for both paths and $\mathbf{F}_1 . d\mathbf{r}$ has a chance to be an exact differential. However, $dW_2$ is different (2 as against 9/5) and so $\mathbf{F}_2 . d\mathbf{r}$ cannot be an exact differential: $\mathbf{F}_2 \cdot d\mathbf{r} = d'W$.

In fact

$$\frac{1}{3}\, d(x^3+y^3) = \frac{1}{3}\left\{\frac{\partial(x^3+y^3)}{\partial x}\right\}_y dx + \frac{1}{3}\left\{\frac{\partial(x^3+y^3)}{\partial y}\right\}_x dy$$

$$= x^2 dx + y^2\, dy$$

$$= \mathbf{F}_1 \cdot d\mathbf{r} .$$

So

$$\mathbf{F}_1 \cdot d\mathbf{r} = d\left(\frac{x^3+y^3}{3}\right) \quad (\equiv df) .$$

There thus exists a function of $x$ and $y$, $f(x,y)$, such that

$$\int_A^B dW_1 = f_B - f_A .$$

In the present case $f_B = f(1,2) = \frac{1}{3}(1+8) = 3$ and $f_A = f(0,0) = 0$ .

Once one has read section 4.1 one has a very simple argument at one's disposal:

$$d'W_j = \mathbf{F}_j \cdot d\mathbf{r} \equiv \mathbf{y}_j \cdot d\mathbf{r}$$

where

$$\mathbf{y}_1 \equiv (x^2, y^2, 0) \qquad \mathbf{y}_2 \equiv (y^2, x^2, 0) .$$

Hence

$$\nabla \times \mathbf{y}_1 = 0 \qquad \nabla \times \mathbf{y}_2 \neq 0$$

so that $d'W_1$ is exact and therefore $F_1$ is conservative, while $d'W_2$ is not exact.

CHAPTER 4.

4.1. The error occurs when the second derivative of $Q$ is taken. As $\int d'Q$ is path-dependent, it is not correct to infer from

$$\left(\frac{\partial Q}{\partial t}\right)_v = \theta(t,v) , \quad \left(\frac{\partial Q}{\partial v}\right)_t = \omega(t,v)$$

that

$$\frac{\partial^2 Q}{\partial t \partial v} = \frac{\partial \theta}{\partial v} = \frac{\partial \omega}{\partial t} .$$

4.2. This may be obtained from books on differential equations.

4.3. The general solution is

$$ax^2 + 2hxy + by^2 = c \quad (a \text{ constant}).$$

4.4. Choose

$$\mathbf{a} = \mathbf{c} = (y,z,x) .$$

Then

$$\mathbf{a}\cdot\underline{\nabla} \times \mathbf{a} = a_x\left(\frac{\partial a_y}{\partial z} - \frac{\mathrm{d}a_z}{\mathrm{d}y}\right) + a_y\left(\frac{\partial a_z}{\partial x} - \frac{\partial a_x}{\partial z}\right) + a_z\left(\frac{\partial a_x}{\partial y} - \frac{\partial a_y}{\partial x}\right)$$

$$= a_x + a_y + a_z = x+y+z .$$

This is in general not zero.

4.5. If $\lambda_1(x_1,\ldots,x_n)$ is one integrating factor, then $\mathrm{d}'Q = \lambda_1 \mathrm{d}\phi_1$. Let $\phi_2 = \phi_2[\phi_1(x_1,\ldots,x_n)]$ be a function of $\phi_1$. Then

$$\mathrm{d}\phi_2 = \frac{\mathrm{d}\phi_2}{\mathrm{d}\phi_1}\mathrm{d}\phi_1$$

so that

$$\mathrm{d}'Q = \lambda_1 \frac{\mathrm{d}\phi_1}{\mathrm{d}\phi_2}\mathrm{d}\phi_2 = \lambda_2 \mathrm{d}\phi_2 .$$

Hence for each function $\phi_2$ one can construct a new integrating factor



$$\lambda_2 = \lambda_1 \frac{\mathrm{d}\phi_1}{\mathrm{d}\phi_2}$$

CHAPTER 5

5.1. (i) $\mathrm{d}U = T\mathrm{d}S - p\mathrm{d}v \Rightarrow \frac{\partial^2 U}{\partial S \partial v} = \left(\frac{\partial T}{\partial v}\right)_S = -\left(\frac{\partial p}{\partial S}\right)_v$

(ii)

$$\mathrm{d}F = \mathrm{d}U - T\mathrm{d}S - S\mathrm{d}T = T\mathrm{d}S - p\mathrm{d}v - T\mathrm{d}S - S\mathrm{d}T = -p\mathrm{d}v - S\mathrm{d}T$$

$$\Rightarrow -\frac{\partial^2 F}{\partial v \partial T} = \left(\frac{\partial p}{\partial T}\right)_v = \left(\frac{\partial S}{\partial v}\right)_T$$

(iii) $\mathrm{d}H = \mathrm{d}U + p\mathrm{d}v + v\mathrm{d}p = T\mathrm{d}S + v\mathrm{d}p$

$$\Rightarrow \left(\frac{\partial T}{\partial p}\right)_S = \left(\frac{\partial v}{\partial S}\right)_p$$

(iv) $\mathrm{d}G = v\mathrm{d}p - S\mathrm{d}T \Rightarrow \left(\frac{\partial v}{\partial T}\right)_p = -\left(\frac{\partial S}{\partial p}\right)_T .$

5.2. (i)

$$[C_v\mathrm{d}T + l_v\mathrm{d}v = C_p\mathrm{d}T + l_p\mathrm{d}p] \Rightarrow [C_p - C_v = l_v\left(\frac{\partial v}{\partial T}\right)_p] .$$

$$[T\mathrm{d}S = C_v\mathrm{d}T + l_v\mathrm{d}v] \Rightarrow [l_v = T\left(\frac{\partial S}{\partial v}\right)_T = T\left(\frac{\partial p}{\partial T}\right)_v]$$

Therefore

$$C_p - C_v = T\left(\frac{\partial v}{\partial T}\right)_p\left(\frac{\partial p}{\partial T}\right)_v = -T\left(\frac{\partial v}{\partial T}\right)_p\left(\frac{\partial p}{\partial v}\right)_T\left(\frac{\partial v}{\partial T}\right)_p = -T(v\alpha_p)^2\left(-\frac{1}{vK_T}\right)$$

$$= Tv\frac{\alpha_p^2}{K_T}$$

(ii) $\frac{K_T}{K_a} = \frac{C_p}{C_v}$ (problem 3.2

Therefore

$$(K_T - K_a)C_p = K_T(C_p - C_v) = Tv\alpha_p^2$$

(iii) Since $C_p dT + l_p dp = dU + p dv$, it follows that

$$C_p = \left(\frac{\partial U}{\partial T}\right)_p + p\left(\frac{\partial v}{\partial T}\right)_p = \left(\frac{\partial H}{\partial T}\right)_p .$$

5.3. By section 4.2 the condition for the existence of an integrating factor of $d'F = \Sigma_1^n Y_j dx_j$ is

$$Y_i\left(\frac{\partial Y_k}{\partial x_j} - \frac{\partial Y_j}{\partial x_k}\right) + Y_j\left(\frac{\partial Y_i}{\partial x_k} - \frac{\partial Y_k}{\partial x_i}\right) + Y_k\left(\frac{\partial Y_j}{\partial x_i} - \frac{\partial Y_i}{\partial x_j}\right) = 0$$

for $(i,j,k) = (1,2,3), (1,2,4), (1,3,4), (2,3,4)$ where $n = 4$ is assumed. This will not in general be satisfied. The equation results from putting two fluids together *conceptually*, and this yields $2\times2 = 4$ independent variables, e.g. $T_1, T_2, v_1, v_2$.

(i) Pressure coupling:

$$p_1 = p_2 = p \quad \text{and} \quad d'Q = C_1 dT_1 + C_2 dT_2 + p dv$$

where $v \equiv v_1 + v_2$. We have only one equation since $n = 3$. With $(i,j,k) = (1,2,3)$ an integrating factor requires

$$C_1\left(\frac{\partial p}{\partial T_2} - \frac{\partial C_2}{\partial v}\right) + C_2\left(\frac{\partial C_1}{\partial v} - \frac{\partial p}{\partial T_1}\right) + p\left(\frac{\partial C_2}{\partial T_1} - \frac{\partial C_1}{\partial T_2}\right) = 0.$$

(ii) Temperature coupling:

$$d'Q = (C_1 + C_2)dT + p_1 dv_1 + p_2 dv_2, \quad C_1 + C_2 \equiv C .$$

For an integrating factor we have

$$C\left(\frac{\partial p_2}{\partial v_1} - \frac{\partial p_1}{\partial v_2}\right) + p_1\left(\frac{\partial C}{\partial v_2} - \frac{\partial p_2}{\partial T}\right) + p_2\left(\frac{\partial p_1}{\partial T} - \frac{\partial C}{\partial v_1}\right) = 0 .$$

5.4. (*a*) A proof is given in books on differential equations.
(*b*) Let us link $P = (p_1,p_2,p_3)$ and $Q = (q_1,q_2,q_3)$, where $q_3 - p_3 \neq 0$ is assumed. Define two intermediate points



$$A = (p_1,r_2,p_3) , \quad B = (q_1,r_2,q_3)$$

The straight line PA lies in the hyperplane $z_1 = p_1$, $z_3 = p_3$ so that $d'Q = dz_1 + z_2 dz_3 = 0$ along it. The same is true for the straight line BQ. For the straight line AB,

$$\int_{AB} d'Q = q_1 - p_1 + r_2(q_3 - p_3) .$$

This is also adiabatic and quasi-static if we choose $r_2$ properly:

$$r_2 = \frac{q_1 - p_1}{q_3 - p_3}$$

A general argument is available in reference R6, p.51, for the case of any Pfaffian form which has no integrating factor.

**5.5.** For the increments of the cycle for which heat is gained, the entropy increase of the fluid is

$$S_+ = \int_+ \frac{d'Q}{T} > \frac{1}{T_1}\int_+ d'Q \equiv \frac{Q_+}{T_1}$$

where + indicates the portion of the cycle during which heat is gained, and $Q_+$ is the total amount of heat gained in these processes. Similarly the heat lost is $Q_-$ and the entropy lost by the fluid in these incremental processes is

$$S_- = \int_- \frac{d'Q}{T} < \frac{1}{T_2}\int_- d'Q \equiv \frac{Q_-}{T_2} .$$

In a complete cycle the entropy of the fluid does not change and the mechanical work done by the fluid is $W = Q_+ - Q_-$. Therefore

$$0 = S_+ - S_- > \frac{Q_+}{T_1} - \frac{Q_-}{T_2} , \quad \text{that is} \quad \frac{Q_-}{Q_+} > \frac{T_2}{T_1} .$$

Thus

$$\eta \equiv \frac{Q_+ - Q_-}{Q_+} = 1 - \frac{Q_-}{Q_+}$$

and so

$$\eta < 1 - \frac{T_2}{T_1} \ .$$

5.6. This is a special case of problem 5.5 in which the heat accepted by the fluid is at the higher temperature, $T_1$ say. The rejected heat is at temperature $T_2$. Hence

$$S_+ = \frac{Q_+}{T_1} \ , \quad S_- = \frac{Q_-}{T_2} \ .$$

The solution of 5.5 goes through with equality signs instead of inequality signs.

5.7. From the solution of problem 3.4

$$d'Q = C_v dT + T \frac{\alpha_p}{K_T} dv$$

and we are given that $C_v > 0$, $K_T > 0$. Thus if an isothermal $T = T_c$ has $\alpha_p = 0$ for a range, then it is also an adiabatic curve for that range. If it is made the lower isothermal in a Carnot cycle, then the fluid takes up heat $Q_h$ from the hot reservoir at temperature $T_h$, and its rejected heat $Q_c$ during isothermal compression at the lower temperature $T_c$ is zero, because the isothermal is also an adiabatic. The work done by the fluid in a cycle is therefore $W = Q_h - Q_c = Q_h$. The cycle has a theoretical efficiency of 100 per cent and this is excluded by the second law of thermodynamics. This range of values can therefore not be quasi-statically and adiabatically linked with other states, otherwise this Carnot cycle could be devised and would contradict the second law.

5.8. From equation (3.12) with the absolute temperature scale

$$\Gamma = \frac{v}{C_v}\left(\frac{\partial p}{\partial T}\right)_v$$

so that

$$\left(\frac{\partial^2 p}{\partial T^2}\right)_v = \left\{\frac{\partial}{\partial T}\left(\frac{C_v \Gamma}{v}\right)\right\}_v \qquad \text{(i)}$$

Also from section 5.4

$$C_v = T\left(\frac{\partial S}{\partial T}\right)_v \ ,$$

so that

$$\left(\frac{\partial C_v}{\partial v}\right)_T = T\,\frac{\partial^2 S}{\partial T \partial v} = T\,\frac{\partial}{\partial T}\left(\frac{\partial S}{\partial v}\right)$$

$$= T\,\frac{\partial}{\partial T}\left(\frac{\partial p}{\partial T}\right)_v = T\left(\frac{\partial^2 p}{\partial T^2}\right)_v$$

where a Maxwell relation of problem 5.1 has been used. It follows that

$$\left(\frac{\partial^2 p}{\partial T^2}\right)_v = \frac{1}{T}\left(\frac{\partial C_v}{\partial v}\right)_T \qquad \text{(ii)}$$

The two relations (i) and (ii) imply

$$\left(\frac{\partial C_v}{\partial v}\right)_T = \frac{T}{v}\left\{\Gamma\left(\frac{\partial C_v}{\partial T}\right)_v + C_v\left(\frac{\partial \Gamma}{\partial T}\right)_v\right\} .$$

Thus one sees that if $C_v$ is independent of $v$ and if $\Gamma$ is independent of $T$ than $C_v$ is a constant. Also if $C_v$ is a constant then $\Gamma$ is independent of $T$.

5.9. We shall write the equation to be proved in the form

$$\frac{1}{T}\left\{\left(\frac{\partial C_v}{\partial T}\right)_S - \frac{C_v}{T}\right\} = v\left(\frac{\partial p}{\partial T}\right)_v \left\{\frac{\partial}{\partial T}\left[\frac{1}{\Gamma T}\right]\right\}_v$$

and show that each side is equal to $(\partial^2 S/\partial T^2)_v$ . The left-hand side is

$$\frac{1}{T}\left\{\frac{\partial}{\partial T}\left\{T\left(\frac{\partial S}{\partial T}\right)_v\right\} - \left(\frac{\partial S}{\partial T}\right)_v\right\} = \left(\frac{\partial^2 S}{\partial T^2}\right)_v \ .$$

The right-hand side can be rewritten by noting that

$$\frac{1}{\Gamma T} = \frac{1}{T}\frac{C_v}{v}\left(\frac{\partial T}{\partial p}\right)_v = \frac{1}{v}\left(\frac{\partial S}{\partial T}\frac{\partial T}{\partial p}\right)_v = \frac{1}{v}\left(\frac{\partial S}{\partial p}\right)_v .$$

It therefore becomes

$$\left\{\left(\frac{\partial p}{\partial T}\right)\frac{\partial}{\partial T}\left(\frac{\partial S}{\partial p}\right)\right\}_v = \left(\frac{\partial^2 S}{\partial T^2}\right)_v .$$

5.10. Since $dH = TdS + vdp$

$$pv = p\left(\frac{\partial H}{\partial p}\right)_S = \left(\frac{\partial H}{\partial \ln p}\right)_S = -\frac{(\partial H/\partial S)_p}{[\partial \ln p/\partial S]_H} = AT .$$

5.11. (*a*) An isolated system does not receive or give out energy. An adiabatic system does not receive or give out heat or radiation. It follows that all isolated systems are also adiabatic and that an adiabatic system need not be an isolated system.

(*b*) The statement is correct. However, it does not tell us anything about a system which is adiabatic but not isolated. For such systems the entropy is also non-decreasing. Hence the statement is weaker than it could be.

(*c*) Since the two states are equilibrium states they are associated with definite entropies. In order to evaluate the entropy change any quasi-static path may be chosen to link them provided only the terminal states are the given states. Hence

$$S_2 - S_1 = \int_1^2 \frac{dQ}{T}$$

is a correct way of doing this, even though $\delta S \geq \delta' Q/T$ is correct for a non-static change.

CHAPTER 6

6.1.

$$C_v dt + l_v dv = dU + pdv = \left(\frac{\partial U}{\partial t}\right)_v dt + \left\{p + \left(\frac{\partial U}{\partial v}\right)_t\right\} dv$$




and therefore

$$\left(\frac{\partial U}{\partial t}\right)_v = C_v , \quad l_v = p + \left(\frac{\partial U}{\partial v}\right)_t .$$

This becomes

$$l_v = p = \frac{Nkt}{v}$$

$$\int_A^B \frac{d'Q}{t} = \int_A^B \frac{C_v}{t}\, dt + Nk\int_A^B \frac{dv}{v} .$$

If $C_v$ is a constant,

$$\int_A^B \frac{d'Q}{t} = C_v \ln\left(\frac{t_B}{t_A}\right) + Nk\ln\left(\frac{v_B}{v_A}\right) = C_v \ln \frac{t_B}{t_A}\left(\frac{v_B}{v_A}\right)^{Nk/C_v} .$$

This is a function only of the end-points A and B, and so is independent of the path. For two variables *any* inexact differential has an integrating factor (see section 4.3) for coefficients which are well behaved.

6.2.

$$TdS = C_p dT + l_p dp$$

and therefore

$$l_p = T\left(\frac{\partial S}{\partial p}\right)_T = -T\left(\frac{\partial v}{\partial T}\right)_p \qquad \text{( by a Maxwell relation )}$$

$$= -Tv\,\alpha_p .$$

Therefore

$$\int \frac{d'Q}{T} = \int_A^B \frac{C_p}{T}\, dT - \int_A^B v\,\alpha_p\, dp .$$

6.3. (*a*)

$$C_p\, dT + l_p\, dp = dU + pdv$$

and therefore

$$C_p = \left(\frac{\partial U}{\partial T}\right)_p + p\left(\frac{\partial v}{\partial T}\right)_p .$$

Also

$$C_v \mathrm{d}T + l_v \mathrm{d}v = \mathrm{d}U + p\mathrm{d}v \Rightarrow C_v = \left(\frac{\partial U}{\partial T}\right)_v .$$

Therefore

$$C_p - C_v = \left(\frac{\partial U}{\partial T}\right)_p - \left(\frac{\partial U}{\partial T}\right)_v + p\left(\frac{\partial v}{\partial T}\right)_p = \left\{\left(\frac{\partial U}{\partial v}\right)_T + p\right\}\left(\frac{\partial v}{\partial T}\right)_p \qquad \text{(i)}$$

For an ideal classical gas this is

$$p\left(\frac{\partial v}{\partial T}\right)_p = p\,\frac{Nk}{p} = Nk .$$

To prove the last step in (i):

$$\mathrm{d}U = \left(\frac{\partial U}{\partial v}\right)_t \mathrm{d}v + \left(\frac{\partial U}{\partial t}\right)_v \mathrm{d}t$$

$$= \left(\frac{\partial U}{\partial v}\right)_t \left\{\left(\frac{\partial v}{\partial p}\right)_t \mathrm{d}p + \left(\frac{\partial v}{\partial t}\right)_p \mathrm{d}t\right\} + \left(\frac{\partial U}{\partial t}\right)_v \mathrm{d}t .$$

But

$$\mathrm{d}U = \left(\frac{\partial U}{\partial t}\right)_p \mathrm{d}t + \left(\frac{\partial U}{\partial p}\right)_t \mathrm{d}p .$$

By comparison

$$\left(\frac{\partial U}{\partial t}\right)_p = \left(\frac{\partial U}{\partial t}\right)_v + \left(\frac{\partial U}{\partial v}\right)_t \left(\frac{\partial v}{\partial t}\right)_p .$$

(*b*) **We have**

$$pv = Nkt, \quad (\partial U/\partial v)_t = 0 \quad \text{(classical ideal gas)} .$$

From the Maxwell relation

$$\left(\frac{\partial S}{\partial v}\right)_T = \left(\frac{\partial p}{\partial T}\right)_v ,$$

therefore

$$\mathrm{d}S(v,T) = Nk\,\frac{\mathrm{d}v}{v}$$



therefore

$$S(v,T) - S(v_0,T) = Nk \ln\frac{v}{v_0} .$$

Also

$$\left(\frac{\partial S}{\partial p}\right)_T = -\left(\frac{\partial v}{\partial T}\right)_p , \qquad \mathrm{d}S(p,T) = -\frac{Nk}{p}\,\mathrm{d}p$$

therefore

$$S(p,T) - S(p_0,T) = -Nk \ln\frac{p}{p_0} .$$

6.4. Heat $Q$ leaks from the reservoir at temperature $T_h$ to the reservoir at temperature $T_c$. The entropy loss is $Q/T_h$ and the gain is $Q/T_c$. The entropy change for the whole system is

$$\Delta S = Q\left(\frac{1}{T_c} - \frac{1}{T_h}\right) = \frac{T_h - T_c}{T_h T_c}\,Q > 0.$$

The system has undergone a non-static adiabatic process, i.e. a *dissipative* process. It is adiabatic because the system *as a whole* has neither gained nor lost heat. The second law states that the entropy increases under these conditions.

6.5. Initial entropy is in the classical limit

$$kN_A\left\{c_A + \ln\left(\frac{v_A}{N_A}\right)\right\} + kN_B\left\{c_B + \left(\ln\frac{v_B}{N_B}\right)\right\} \equiv S_{\text{initial}} \equiv S_i .$$

The factor $c_A = c_B$ depends on the masses of the particles as can be deduced from problem 12.4 by forming $(U-F)/T = (s+3/2-\gamma_N)Nk$ and noting that $\gamma_N \propto \ln(N/v)$.

(i) <u>A and B particles distinguishable</u>

$$S_{\text{final}} \equiv S_{f1} = kN_A\left\{c + \ln\left(\frac{v_A+v_B}{N_A}\right)\right\} + kN_B\left\{c + \ln\left(\frac{v_A+v_B}{N_B}\right)\right\}$$

$$S_1 \equiv S_{f1} - S_i = k\left\{\ln\left(\frac{v_A+v_B}{N_A}\,\frac{N_A}{v_A}\right)\right\}N_A + kN_B \ln\left(\frac{v_A+v_B}{N_B}\,\frac{N_B}{v_B}\right)$$

$$= kN\{-q\ln p - (1-q)\ln(1-p)\} ,$$

where $$p \equiv \frac{v_A}{v}, \quad q \equiv \frac{N_A}{N}, \quad 1-q = \frac{N_B}{N}, \quad 1-p = \frac{v_B}{v}.$$

(ii) <u>A and B particles indistinguishable</u>

$$S_{final} \equiv S_{f2} = (N_A + N_B)\, k\left\{c + \ln\left(\frac{v_A+v_B}{N_A+N_B}\right)\right\}$$

$$S_2 \equiv S_{f2} - S_i = kN\left\{\ln\left(\frac{v}{N}\right) - q\ln\left(\frac{v_A}{N_A}\right) - (1-q)\ln\left(\frac{v_B}{N_B}\right)\right\}$$

$$= kN\left\{q\ln\left(\frac{v}{N}\right) + (1-q)\ln\frac{v}{N} - q\ln\left(\frac{v_A}{N_A}\right) - (1-q)\ln\left(\frac{v_B}{N_B}\right)\right\}$$

$$= kN\left\{q\ln\left(\frac{q}{p}\right) + (1-q)\ln\left(\frac{1-q}{1-p}\right)\right\}.$$

(ii)

$$\left\{\frac{1}{kN}\frac{\partial S_1}{\partial q} = -\ln p + \ln(1-p) = 0\right\} \Rightarrow \left\{p = \tfrac{1}{2}\right\}$$

$$\left\{\frac{1}{kN}\frac{\partial S_1}{\partial p} = -\frac{q}{p} + \frac{1-q}{1-p} = 0\right\} \Rightarrow \left\{p = q\right\}$$

$$\Rightarrow p = q = \tfrac{1}{2}, \quad \text{i.e. } v_A = v_B;\ N_A = N_B$$

This is a minimum and $S_1 \geq S_1\ (p=q=\frac{1}{2}) = kN \ln 2$.

$$\left\{\frac{1}{kN}\frac{\partial S_2}{\partial q} = \ln\left(\frac{q}{p}\right) + 1 - \ln\left(\frac{1-q}{1-p}\right) - 1\right\} \Rightarrow \left\{\frac{1-q}{q} = \frac{1-p}{p}\right\} \Rightarrow \{p = q\}.$$

$$\left\{\frac{1}{kN}\frac{\partial S_2}{\partial p} = -\frac{q}{p} + \frac{1-q}{1-p}\right\} \Rightarrow \left\{\frac{1-q}{q} = \frac{1-p}{p}\right\}$$

$$\Rightarrow p = q, \quad \text{i.e. } \frac{v_A}{N_A} = \frac{v}{N} = \frac{v_B}{N_B} \qquad S_2 \geq 0.$$

(iii) Entropy has a meaning only against a background of specified measurements which may enable one to distinguish between the particles, or may not. A limiting procedure is not possible for classical gases.

6.6. (*a*) See equation (6.2).

(*b*) If $pv = gU$, the result of (*a*) is

$$\frac{gU}{v} = T\frac{g}{v}\left(\frac{\partial U}{\partial T}\right)_v - \left(\frac{\partial U}{\partial v}\right)_T$$

that is

$$U = T\left(\frac{\partial U}{\partial T}\right)_v - \frac{v}{g}\left(\frac{\partial U}{\partial v}\right)_T.$$

This is satisfied by $Uv^g = f(Tv^g)$ since the right-hand side of the above relation becomes then

$$Tv^{-g}f'v^g - \frac{v}{g}\left\{-\frac{g}{v^{g+1}}f(Tv^g) + v^{-g}f'\,Tg\,v^{g-1}\right\} = U.$$

(*c*) $dS = T^{-1}(dU + pdv)$

therefore

$$Tv^g dS = v^g\left(dU + \frac{gU}{v}dv\right) = d(Uv^g) = d\{f(Tv^g)\}.$$

Hence $S \equiv h(z)$, $z \equiv Tv^g$, provided the function $h$ satisfies

$$z\,dh = df,$$

that is

$$h' \equiv \frac{dh}{dz} = z^{-1}f'.$$

(*d*) In a quasi-static adiabatic change the entropy is constant. Hence $Tv^g \equiv z$ is constant. This implies the constancy of $Uv^g = (pv/g)v^g$, i.e. of $pv^{1+g}$. Dividing these two expressions, we also have as a constant

$$\frac{Tv^g}{pv^{1+g}}, \quad \text{i.e. } \frac{T^{1/g}v}{p^{1/1+g}v}, \quad \text{i.e. } \frac{T^{1+g}}{p^g}.$$

6.7. (*a*) In problem 5.1 it has been shown that

(i) $$dF = -SdT - pdv$$

so that

$$\left(\frac{\partial F}{\partial T}\right)_v = -S = \frac{F-U}{T}.$$

As $T \to 0$, $F \to U$ for finite $S$ so that the right-hand side becomes indeterminate. Hence by L'Hospital's rule,

$$\lim_{T\to 0}\left(\frac{\partial F}{\partial T}\right)_v = \lim_{T\to 0}\left\{\left(\frac{\partial F}{\partial T}\right)_v - \left(\frac{\partial U}{\partial T}\right)_v\right\}$$

i.e.

$$\{S_{T=0} = S_{T=0} + C_{V,T=0}\} \Rightarrow \{C_{v,T=0} = 0\} .$$

(ii) $dG = -SdT + vdp$

$$-\left(\frac{\partial G}{\partial T}\right)_p = S = \frac{H-G}{T}$$

therefore

$$\left\{\lim_{T\to 0}\left\{-\left(\frac{\partial G}{\partial T}\right)_p\right\} = \lim_{T\to 0}\left[C_p - \left(\frac{\partial G}{\partial T}\right)_p\right]\right\} \Rightarrow \left\{C_{p,T=0} = 0\right\}$$

(iii) If $x$ and $y$ be two independent variables, then

$$T\left(\frac{\partial S}{\partial x}\right)_y = C_v\left(\frac{\partial T}{\partial x}\right)_y + l_v\left(\frac{\partial v}{\partial x}\right)_y = C_p\left(\frac{\partial T}{\partial x}\right)_y + l_p\left(\frac{\partial p}{\partial x}\right)_y = m_v\left(\frac{\partial v}{\partial x}\right)_y + m_p\left(\frac{\partial p}{\partial x}\right)_y .$$

Consider as $T \to 0$ the following cases

| | | | | | | |
|---|---|---|---|---|---|---|
| $(x,y) =$ | $T,v$ | $T,p$ | $v,T$ | $p,T$ | $v,p$ | $p,v$ |
| Result | $C_v\to 0$ | $C_p\to 0$ | $l_v\to 0$ | $l_p\to 0$ | $m_v\to 0$ | $m_p\to 0$ |

(*b*) Ideal classical gas. By problem 6.3 $C_p - C_v$ is a constant, and this cannot tend to zero as $T \to 0$.

(*c*) If $pv = gU$, then

$$\left(\frac{\partial p}{\partial T}\right)_v = \frac{g}{v}C_v .$$

If this tends to zero as $T \to 0$, then from a Maxwell relation one obtains something like equation (6.19):

$$\left(\frac{\partial S}{\partial v}\right)_T \to 0$$

SOLUTIONS OF PROBLEMS

A more general proof would be desirable.

6.9. (i) By (6.15)

$$U = f(T) - \frac{a}{v} + B_1$$

where $df/dT = C_v$. Hence the result.

(ii) From

$$T\left(\frac{\partial S}{\partial T}\right)_v = C_v$$

and

$$T\left(\frac{\partial S}{\partial v}\right)_T = p + \left(\frac{\partial U}{\partial v}\right)_T = p + \frac{a}{v^2} = \frac{cT}{v-b}$$

one finds

$$S = \int \frac{C_v}{T}\,dT + f_1(v)$$

$$S = c\ln(v-b) + f_2(T)$$

whence the results (ii) and (iii) follow.

(iv) For given $T$ the work required is $F(v_2) - F(v_1)$.

CHAPTER 7

7.1. The method follows that of section 7.3 where several of the results are proved.

7.2. For a closed system one can use relations (7.11) and (7.11'). For example if $\delta T = \delta v = 0$ then $\delta(U-TS) \le 0$ for positive temperatures and $\delta(U-TS) \ge 0$ for negative temperatures. Thus $\delta F \le 0$ and $\delta F \ge 0$ in the two cases. The other cases are discussed similarly.

7.3. (*a*) The work done on the system in compressing at pressure $p_1$ is

$$\int_0^{v_1} p_1 \mathrm{d}v_1 = p_1 v_1 .$$

The work done by the system in expanding against pressure $p_2$ is

$$\int_0^{v_2} p_2 \mathrm{d}v_2 = p_2 v_2 .$$

Hence work done by the system is $p_2v_2 - p_1v_1$ . This must be the drop in its internal energy:

$$-(U_2 - U_1) = p_2v_2 - p_1v_1$$

that is

$$U_1 + p_1v_1 = U_2 + p_2v_2$$

that is

$$H_1 = H_2 .$$

(*b*) As the enthalpy $H$ is constant and the pressure is changed, the coefficient to discuss is

$$\mu \equiv \left(\frac{\partial T}{\partial p}\right)_H .$$

Now

$$\mathrm{d}H = T\mathrm{d}S + v\mathrm{d}p = T\left(\frac{\partial S}{\partial p}\right)_T \mathrm{d}p + T\left(\frac{\partial S}{\partial T}\right)_p \mathrm{d}T + v\mathrm{d}p$$

$$= \left\{T\left(\frac{\partial S}{\partial p}\right)_T + v\right\}\mathrm{d}p + T\left(\frac{\partial S}{\partial T}\right)_p \mathrm{d}T .$$

Therefore

$$\mu = -\frac{T(\partial S/\partial p)_T + v}{T(\partial S/\partial T)_p} = \frac{1}{C_p}\left\{T\left(\frac{\partial v}{\partial T}\right)_p - v\right\}$$





by a Maxwell relation. Hence

$$\mu = \frac{v}{C_p}(\alpha_p T - 1) .$$

(*c*) If $\mu = 0$, $\alpha_p = 1/T$,

that is

$$\left(\frac{\partial v}{\partial T}\right)_p = \frac{v}{T}$$

that is

$$\ln v = \ln T - \ln f(p)$$

where $f$ is some function of pressure. Hence the required condition is that there exists a function $f$ of pressure such that

$$f(p)v = AT$$

where $A$ is a constant. In particular, for an ideal classical gas $pv = AT$ and hence $\mu = 0$.

7.4. By the Gibbs-Duhem relation if $n$ intensive variables are present and $n$-1 are kept constant then the $n$th is also constant. Hence for $n = 3$

$$\left(\ \right)_{\mu,T} = \left(\ \right)_{p,\mu,T} = \left(\ \right)_{p,T} = \left(\ \right)_{p,\mu} . \qquad \text{(i)}$$

If all $n$ intensive variables are kept constant, the size of the system is not fixed, but the thermodynamic states are otherwise specified. Hence the ratio of any extensive variables, $E_r$ and $E_s$ say, is fixed since such a ratio is independent of the size of the system and has the value $k_{rs}$ (say):

$$E_r = k_{rs} E_s \qquad (k_{rs} \text{ a constant})$$

$$\frac{\partial E_r}{\partial E_s} = \frac{E_r}{E_s} .$$

By a less direct method, proceed in two steps:

(1) $dN = \left(\frac{\partial N}{\partial v}\right)_{T,p} dv + \left(\frac{\partial N}{\partial T}\right)_{v,P} dT + \left(\frac{\partial N}{\partial p}\right)_{v,T} dp$

$$\left(dp = \frac{S}{v}\,dT + \frac{N}{v}\,d\mu \text{ (Gibbs-Duhem)}\right)$$

$$= \left(\frac{\partial N}{\partial v}\right)_{T,p} dv + \left\{\left(\frac{\partial N}{\partial T}\right)_{v,p} + \left(\frac{\partial N}{\partial p}\right)_{v,T} \frac{S}{v}\right\} dT + \left(\frac{\partial N}{\partial p}\right)_{v,T} \frac{N}{v}\,d\mu$$

Thus a special case of (i) is found:

$$\left(\frac{\partial N}{\partial v}\right)_{T,\mu} = \left(\frac{\partial N}{\partial v}\right)_{T,p} .$$

(2) In $dG = -SdT + vdp + \mu dN$ consider second derivatives:

$$\left(\frac{\partial v}{\partial N}\right)_{T,p} = \left(\frac{\partial \mu}{\partial p}\right)_{T,N} = \frac{v}{N} \text{ (Gibbs-Duhem)} .$$

7.5. (*a*) From the equation of state

$$p = \frac{At}{v-b} - \frac{a}{v^2}$$

$$\left(\frac{\partial p}{\partial v}\right)_t = -\frac{At}{(v-b)^2} + \frac{2a}{v^3} . \qquad \text{(i)}$$

Thus for extrema, labelled by $i$,

$$\frac{v_i^3}{2a} = \frac{(v_i-b)^2}{At} = \frac{(v_i-b)^2}{(p_i+a/v_i^2)(v_i-b)} .$$

Hence the required curve on the $(p,v)$-diagram is given by

$$p_i = \frac{a}{v_i^2} - \frac{2ab}{v_i^3} .$$

To check that these extrema are maxima, note that

$$\left(\frac{\partial^2 p}{\partial v^2}\right)_t = \frac{2At}{(v-b)^3} - \frac{6a}{v^4}$$

which at an extremum becomes



$$\frac{2}{v_i-b}\,\frac{2a}{v_i^3} - \frac{6a}{v_i^4} \doteqdot -\frac{2a}{v_i^4} .$$

(*b*) This last curve has a maximum given by

$$-\frac{2a}{v_i^3} + \frac{6ab}{v_i^4} = \frac{2a}{v_i^4}(3b - v_i) = 0 .$$

To check that this is a maximum, observe that

$$\frac{d^2 p_i}{dv_i^2} = \frac{6a}{v_i^4} - \frac{24ab}{v_i^5} = \frac{6a}{v_i^4}\left(1 - \frac{4b}{v_i}\right) .$$

At the point at which $dp_i/dv_i$ is zero, this quantity is negative as required. Hence, at this maximum, $v_c = 3b$ . Also $p_i = p_c$ at $v_i = 3b$, so that

$$p_c = \frac{a}{9b^2} - \frac{2ab}{27b^3} = \frac{a}{27b^2} .$$

The van der Waals equation at the critical point furnishes now a value of $t_c$:

$$At_c = \left(\frac{a}{27b^2} + \frac{a}{9b^2}\right)(3b - b) = \frac{8a}{27b} .$$

(*c*) An easy algebraic result.

(*d*) We write

$$pv = At\left(1 + \frac{a}{pv^2}\right)^{-1}\left(1 - \frac{b}{v}\right)^{-1} \doteqdot At - \frac{aAt}{pv^2} + \frac{bAt}{v} .$$

For small $a,b$ the approximate equation is $pv = At$. Using this in the two correction terms, we have

$$pv \doteqdot At - \frac{ap}{At} + bp$$

that is

$$E_2 = b - \frac{a}{At} .$$

(*e*) $E_2 = 0$ if $t = t_B = a/Ab$. Hence

$$\frac{t_B}{t_c} = \frac{a}{Ab}\,\frac{27Ab}{8a} = \frac{27}{8} = 3\cdot375$$

(*f*) Recall from equation (6.2) that

$$\left(\frac{\partial U}{\partial v}\right)_T + p = T\left(\frac{\partial p}{\partial T}\right)_v .$$

Differentiating with respect to temperature at constant volume yields

$$\frac{\partial^2 U}{\partial v \partial T} + \left(\frac{\partial p}{\partial T}\right)_v = T\left(\frac{\partial^2 p}{\partial T^2}\right)_v + \left(\frac{\partial p}{\partial T}\right)_v$$

that is

$$\left(\frac{\partial C_v}{\partial v}\right)_T = T\left(\frac{\partial^2 p}{\partial T^2}\right)_v .$$

The right-hand side vanishes for a van der Waals gas.

(*g*) Observe first that

$$\frac{\alpha_p}{K_t} = -\frac{(1/v)(\partial v/\partial t)_p}{(1/v)(\partial v/\partial p)_t} = \left(\frac{\partial p}{\partial t}\right)_v = \frac{A}{v-b} . \qquad \text{(ii)}$$

From (i) above

$$K_t = \left\{\frac{Avt}{(v-b)^2} - \frac{2a}{v^2}\right\}^{-1} , \qquad \alpha_p = \frac{AK_t}{v-b} .$$

On putting $v = v_c$, and keeping $t$ arbitrary,

$$K_t^{-1} = \frac{3A}{4b}\left(t - \frac{8a}{27Ab}\right) = \frac{3A}{4b}\,(t - t_c) . \text{ (iii)}$$

This is the first result. The second result follows from (ii) and (iii):

$$\alpha_p = \frac{AK_t}{2b} = \frac{A}{2b}\,\frac{4b}{3A(t-t_c)} = \frac{2}{3(t-t_c)} .$$



(*h*) By problem 7.3 part (*b*) the coefficient is

$$\frac{1}{C_p}\left[T\left(\frac{\partial v}{\partial T}\right)_p - v\right] .$$

Now $AT = pv - bp + \frac{a}{v} - \frac{ab}{v^2}$

so that

$$A = \left(p - \frac{a}{v^2} + \frac{2ab}{v^3}\right)\left(\frac{\partial v}{\partial T}\right)_p .$$

It follows that the coefficient is

$$\frac{1}{C_p}\left[AT/\left(p - \frac{a}{v^2} + \frac{2ab}{v^3}\right) - v\right]$$

$$= (1/C_p)\left[p - \frac{a}{v^2} + \frac{2ab}{v^3}\right]^{-1}\left[pv - bp + \frac{a}{v} - \frac{ab}{v^2} - v\left(p - \frac{a}{v^2} + \frac{2ab}{v^3}\right)\right]$$

$$= (1/C_p)\left[p - \frac{a}{v^2} + \frac{2ab}{v^3}\right]^{-1}\left[\frac{2a}{v} - \frac{3ab}{v^2} - bp\right] .$$

Hence it vanishes at pairs of values $p_i, v_i$ related by

$$p_i = \frac{2a}{bv_i} - \frac{3a}{v_i^2} .$$

**7.6.** (*a*)
$$\left(\frac{\partial p}{\partial T}\right)_v = \left(\frac{\partial S}{\partial v}\right)_T = \frac{1}{T}\left(\frac{\partial Q}{\partial v}\right)_T = \frac{1}{T}\,\frac{\Lambda^{1\to 2}}{v_2 - v_1}$$

(*b*) Take $\Lambda^{1\to 2} > 0$. Then we have the results given in the table

| | State 1 | State 2 | | $dp/dT$ |
|---|---|---|---|---|
| Vaporization | Liquid | Vapour | $v_2 > v_1$ | $> 0$ |
| Melting (normal) | Solid | Liquid | $v_2 > v_1$ | $> 0$ |
| Melting (ice) | Solid | Liquid | $v_2 < v_1$ | $< 0$ |

(c) We have $\delta T = [T(v_2-v_1)/\Lambda^{1\to 2}]\delta p$ .
One has to take $pv$ as an energy per unit mass if one takes $\Lambda$ as an energy per unit mass. (Alternatively both may be taken simply as energies.) We shall work in SI units. Then, using 1 cal = 4·19 J,

$$v_2-v_1 = -0\cdot 091 \text{ cm}^3\text{g}^{-1} = -0\cdot 091\times 10^{-3} \text{ m}^3 \text{ kg}^{-1}$$

$$\Lambda = 335\cdot 2 \text{ Jg}^{-1} = 335\cdot 2 \times 10^3 \text{ J kg}^{-1}$$

$$\delta p = 1 \text{ atmosphere} = 1\cdot 013 \times 10^5 \text{ N m}^{-2}$$

$$T = 273 \text{ K}.$$

It follows that

$$\delta T = -\frac{273\times 0\cdot 091\times 10^{-3}}{335\cdot 2\times 10^3} \times 1\cdot 013 \times 10^5 = -0\cdot 0075 \text{ K}.$$

In the c.g.s system of units we have with
$\Lambda = 80 \times 4\cdot 19 \times 10^7$ erg $g^{-1}$

$$\delta T = -\frac{273\times 0\cdot 091}{80\times 4\cdot 19\times 10^7} \times 1\cdot 013 \times 10^6 = -0\cdot 0075 \text{ K}.$$

7.8. The equal-area rule in the reduced variables of (7.22) yields:

$$0 = \int_{\tau=\text{const.}} \phi \, d\pi = \int_{\tau=\text{const.}} \phi\left(\frac{\partial\pi}{\partial\phi}\right)_\tau d\phi$$

$$= \int \phi\left[-\frac{24\tau}{(3\phi-1)^2} + \frac{6}{\phi^3}\right] d\phi$$

$$= -24\tau \int \frac{\phi d\phi}{(3\phi-1)^2} + 6\int \phi^{-2}\, d\phi$$

$$= -\frac{24\tau}{9}\left[-\frac{1}{3\phi-1} + \ln(3\phi-1)\right] - \frac{6}{\phi} - B$$



$$= \frac{8}{3}\frac{T+1}{2+3\phi} - \frac{6}{\phi+1} - \frac{8}{3}(T+1)\ln(2+3\phi) - B$$

$$= \frac{4}{3}\frac{1+T}{1+(3/2)\phi} - \frac{6}{1+\phi} - \frac{8}{3}(1+T)\left\{\ln 2+\ln\left(1+\frac{3}{2}\phi\right)\right\} - B$$

$$= \frac{4}{3}(1+T)\left(1 - \frac{3}{2}\phi + \frac{9}{4}\phi^2 - \frac{27}{8}\phi^3 + \frac{81}{16}\phi^4 - \dots\right) - B$$

$$- 6(1 - \phi + \phi^2 - \phi^3 + \phi^4 - \dots)$$

$$- \frac{8}{3}(1+T)\left(\ln 2 + \frac{3}{2}\phi - \frac{9}{8}\phi^2 + \frac{27}{24}\phi^3 - \frac{81}{64}\phi^4 + \dots\right)$$

$$= (1+T)\left(-\frac{8}{3}\ln 2 + \frac{4}{3} - 6\phi + 6\phi^2 - \frac{15}{2}\phi^3 + \frac{81}{8}\phi^4 + \dots\right)$$

$$- 6 + 6\phi - 6\phi^2 + 6\phi^3 - 6\phi^4 - B$$

$$= -\frac{14}{3} - \frac{8}{3}\ln 2 - \frac{3}{2}\phi^3 + \frac{33}{8}\phi^4 \dots$$

$$+ T\left(\frac{4}{3} - \frac{8}{3}\ln 2 - 6\phi + 6\phi^2 + \dots\right) - B .$$

Here $B$ is a constant of integration and we have used

$$(1+x)^{-1} = 1 - x + x^2 - x^3 + x^4 \dots$$

$$\ln(1+x) = x - \tfrac{1}{2}x^2 + \tfrac{1}{3}x^3 - \tfrac{1}{4}x^4 + \dots .$$

If the values of $\phi$ at the ends of the range of integration are $\phi_1$, $\phi_2$ then, putting $\phi = \phi_1, \phi_2$ and taking the difference between the two equations, yields the result stated in problem 7.8 . The expression vanishes for $\phi_1 = \phi_2$; and if $\phi_1 \neq \phi_2$ it vanishes for $\phi_1 + \phi_2 = 0$ provided $3/2\ \phi_1^2 = -6T$, which is possible since $T < 0$. This condition also holds at the critical point when $\phi_1 \to 0$ and $T \to 0$.

7.9. The simplest solution is via the Jacobian matrix

$$\frac{\partial(u_1,u_2,\ldots,u_n)}{\partial(y_1,y_2,\ldots,y_n)} \equiv \begin{pmatrix} \partial u_1/\partial y_1 & \cdots & \partial u_1/\partial y_n \\ \vdots & & \vdots \\ \partial u_n/\partial y_1 & \cdots & \partial u_n/\partial y_n \end{pmatrix} .$$

Let $u_i$ be $n$ functions each of which can depend on the $n$ independent variables $y_j$, which themselves can each depend on the $n$ independent variables $x_k$. Then

$$\frac{\partial u_i}{\partial x_j} = \sum_{k=1}^{n} \frac{\partial u_i}{\partial y_k}\frac{\partial y_k}{\partial x_j} \qquad (i,j = 1,2,\ldots,n) .$$

These $n^2$ equations are equivalent to the single matrix equation

$$\frac{\partial(u_1\ldots)}{\partial(x_1\ldots)} = \frac{\partial(u_1\ldots)}{\partial(y_1\ldots)}\frac{\partial(y_1\ldots)}{\partial(x_1\ldots)} . \qquad \text{(i)}$$

Note that (i) with $n = 2$ and $u_2 = x_2$ implies the following relation for the determinant

$$\frac{\partial(u_1,u_2)}{\partial(x_1,u_2)} = \begin{vmatrix} (\partial u_1/\partial x_1)_{x_2} & (\partial u_1/\partial u_2)_{x_1} \\ (\partial u_2/\partial x_1)_{u_2} & (\partial u_2/\partial u_2)_{x_1} \end{vmatrix} = \begin{vmatrix} (\partial u_1/\partial x_1)_{x_2} & (\partial u_1/\partial u_2)_{x_1} \\ 0 & 1 \end{vmatrix}$$

$$= (\partial u_1/\partial x_1)_{x_2} . \qquad \text{(ii)}$$

With $u_1 = p$, $u_2 = T$; $y_1 = v$, $y_2 = T$; $x_1 = v$, $x_2 = S$ one finds from (i) and (ii)

$$\frac{\partial(p,T)}{\partial(v,S)} = \frac{\partial(p,T)}{\partial(v,T)}\frac{\partial(v,T)}{\partial(v,S)} = \left(\frac{\partial p}{\partial v}\right)_T \left(\frac{\partial T}{\partial S}\right)_v = -\frac{1}{vK_T}\frac{T}{C_v} . \qquad \text{(iii)}$$

This is the required result; the last line is enjoyably direct.





The following solution does not rely on Jacobians.

Observe that

$$dp = \left(\frac{\partial p}{\partial v}\right)_S dv + \left(\frac{\partial p}{\partial S}\right)_v \left\{\left(\frac{\partial S}{\partial v}\right)_T dv + \left(\frac{\partial S}{\partial T}\right)_v dT\right\}$$

so that

$$\left(\frac{\partial p}{\partial v}\right)_T = \left(\frac{\partial p}{\partial v}\right)_S + \left(\frac{\partial p}{\partial S}\right)_v \left(\frac{\partial S}{\partial v}\right)_T .$$

The left-hand side of the equation to be proved is therefore

$$-\left(\frac{\partial T}{\partial S}\right)_v \left(\frac{\partial p}{\partial v}\right)_T + \left(\frac{\partial T}{\partial S}\right)_v \left(\frac{\partial p}{\partial S}\right)_v \left(\frac{\partial S}{\partial v}\right)_T - \left(\frac{\partial p}{\partial S}\right)_v^2$$

$$= -\left(\frac{\partial T}{\partial S}\right)_v \left(\frac{\partial p}{\partial v}\right)_T \left[1 - \left(\frac{\partial S}{\partial p}\right)_T \left(\frac{\partial p}{\partial S}\right)_v + \left(\frac{\partial S}{\partial T}\right)_v \left(\frac{\partial v}{\partial p}\right)_T \left(\frac{\partial p}{\partial S}\right)_v^2\right] .$$

The last two terms cancel since the last term is

$$\left(\frac{\partial v}{\partial p}\right)_T \left(\frac{\partial p}{\partial T}\right)_v \left(\frac{\partial p}{\partial S}\right)_v = -\left(\frac{\partial v}{\partial T}\right)_p \left(\frac{\partial p}{\partial S}\right)_v = \left(\frac{\partial S}{\partial p}\right)_T \left(\frac{\partial p}{\partial S}\right)_v$$

where a Maxwell relation has been used in the last step. Hence we have obtained the right-hand side as

$$-\left(\frac{\partial T}{\partial S}\right)_v \left(\frac{\partial p}{\partial v}\right)_T$$

which is $T/v\, C_v\, K_T$ . This is unduly long compared with (iii).

7.10. Imagine $\dot{x}$ to be plotted as a function of $x$. The curve, being a cubic with three positive roots of $\dot{x} = 0$, intersects the $\dot{x} = 0$ axis three times. If $a_3 < 0$, then

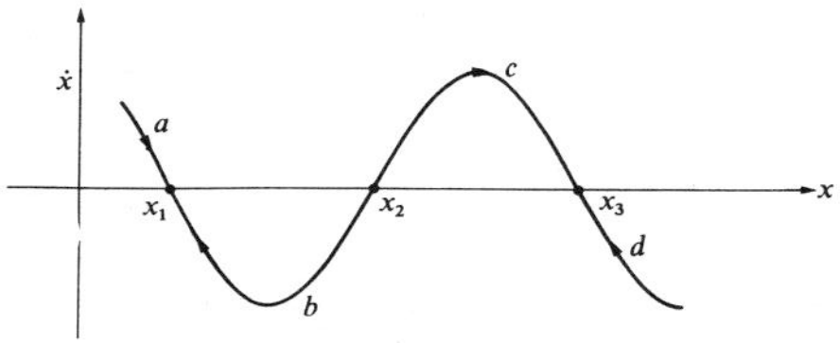


Fig.S7.1.

$\dot{x} < 0$ for large $x$, (Fig, S7.1). We now attach arrows to indicate which way $x$ will tend to change in each section of the curve labelled a, b, c, d. On a and c $x$ will tend to increase; on b and d $x$ will tend to decrease. Hence after a small departure from the root $x = x_1$ $x$ will tend to return to $x_1$. The same applies to $x_3$, and these are stable roots. Similarly, $x_2$ is an unstable root. One can show similarly that only one root is stable if $a_3 > 0$.

7.11. From (3.1) and a Maxwell relation

$$TdS = C_p dT + T\left(\frac{\partial S}{\partial p}\right)_T dp = C_p dT - T\left(\frac{\partial v}{\partial T}\right)_p dp$$

$$= C_B dT + T\left(\frac{\partial M}{\partial T}\right)_B dB$$

whence

$$\left(\frac{\partial T}{\partial B}\right)_S = -\frac{T}{C_B}\left(\frac{\partial M}{\partial T}\right)_B .$$



We have

$$\left(\frac{\partial T}{\partial B}\right)_S \left(\frac{\partial S}{\partial T}\right)_B \left(\frac{\partial B}{\partial S}\right)_T = -1 .$$

The second factor is liable to be positive as entropy increases with temperature. The third factor is liable to be negative since the ordering effect of a magnetic field should decrease the entropy. This makes $(\partial T/\partial B)_S$ liable to be positive, as required.

## CHAPTER 8

8.1. Since

$$p = \frac{AT}{v-b} - \frac{a}{v^2}, \quad A \equiv NkT$$

it follows that for fixed $T$

$$dp = -\frac{AT}{(v-b)^2} dv + \frac{2a}{v^3} dv .$$

Now from

$$G_2 - G_1 = \int v\,dp = NkT \ln \frac{f_2}{f_1} ,$$

$$-NkT \ln\frac{f_2}{f_1} = -\int_1^2 v\,dp = \int \left\{\frac{ATv}{(v-b)^2} - \frac{2a}{v^2}\right\} dv$$

$$= AT\left|\ln(v-b) - \frac{b}{v-b}\right|_1^2 + \left|\frac{2a}{v}\right|_1^2$$

$$= AT\ln\frac{v_2-b}{v_1-b} - bAT\left(\frac{1}{v_2-b} - \frac{1}{v_1-b}\right) + 2a\left(\frac{1}{v_2} - \frac{1}{v_1}\right)$$

therefore

$$\ln f_2 = \ln f_1 - \ln\frac{v_2-b}{v_1-b} + b\left(\frac{1}{v_2-b} - \frac{1}{v_1-b}\right) - \frac{2a}{NkT}\left(\frac{1}{v_2} - \frac{1}{v_1}\right) .$$

Take state 1 to be an ideal gas state (i.e. in the limit $p_1 \to 0$) distinguished by subscript 0. Then dropping subscripts 2 for the general state,

$$\ln f = \ln p_0 + \ln\frac{v_0-b}{v-b} + b\left(\frac{1}{v-b} - \frac{1}{v_0-b}\right) - \frac{2a}{NkT}\left(\frac{1}{v} - \frac{1}{v_0}\right) .$$

Since $v_0 \gg b$,

$$p_0v_0 \sim NkT, \quad p_0(v_0-b) \sim NkT, \quad \frac{b}{v_0-b} \sim 0, \quad \frac{1}{v_0} \sim 0$$

$$\ln f \doteqdot \ln\frac{NkT}{v_0-b} + \ln\frac{v_0-b}{v-b} + \frac{b}{v-b} - \frac{2a}{NkTv}$$

$$\ln f \doteqdot \ln\frac{NkT}{v-b} + \frac{b}{v-b} - \frac{2a}{NkTv} .$$

In the ideal gas case one sees that $f = p$.

8.2. From $p = -(\partial F/\partial v)_{T,N}$ and $\mu = (\partial F/\partial N)_{T,V}$ one finds with $f \equiv f(N,T)$

$$F = -NkT\ln(v-bN) - aN^2/v + f$$

$$\mu = -kT\ln(v-bN) + \frac{bNkT}{v-bN} - \frac{2aN}{v} + \frac{\partial f}{\partial N} .$$

Taking $T$, $v$, and $N$ as independent variables we write from the Gibbs-Duhem equation $S\mathrm{d}T = v\mathrm{d}p - N\mathrm{d}\mu$ . On substituting for d$p$ and d$\mu$ from the equations for $p$ and $\mu$ in terms of d$T$, d$V$ and d$N$, this yields a relation of the form

$$S\mathrm{d}T = \alpha\mathrm{d}T + \beta\mathrm{d}V + \gamma\mathrm{d}N . \qquad \text{(i)}$$

In fact, the coefficient $\beta$ vanishes identically. The condition for $\gamma$ to vanish is found after some algebra to be

$$\left(\frac{\partial^2 f}{\partial N^2}\right)_T = \frac{kT}{N}$$

whence the existence of functions $\phi(T)$, $h(T)$ is inferred such that

$$f = NkT(\ln N-1) - N\phi + h. \qquad \text{(ii)}$$



SOLUTIONS OF PROBLEMS

One can substitute this expression into $F$ and $\mu$ and calculate $\mu N - F = G - F = pv$. The right-hand side is known and this comparison shows that $h(T) = 0$. The function $\alpha$ in (i) is of course $S$ but involves $f$ which can be eliminated using (ii) in favour of $\phi(T)$. Hence the results stated in the problem are obtained: $\mu N$ is $G$, $pv$ is obtained from $p$, $F$ and $S$ yield $U = F + TS$, and $C_v$ is found as $(\partial U/\partial T)_{v,N}$.

CHAPTER 9

9.1. Consider

$$f = -k\sum_i (p_i\ln p_i - \alpha p_i + \beta p_i x_i)$$

where $\alpha,\beta$ are Lagrangian multipliers. We have

$$\frac{\partial f}{\partial p_j} = -k(\ln p_j + 1 - \alpha + \beta x_j) .$$

It follows that

$$p_j = \frac{1}{Z(x)}\exp(-\beta x_j) \qquad (j = 1,2,\ldots)$$

where

$$Z(x) \equiv \sum_i \exp(-\beta x_i) .$$

Also

$$x_0 = \sum p_i x_i = \frac{1}{Z(x)}\sum_i x_i\exp(-\beta x_i) = -\left\{\frac{\partial\ln Z(x)}{\partial\beta}\right\}_{x_1,x_2,\ldots} .$$

Lastly

$$S_2 \equiv S_{\max} = +k\sum_i p_i\{\beta x_i + \ln Z(x)\}$$

$$= k\beta x_0 + k\ln Z .$$

9.2. Proceeding as in the preceding problem with $\mathbf{i} = (i_1,i_2)$

$$f = -k\sum_{\mathbf{i}} (p_{\mathbf{i}}\ln p_{\mathbf{i}} - \alpha p_{\mathbf{i}} + \beta p_{\mathbf{i}}x_{i_1} + \gamma p_{\mathbf{i}}y_{i_2}) ,$$

$$\frac{\partial f}{\partial p_j} = -k(\ln p_j + 1 - \alpha + \beta x_{j_1} + \gamma y_{j_2}).$$

Hence

$$p_j = \frac{1}{\Xi}\exp(-\beta x_{j_1} - \gamma y_{j_2})$$

where

$$\Xi \equiv \sum_i \exp(-\beta x_{i_1} - \gamma y_{i_2}) .$$

Also

$$x_0 = \sum_i p_i x_{i_1} = -\left(\frac{\partial \ln \Xi}{\partial \beta}\right)_{x_i, y_i, \gamma} , \qquad y_0 = -\left(\frac{\partial \ln \Xi}{\partial \gamma}\right)_{x_i, y_i, \beta} .$$

Lastly

$$S_3 \equiv S_{max} = k \sum_i p_i(\beta x_{i_1} + \gamma y_{i_2} + \ln \Xi) = k\beta x_0 + k\gamma y_0 + k \ln \Xi .$$

9.3. In the case of $S_2, x_0$ is now the internal energy $U$, and we write

$$S_2 = k\beta U + k \ln Z = \frac{U-F}{T} .$$

In the case of $S_3$, $y_0$ is the mean number of particles $n$, and we write

$$S_3 = k\beta U + k\gamma n + k \ln \Xi = \frac{U - \mu n + pv}{T} .$$

The identifications proposed in the problem follow.

9.4. (*a*) The entropy is

$$S = -k \int p(x) \ln p(x) \mathrm{d}x$$

$$= -k \int p(x) \left(-\frac{\ln 2\pi\sigma^2}{2} - \frac{x^2}{2\sigma^2}\right) \mathrm{d}x$$

$$= \frac{k}{2}\ln(2\pi\sigma^2) + \frac{k}{2\sigma^2}\int x^2 p(x)\mathrm{d}x$$



$$= \tfrac{1}{2}k \ln(2\pi e\sigma^2) .$$

(*b*) We must maximize with respect to arbitrary variations of $p(x)$ in the integrand of

$$f\{p(x)\} \equiv -k\int_{-\infty}^{\infty} p(x)\ln p(x)\mathrm{d}x - \alpha\int_{-\infty}^{\infty} p(x)\mathrm{d}x - \beta\int_{-\infty}^{\infty} x^2 p(x)\mathrm{d}x$$

where $\alpha, \beta$ are Lagrangian multipliers. Hence

$$-k - k\ln p(x) - \alpha - \beta x^2 = 0 .$$

It follows that

$$p(x) = a\exp(-\beta x^2)$$

for the maximum. The normalization yields

$$a\int_{-\infty}^{\infty}\exp(-\beta x^2)\mathrm{d}x = a\left(\frac{\pi}{\beta}\right)^{1/2} = 1 .$$

Also

$$\sigma^2 = a\int_{-\infty}^{\infty} x^2\exp(-\beta x^2)\mathrm{d}x = \frac{a}{2}\left(\frac{\pi}{\beta^3}\right)^{1/2} = \frac{1}{2\beta} .$$

Hence

$$p(x) = \frac{1}{(2\pi\sigma^2)^{1/2}}\exp\left(-\frac{x^2}{2\sigma^2}\right) ,$$

under the conditions stated. Note that relevant formulae for integrals are given on p. 436

9.5. (*a*) In any State of the ensemble, let us give systems with the same energy the same letter, and systems with different energies different letters. Hence $G(n_1, n_2, \ldots)$ gives the number of distinguishable arrangements of $N$ letters, $n_1$ of one type, and $n_2$ of another type, etc., and is equal to $N!/n_1!n_2!\ldots$ if the degeneracies are neglected. As a result of the degeneracy $g_j$, each of the $G$ arrangements gives rise to a number of further arrangements equal to the number of ways of assigning $n_j$ systems to $g_j$ states, i.e. to $g_j^{n_j}$ arrangements. The result of part ($a$) is thus obtained.

($b$) This matter is discussed in many books.

($c$) From part ($b$) and the definition of $\Psi(n)$, we have

$$\Psi(n) \doteqdot \frac{d}{dn}\{n \ln n - n + \tfrac{1}{2}\ln(2\pi n)\} = \ln n + 1 - 1 + 1/2n$$

$$\doteqdot \ln n + \ln(1+1/2n) = \ln(n+\tfrac{1}{2}) \ .$$

From tables we find $\Psi(3) = 1{\cdot}2561$, $\ln(3{\cdot}5) = 1{\cdot}2528$, so that the error in $\Psi(3) \doteqdot \ln(3{\cdot}5)$ is $0{\cdot}0033/1{\cdot}2561 = 0{\cdot}26$ per cent. It is less for $n > 3$.

($d$) For the most probable State of the ensemble, we have to consider

$$f \equiv \ln G(n_1, n_2, \ldots) - \alpha \sum_j n_j - \beta \sum_j E_j n_j$$

where $\alpha$ and $\beta$ are undetermined multipliers, to take account of the conditions that $N$ and $E_0$ are given, i.e.

$$\sum_j n_j = N, \quad \sum_j E_j n_j = E_0 \ .$$

Instead of maximizing $G$ with respect to each $n_j$, it is more convenient to maximize $\ln G$. We have

$$\frac{\partial f}{\partial n_j} = \frac{\partial}{\partial n_j}(-\ln n_j! + n_j \ln g_j - \alpha n_j - \beta E_j n_j)$$

$$= -\Psi(n_j) + \ln g_j - \alpha - \beta E_j = 0 \ .$$

Hence, if $(n_1^*, n_2^*, \ldots)$ is the most probable state,

$$\Psi(n_j^*) = \ln[g_j \exp(-\alpha - \beta E_j)] \ ,$$

i.e.

$$n_j^* = g_j \exp(-\alpha - \beta E_j) - \tfrac{1}{2} \ .$$



If there are $M$ energy levels, $\alpha$ may be identified by summing over the energy levels to find

$$N + \tfrac{1}{2}M = \exp(-\alpha) \sum_j g_j \exp(-\beta E_j) = Z \exp(-\alpha)$$

where the canonical partition function has been introduced. It follows that

$$\frac{n_j^*}{N} = \left(1 + \frac{M}{2N}\right) \frac{g_j \exp(-\beta E_j)}{Z} - \frac{1}{2N} \ .$$

When the terms in $1/N$ are collected together, the desired result is obtained.

($e$) The most probable State of the ensemble leads to the most probable values $n_j^*$. The canonical ensemble of problem 9.1 yields the mean values of the $n_j$ as averaged over all States of the ensemble. In this average the most probable State makes a dominant contribution, but less probable States will also contribute, and hence the different results which are, strictly speaking, obtained by the two methods.

The present method requires the $n_j^*$ to be large enough for the assumption of continuity to be justifiable. This condition is difficult to fulfil unless $N \to \infty$. Although $M$ is often also infinitely large the usual result follows only if $M/N \to 0$. These difficulties are often overlooked. It is particularly easy to overlook them if the coarser approximation $\Psi(n) \sim \ln n$ is used, which yields at once

$$\frac{n_j^*}{N} = \frac{g_j \exp(-\beta E_j)}{Z} \ .$$

9.6. ($a$) Let $u$ refer to the upper level, and $\ell$ to the lower level. Then

$$n_\ell + n_u = N, \quad n_\ell - n_u = n$$

whence

$$n_\ell = \tfrac{1}{2}(N+n), \quad n_u = \tfrac{1}{2}(N-n) \ .$$

Since $N$ is even, it follows that $n$ is even.

The number of ways of realizing a State $n$ is, from problem 9.5($a$) with $g_1 = g_2 = 1$,

$$G(n) = \frac{N!}{\{\frac{1}{2}(N+n)\}!\{\frac{1}{2}(N-n)\}!} .$$

Hence using problem 9.5($b$)

$$\ln G(n) = N\ln N - N + \tfrac{1}{2}\ln(2\pi N) - \tfrac{1}{2}(N+n)\ln\{\tfrac{1}{2}(N+n)\} + \tfrac{1}{2}(N+n)$$

$$-\tfrac{1}{2}\ln\{\pi(N+n)\} - \tfrac{1}{2}(N-n)\ln\{\tfrac{1}{2}(N-n)\} + \tfrac{1}{2}(N-n) - \tfrac{1}{2}\ln\{\pi(N-n)\}$$

$$= N\ln 2 + \tfrac{1}{2}\ln\left(\frac{2}{\pi N}\right) - \tfrac{1}{2}(N+n+1)\ln\left(1+\frac{n}{N}\right) - \tfrac{1}{2}(N-n+1)\ln\left(1-\frac{n}{N}\right) .$$

Using $\ln(1+x) = x - \frac{1}{2}x^2$ for $x \ll 1$, we find

$$\ln G(n) = N\ln 2 + \tfrac{1}{2}\ln\left(\frac{2}{\pi N}\right) - \frac{n^2}{N} + \frac{n^2(N+1)}{2N^2}$$

whence the result follows.

($b$) The number of States of the ensemble is $2^N$ since each of $N$ systems can be in one of two states. Hence

$$P(n) = \frac{G(n)}{2^N} .$$

To test the normalization, put $y = \frac{1}{2}(N+n)$, when $\frac{1}{2}(N-n) = N-y$. Then $y = 0$ at $n = -N$ and $y = N$ at $n = N$, and the corresponding values are given below:

| | | | | | | | |
|---|---|---|---|---|---|---|---|
| $y$ | 0 | 1 | ... | $\frac{1}{2}N$ | $\frac{1}{2}N+1$ | ... | $N$ |
| $n$ | $-N$ | $-N+2$ | | 0 | 2 | | $N$ |

Observe also that

$$(a+b)^N = \sum_{y=0}^{N} \frac{N!}{y!(N-y)!} a^y b^{N-y}$$

whence

$$\sum_{y=0}^{N} \frac{N!}{y!(N-y)!} = 2^N .$$



Using all these results, we can verify the correct normalization of $P(n)$:

$$\sum_{\substack{n=-N \\ (\text{even } n)}}^{N} P(n) = 2^{-N} \sum_{\substack{n=-N \\ (\text{even } n)}}^{N} G(n) = 2^{-N} \sum_{y=0}^{N} \frac{N!}{y!(N-y)!} = 1 .$$

Hence the value of $A$ introduced in the problem is unity.

One may be tempted to check the normalization by evaluating the integral

$$\int_{-\infty}^{\infty} P(n)\,dn = \left(\frac{2}{\pi N}\right)^{1/2} \int_{-\infty}^{\infty} \exp\left(-\frac{n^2}{2N}\right) dn = \left(\frac{2}{\pi N}\right)^{1/2} (2N\pi)^{1/2} = 2.$$

This suggests that the normalization is incorrect. However, an approximate formula for $n \ll N$ has here been used for $-N \le n \le N$ as $n$ and $N \to \infty$, and the fact that $n$ takes only every other integral value has been ignored. This supplies a correction factor of $\frac{1}{2}$.

($c$) The symmetry between the two states of the system gives $G(n) = G(-n)$, whence the mean value of $n$ is

$$\sum_{\substack{n=-N \\ (\text{even } n)}}^{N} nP(n) = 2^{-N} \sum nG(n) = 0.$$

Direct algebraic study of the original expression for $G(n)$ in terms of factorials gives the most probable value of $n$ as $n = 0$. This is also the value obtained from the expression given in part ($b$). Hence the mean and the most probable values coincide in this case.

($d$)

$$G_T = 2^N, \quad G(0) = \frac{N!}{\{(\frac{1}{2}N)!\}^2}$$

and one has $G(0) \sim 2^N \surd(2/\pi N)$ so that

$$f_1 \equiv \frac{G_T - G(0)}{G_T} = 1 - \left(\frac{2}{\pi N}\right)^{\frac{1}{2}} \to 1 .$$

Also

$$S_T = kN\ln 2, \quad S(0) \equiv k\ln G(0).$$

Hence

$$f_2 \equiv \frac{S_T - S(0)}{S_T} = \frac{\ln(\frac{1}{2}\pi N)}{2N \ln 2} \to 0.$$

Thus the value of $f$, which gives the fractional error in replacing all States by the most probable States in a calculation, is large in a calculation of the number of states, but small for the entropy. This circumstance is typical and shows that the entropy is a very *insensitive* function.

The probability $P(n)$ has a maximum at $n = 0$. This is not, however, a sufficient condition for an average over the ensemble to be approximated well by the most probable States. One must show that this maximum is also very steep. As a measure of the 'steepness' one could take $G(0)$ divided by the standard deviation. This ratio is

$$\frac{2^N (2/\pi N)^{1/2}}{N^{1/2}} = \left(\frac{2}{\pi}\right)^{1/2} \frac{2^N}{N}$$

and it indicates infinite 'steepness' as $N \to \infty$. The replacement of all States by the most probable States is therefore justifiable in this case.

(*a*) The independence of the probability distributions $p_i(x_i)$ implies that the probability of finding a state $x_1, x_2, \ldots, x_N$ of the system is

$$P(x_1, x_2, \ldots, x_N) = \prod_{i=1}^{N} p_i(x_i) .$$

Hence the entropy is

$$S = -k \int \ldots \int p_1 \ldots p_N \ln(p_1 \ldots p_N) \mathrm{d}x_1 \ldots \mathrm{d}x_N$$

(*b*) The following constraints apply to the $p_i$'s:



$$\int_{-\infty}^{\infty} p_i(V_i)\mathrm{d}V_i = 1 \qquad \int_{-\infty}^{\infty} p_i(\tfrac{1}{2}mV_i^2)\mathrm{d}V_i = \tfrac{1}{2}kT \qquad (i = 1,2,3) .$$

Maximizing the entropy subject to these constraints, consider

$$f = -k \sum_{i=1}^{3} \left(\int p_i \ln p_i \mathrm{d}V_i + \alpha_i' \int p_i \mathrm{d}V_i + \beta_i \int p_i V_i^2 \mathrm{d}V_i\right) .$$

Hence

$$\frac{\partial f}{\partial p_j} = -k \int (\ln p_j + \alpha_j' + \beta_j V_j^2 + 1)\mathrm{d}V_j = 0$$

whence

$$p_j = \exp(-\beta_j V_j^2 - \alpha_j), \qquad \alpha_j \equiv \alpha_j' + 1 .$$

To identify the Lagrangian multipliers observe, using the integrals given on p. 456

$$\int_{-\infty}^{\infty} \exp(-\beta_j V_j^2 - \alpha_j)\mathrm{d}V_j = 1 \quad \text{i.e.} \quad \left(\frac{\pi}{\beta_j}\right)^{1/2} \exp(-\alpha_j) = 1$$

$$\int_{-\infty}^{\infty} V_j^2 \exp(-\beta_j V_j^2 - \alpha_j)\mathrm{d}V_j = \frac{kT}{m} .$$

This yields

$$\left(\frac{\pi}{\beta_j}\right)^{1/2} \exp(-\alpha_j) = 1 = \frac{m}{2kT\beta_j} \left(\frac{\pi}{\beta_j}\right)^{1/2} \exp(-\alpha_j),$$

so that

$$\exp(-\alpha_j) = \left(\frac{m}{2\pi kT}\right)^{1/2} \qquad \beta_j = \frac{m}{2kT}$$

and the stated expression for $p_i$ follows.

(*c*) The Jacobian for the transformation to polar coordinates $V_1 = V \sin\theta \cos\phi$, $V_2 = V \sin\theta \sin\phi$, $V_3 = V \cos\theta$ is $J = V^2 \sin\theta$, so that $\mathrm{d}V_1 \mathrm{d}V_2 \mathrm{d}V_3$ is after $\theta$- and $\phi$-integrations $V^2 \mathrm{d}V \int_0^{2\pi} \mathrm{d}\phi \int_0^{\pi} \sin\theta \, \mathrm{d}\theta = 4\pi V^2 \mathrm{d}V$.

To obtain the expression for $J$, note that it is

$$\frac{\partial(V_1,V_2,V_3)}{\partial(\theta,\phi,V)} \equiv \begin{vmatrix} \partial V_1/\partial\theta & \partial V_1/\partial\phi & \partial V_1/\partial V \\ \partial V_2/\partial\theta & \partial V_2/\partial\phi & \partial V_2/\partial V \\ \partial V_3/\partial\theta & \partial V_3/\partial\phi & \partial V_3/\partial V \end{vmatrix} = \begin{vmatrix} V\cos\theta\cos\phi & -V\sin\theta\sin\phi & \sin\theta\cos\phi \\ V\cos\theta\sin\phi & V\sin\theta\cos\phi & \sin\theta\sin\phi \\ -V\sin\theta & 0 & \cos\theta \end{vmatrix}.$$

Also one has to use with $f(V_1,V_2,V_3) \equiv g(\theta,\phi,V)$

$$f(V_1,V_2,V_3)\mathrm{d}V_1\mathrm{d}V_2\mathrm{d}V_3 = g(\theta,\phi,V)J\mathrm{d}\theta\mathrm{d}\phi\mathrm{d}V \ .$$

(*d*) The main assumption is that $p_1(V_1), p_2(V_2), p_3(V_3)$ are independent probability distributions and that the point particles have only point interactions. In a dense gas the interactions cannot be approximated in this way. It has also been assumed that one particle can be treated separately from the rest. This is invalid for a system of indistinguishable particles when exchange effects must be expected.

(*e*) We have

$$\bar{e} = \int_0^\infty \tfrac{1}{2} mV^2 \, P(V)\mathrm{d}V$$

$$= \int_0^\infty e \, \frac{2\sqrt{e}}{\sqrt{\pi}(kT)^{3/2}} \exp\left(-\frac{e}{kT}\right)\mathrm{d}e$$

$$= \frac{2}{kT\sqrt{(\pi kT)}} \int_0^\infty e^{3/2} \exp\left(-\frac{e}{kT}\right) \mathrm{d}e$$

$$= \frac{2}{\sqrt{\pi}} kT \int_0^\infty x^{3/2} \exp(-x)\mathrm{d}x$$

$$= \frac{3}{2} kT.$$

9.8. (*a*) Using (9.11)

$$\frac{\partial \ln Q}{\partial \beta_r} = -\frac{\Sigma_{j_1}\ldots\Sigma_{j_t} X_{rj_r} \exp(-\beta_1 X_{1j_1} - \ldots)}{\Sigma_{j_1}\ldots\Sigma_{j_t} \exp(-\beta_1 X_{1j_1} - \ldots)} = -\bar{X}_r \ .$$

Note that the denominator in the centre of this expression is just $Q$.

(*b*)

$$\frac{\partial^2 \ln Q}{\partial\beta_r \partial\beta_s} = Q^{-2}\{Q \,\Sigma_{j_1}..\Sigma_{j_t} X_{rj_r} X_{sj_s} \exp(-\beta_1 X_{1j_1}\ldots) -$$

$$[\Sigma_{j_1}\ldots\Sigma_{j_t} X_{rj_r} \exp(-\beta_1 X_{1j_1}\ldots)][\Sigma_{j_1}\ldots\Sigma_{j_t} X_{sj_s} \exp(-\beta_1 X_{1j_1} - \ldots)]$$

$$= \overline{X_r X_s} - \bar{X}_r \bar{X}_s$$

$$= -\frac{\partial \bar{X}_r}{\partial \beta_s} \text{ by } (a) \ .$$

Note also that

$$\overline{\Delta X_r \Delta X_s} = \overline{(X_{rj_r} - \bar{X}_r)(X_{sj_s} - \bar{X}_s)} = \overline{X_r X_s} - \bar{X}_r\bar{X}_s - \bar{X}_r\bar{X}_s + \bar{X}_r\bar{X}_s \ .$$

This proves all parts of (*b*)

(*c*) $-\overline{\ln p_{\mathbf{i}}} = \ln Q + \sum_j \beta_j \bar{X}_j$

by (9.10), and the result is $\bar{S}/k$ by (9.13).

(*d*) $\Delta(-\ln p_{\mathbf{i}}) \equiv -\ln p_{\mathbf{i}} + \overline{\ln p_{\mathbf{i}}} = \ln Q + \Sigma\, \beta_j X_{ji_j} - \ln Q - \sum_j \beta_j \bar{X}_j$,

where (9.10) has been used.

(*e*) $\overline{\Delta(-\ln p_{\mathbf{i}})\Delta(-\ln p_{\mathbf{j}})} = \overline{\sum_{r,s} \beta_r \Delta X_r \beta_s \Delta X_s} = \sum_{r,s} \beta_r\beta_s \overline{\Delta X_r \Delta X_s}$

$$= -\sum_{r,s} \beta_r\beta_s \frac{\partial \bar{X}_r}{\partial \beta_s} \ ,$$

where a result of part (*b*) has been used.

9.9. (*a*) Using (9.13)

$$\bar{S} = k \sum_{r=1}^{t} \beta_r \bar{X}_r + k \ln Q$$

where by (9.12) $Q = Q(\beta_1,\ldots,\beta_t)$. It follows that

$$\left(\frac{\partial \bar{S}}{\partial \bar{X}_s}\right)_{\bar{X}} = k\beta_S \ .$$

(*b*) If $(\ )_\beta$ means that the 'other' $\beta$'s are kept constant, then by (*a*)

$$\left(\frac{\partial \bar{S}}{\partial \beta_s}\right)_\beta = \sum_r \left(\frac{\partial \bar{S}}{\partial \bar{X}_r}\right)_{\bar{X}} \left(\frac{\partial \bar{X}_r}{\partial \beta_s}\right)_\beta = k \sum_r \beta_r \left(\frac{\partial \bar{X}_r}{\partial \beta_s}\right)_\beta$$

(*c*) From $\gamma_1 = 1/kT$, $\beta_r = -\gamma_1 \gamma_r (r = 2,3,\ldots)$ it follows that $d\gamma_1/dT = -1/kT^2 = -\gamma_1/T$, so that

$$-\gamma_1 \left(\frac{\partial \bar{S}}{\partial \gamma_1}\right)_\gamma = -\gamma_1 \left(\frac{\partial \bar{S}}{\partial T}\right)_\gamma \frac{dT}{d\gamma_1} = T\left(\frac{\partial \bar{S}}{\partial T}\right)_\gamma .$$

Also

$$\gamma_1 \left(\frac{\partial \beta r}{\partial \gamma_1}\right)_\gamma = -\gamma_1 \gamma_r = \beta_r .$$

(*d*) We have, using results from parts (*b*) and (*c*),

$$C_\gamma \equiv T\left(\frac{\partial \bar{S}}{\partial T}\right)_\gamma = -\gamma_1 \left(\frac{\partial \bar{S}}{\partial \gamma_1}\right)_\gamma = -\gamma_1 \sum_s \left(\frac{\partial \bar{S}}{\partial \beta_s}\right)_\beta \left(\frac{\partial \beta_s}{\partial \gamma_1}\right)_\gamma$$

$$= -k \sum_{r,s} \beta_r \beta_s \left(\frac{\partial \bar{X}_r}{\partial \beta_s}\right)_\beta$$

Using problem 9.8 one finds the required result:

$$C_\gamma = k\, \overline{\Delta(-\ln p_i)\Delta(-\ln p_j)} .$$

9.10. Let

$$F = -k\left(\int_{-\infty}^{\infty} p \ln p \, dx + \lambda M_\nu^\nu + \mu \int_{-\infty}^{\infty} p \, dx\right)$$

where $\lambda$ and $\mu$ are Lagrangian multipliers. The maximization of $F$ leads then to the required distribution $p$ of maximum entropy. One finds

$$\frac{\partial F}{\partial p} = -k \int_{-\infty}^{\infty} (\ln p + 1 + \lambda |x-a|^\nu + \mu)\, dx = 0.$$

The distribution is

$$p = \exp(-1-\mu-\lambda|x-a|^\nu) . \qquad \text{(i)}$$

Normalization yields

$$\exp(1+\mu) = \int_{-\infty}^{\infty} \exp[-\lambda|x-a|^\nu]\, dx = 2\int_a^\infty \exp\{-\lambda(x-a)^\nu\}\, dx$$

$$= 2\lambda^{-1/\nu}\Gamma\left(1+\frac{1}{\nu}\right) , \qquad \text{(ii)}$$

which is obtained by changing the variable of integration from $x$ to $y = \lambda(x-a)^\nu$. A second relation for $\lambda$ is obtained by evaluating

$$M_\nu^\nu = 2\int_a^\infty (x-a)^\nu \exp\{-1-\mu-\lambda(x-a)^\nu\}\, dx = (\nu\lambda)^{-1} , \qquad \text{(iii)}$$

where (ii) has been used. From (i) to (iii) the required distribution is

$$p(x) = B_\nu^{-1} \exp\left(-\frac{|x-a|^\nu}{\nu M_\nu^\nu}\right)$$

$$B_\nu = 2\,\nu^{1/\nu} M_\nu \Gamma\left(1+\frac{1}{\nu}\right)$$

The entropy of $p(x)$, i.e. the maximum entropy of all normalized distributions with given value of $M_\nu^\nu$, is

$$S = -k\int_0^\infty p(x)\left(-\ln B_\nu - \frac{|x-a|^\nu}{\nu M_\nu^\nu}\right) dx = k \ln B_\nu + \frac{k}{\nu M_\nu^\nu}\int_0^\infty |x-a|^\nu p\, dx$$

$$= k \ln B_\nu + \frac{k}{\nu}$$

Thus all entropies of such distributions satisfy

$$S \le \frac{k}{\nu} + k \ln B_\nu$$

that is

$$B_\nu \ge \exp\left(\frac{S}{k} - \frac{1}{\nu}\right) .$$

## CHAPTER 10

10.1. Equation (10.1) can be written in the stated form if (10.2) is used. All sums over $n_j$, except the sum over $n_k$ itself, occur in both numerator and denominator, and cancel. Hence

$$\bar{n}_k = \frac{\Sigma_{n_k} n_k t_k^{n_k}}{\Sigma_{n_k} t_k^{n_k}} = t_k \frac{d}{dt_k}\left\{\ln\left(\sum_{n_k} t_k^{n_k}\right)\right\}$$

$$= t_k \frac{d}{dt_k}\left\{\ln(1 \pm t_k)^{\pm 1}\right\}$$

$$= \pm\, t_k \frac{d}{dt_k}\left\{\ln(1 \pm t_k)\right\}$$

$$= \pm\, t_k \frac{1}{1 \pm t_k}\,(\pm 1)$$

$$= \frac{1}{t_k^{-1} \pm 1} \quad .$$

10.2. We have

$$n(\mu - 2kT) = \frac{1}{e^{-2}+1} \sim 0\cdot 88$$

$$n(\mu + 2kT) = \frac{1}{e^{2}+1} \sim 0\cdot 12$$

so that the slope is

$$-\frac{0\cdot 76}{4kT} = -0\cdot 19/kT$$

10.3. The required sum is with $x \equiv \exp\lambda$

$$\frac{1}{x+1} + \frac{1}{x^{-1}+1} = 1$$

10.4. Let $\eta = \pm 1$, the top sign being for fermions and the bottom sign for bosons. Then by (10.9)

$$\frac{1-\eta n}{n} = \exp\left(\frac{e-\mu}{kT}\right) \quad (\equiv x) \quad .$$

The energy conservation condition $e_1 + e_2 = e_{1'} + e_{2'}$ implies $x_1 x_2 = x_{1'} x_{2'}$, i.e.

$$\frac{(1-\eta n_1)(1-\eta n_2))}{n_1 n_2} = \frac{(1-\eta n_{1'})(1-\eta n_{2'})}{n_{1'}\, n_{2'}}$$



so that

$$n_1 n_2 (1 - \eta n_{1'})(1 - \eta n_{2'}) = n_{1'} n_{2'} (1-\eta n_1)(1-\eta n_2) .$$

The left-hand side can be taken to refer to the forward reaction $1+2 \to 1'+2'$, and yields for $\eta = 0$ (Boltzmann or classical case) a reaction rate proportional to $n_1 n_2$. The right-hand side can be taken to refer to the reverse rate $1' + 2' \to 1 + 2$ and is for $\eta = 0$ proportional to $n_1' n_2'$. This is in agreement with detailed balance for Boltzmann statistics. For quantum statistics the final state contributes a factor $1-n$ for fermions and $1+n$ for bosons. Thus mean occupation of the final state discourages a process for fermions, and makes it impossible if $n = 1$. This extreme case is a direct consequence of the *Pauli exclusion principle*. Mean occupation of the final state *increases* the rate into that state for bosons.

10.5. The net transition rate - forward minus reverse - from group $i$ to group $j$ is

$$u_{ij} = \sum_{I(\varepsilon i)} \sum_{J(\varepsilon j)} (p_I\, S_{IJ}\, q_J - p_J\, S_{JI}\, q_I) \quad . \qquad \text{(i)}$$

In equilibrium each individual net transition rate is zero. Now for *fermions*

$$p_{I0} \doteq \left\{\exp\left(\frac{E_I - \mu_{i0}}{kT}\right) + 1\right\}^{-1}$$

where $\mu_{i0}$ is the chemical potential for the states of group $i$ when they are in thermal equilibrium with each other. Thus

$$\left(\frac{p_I q_J}{q_I p_J}\right)_0 = \exp\left(\frac{E_J - E_I}{kT}\right)\exp\left(\frac{\mu_{i0} - \mu_{j0}}{kT}\right)$$

The second factor is unity since for equilibrium between the groups: $\mu_{i0} = \mu_{j0}$. Detailed balance shows that for all $I$ and $J$

$$\left(\frac{S_{JI}}{S_{IJ}}\right)_0 = \exp\left(\frac{E_J - E_I}{kT}\right) . \qquad \text{(ii)}$$

For a Maxwell-Boltzmann distribution one has to recall that all levels are rather poorly populated so that one can substitute in (i) for $\lambda \equiv \exp(\mu/kT)$

$$p_{I0} = \lambda \exp\left(-\frac{E_I}{kT}\right) \ll 1, \quad q_{I0} = 1 - p_{I0} \sim 1 .$$

One then regains (ii).

10.6. The net transition rate is given again by the expression (i) for $u_{ij}$ in the preceding solution. One finds also

$$\left(\frac{S_{JI}}{S_{IJ}}\right)_0 = N_{\nu 0}\left(\frac{B_{JI}}{A_{IJ}}\right)_0 = \exp\left(\frac{E_J - E_I}{kT}\right) .$$

As $T \to \infty$ the right-hand side approaches unity, so that

$$N_{\nu 0} \to \left(\frac{A_{IJ}}{B_{JI}}\right)_0$$

which is a finite limit. This is in disagreement with the Bose distribution which shows that the mean number of photons goes to infinity with the temperature.

10.7. The formula (ii) of solution 10.5 now yields

$$\left(\frac{S_{JI}}{S_{IJ}}\right)_0 = \left(\frac{B_{JI}N}{A_{IJ} + B_{IJ}N}\right)_0$$

$$= \exp\left(\frac{E_J - E_I}{kT}\right) = \exp\left(-\frac{h\nu}{kT}\right) .$$

It follows that

$$N_{\nu 0} = \frac{(A_{IJ}/B_{IJ})_0}{(B_{JI}/B_{IJ})_0 \exp(h\nu/kT) - 1}$$

As $T \to \infty$, $N_{\nu 0} \to \infty$, as required, provided that

$$(B_{IJ})_0 = (B_{JI})_0 .$$





The Bose formula has unity in the numerator so that all three coefficients must be equal.

## CHAPTER 11.

11.1.

$$U = \sum \bar{n}_j e_j = \sum_j \frac{e_j}{t_j^{-1} \pm 1}$$

$$= A \int \frac{e^s e \, de}{\exp(x-\gamma) \pm 1} = A(kT)^{s+2} \int \frac{x^{s+1} dx}{\exp(x-\gamma) \pm 1}$$

$$= A(kT)^{s+2}\Gamma(s+2) I(\gamma, s+1, \pm)$$

11.2.

$$TS = U - F = A(kT)^{s+2}\Gamma(s+2) I(\gamma, s+1, \pm)$$

$$- \mu N + A(kT)^{s+2}\Gamma(s+1) I(\gamma, s+1, \pm)$$

$$= (s+2)A(kT)^{s+2}\Gamma(s+1) I(\gamma, s+1, \pm) - \mu N$$

$$\frac{S}{k} = (s+2)A(kT)^{s+1}\Gamma(s+1) I(\gamma, s+1, \pm) - \gamma N$$

11.3. We have

$$G = \mu N = U - TS + pv$$

so that

$$pv = \mu N + TS - U$$

$$= \mu N + (s+2)A(kT)^{s+2}\Gamma(s+1) I(\gamma, s+1, \pm) - \mu N$$

$$- A(kT)^{s+2}\Gamma(s+2) I(\gamma, s+1, \pm)$$

$$= A(kT)^{s+2}\Gamma(s+1) I(\gamma, s+1, \pm)$$

11.4. We have

$$N = A(kT)^{s+1}\Gamma(s+1) I(\gamma, s, \pm) .$$

Applying $(\partial/\partial T)_{v,N}$ ,

$$0 = (s+1)\frac{N}{T} + N\frac{I(\gamma,s-1,\pm)}{I(\gamma,s,\pm)}\left(\frac{\partial\gamma}{\partial T}\right)_{v,N}$$

Hence

$$\left(\frac{\partial\gamma}{\partial T}\right)_{v,N} = -\frac{s+1}{T}\frac{I(\gamma,s,\pm)}{I(\gamma,s-1,\pm)} .$$

Similarly, applying $(\partial/\partial v)_{T,N}$,

$$0 = \frac{N}{v} + N\frac{I(\gamma,s-1,\pm)}{I(\gamma,s,\pm)}\left(\frac{\partial\gamma}{\partial v}\right)_{T,N} .$$

Hence

$$\left(\frac{\partial\gamma}{\partial v}\right)_{T,N} = -\frac{1}{v}\frac{I(\gamma,s-1,\pm)}{I(\gamma,s,\pm)} .$$

Lastly, applying $(\partial/\partial T)_{p,N}$ ,

$$0 = \frac{s+1}{T}N + \frac{N}{v}\left(\frac{\partial v}{\partial T}\right)_{p,N} + A(kT)^{s+1}\Gamma(s+1)I(\gamma,s-1,\pm)\left(\frac{\partial\gamma}{\partial T}\right)_{p,N} .$$

Hence

$$\left(\frac{\partial\gamma}{\partial T}\right)_{p,N} = -\left\{\frac{s+1}{T} + \frac{1}{v}\left(\frac{\partial v}{\partial T}\right)_{p,N}\right\}\frac{I(\gamma,s,\pm)}{I(\gamma,s-1,\pm)} .$$

11.5. By (11.7) one has for an average total number $N$ of indistinguishable particles

$$F_N = \mu N - A(kT)^{s+2}\Gamma(s+1)I(\gamma_N,s+1,\pm)$$

$$= -kTN\left\{\frac{I(\gamma,s+1,\pm)}{I(\gamma,s,\pm)} - \gamma\right\} .$$

In the classical approximation (11.6) shows that

$$\gamma = \ln\frac{N}{A(kT)^{s+1}\Gamma(s+1)}$$

so that, if e is the base of the natural logarithms,



$$-\frac{F_N}{kT} = N + \ln\left\{\frac{A(kT)^{s+1}\Gamma(s+1)}{N}\right\}^N$$

$$= \ln\left\{\frac{eA(kT)^{s+1}\Gamma(s+1)}{N}\right\}^N .$$

Hence

$$\exp\left(-\frac{F_N}{kT}\right) = \frac{\{A(kT)^{s+1}\Gamma(s+1)e\}^N}{N^N} .$$

The last factor on the right is by Stirling's approximation $N!$ for large $N$ since

$$\ln N! \doteqdot N\ln N - N \doteqdot N\ln N = \ln(N^N) .$$

It follows that

$$\exp\left(-\frac{F_N}{kT}\right) = \frac{\exp(-NF_1/kT)}{N!}$$

If the same system at the same volume and temperature is represented by the canonical ensemble with $N$ now the total *fixed* number of particles, one should expect a similar free energy. For the canonical ensemble the partition function is

$$Z_N = \exp\left(-\frac{F_N}{kT}\right) = \frac{1}{N!}\exp\left(-\frac{NF_1}{kT}\right)$$

whence

$$Z_N = Z_1^N/N!$$

11.6. We start (using an obvious, abbreviated notation) with (11.20):

$$U = \frac{(s+1)NkT\,I(s+1)}{I(s)}$$

Note that for a fixed number of particles $d'Q = dU + pdv = C_v dT + l_v dv$ so that

$$C_v = \left(\frac{\partial U}{\partial T}\right)_v = (s+1)Nk\frac{I(s+1)}{I(s)} + (s+1)NkT\frac{I(s)^2 - I(s+1)I(s-1)}{I(s)^2}\left(\frac{\partial\gamma}{\partial T}\right)_v .$$

Since by (11.16)

$$\left(\frac{\partial \gamma}{\partial T}\right)_v = -\frac{s+1}{T}\frac{I(s)}{I(s-1)},$$

one finds

$$C_v = \frac{U}{T} - (s+1)^2 Nk \frac{I(s)^2 - I(s+1)I(s-1)}{I(s)^2}\frac{I(s)}{I(s-1)}$$

$$= \frac{U}{T} - (s+1)\frac{U}{T}\frac{I(s)}{I(s+1)}\frac{I(s)^2 - I(s+1)I(s-1)}{I(s)^2}\frac{I(s)}{I(s-1)}$$

$$= \frac{U}{T}\left\{1 - (s+1)\frac{I(s)^2}{I(s+1)I(s-1)} + (s+1)\right\}.$$

This is (11.22).

11.7. The partition function for one particle is

$$Z = z\iint \exp\left\{-\left(\frac{p^2}{2mkT} + \frac{mgq}{kT}\right)\right\} dp_1\, dp_2\, dp_3\, dq_1\, dq_2\, dq_3$$

$$= zA\left\{\int_0^\infty \exp\left(-\frac{mgq_3}{kT}\right)dq_3\right\}\left\{\int_0^\infty 4\pi \int p^2 \exp\left(-\frac{p^2}{2mkT}\right) dp\right\}$$

where $A$ is the cross-sectional area of the cylinder. Let

$$x \equiv \frac{mgq_3}{kT}, \quad a \equiv \frac{1}{2mkT}.$$

Then the integrals are

$$\frac{kT}{mg}\int_0^\infty e^{-x}dx \quad \text{and} \quad 4\pi\int_0^\infty p^2 e^{-ap^2}\, dp$$

and they yield respectively

$$\frac{kT}{mg} \quad \text{and} \quad 4\pi \times \frac{\sqrt{\pi}}{4a^{3/2}} = (2\pi mkT)^{3/2}.$$

It follows that

$$Z = zA\frac{(2\pi)^{3/2}\sqrt{m}}{g}(kT)^{5/2}.$$

SOLUTIONS OF PROBLEMS

The internal energy per particle is, with $\beta \equiv 1/kT$,

$$-\frac{\partial \ln Z}{\partial \beta} = -\frac{\partial}{\partial \beta}\left(-\frac{5}{2}\ln \beta + \text{constant}\right)$$

$$= \frac{5}{2}kT.$$

11.8. From

$$E = c(p_1^2 + p_2^2 + p_3^2 + m_0^2c^2)^{\frac{1}{2}}$$

$$\frac{1}{2}p_j\frac{\partial E}{\partial p_j} = \frac{c^2p_j^2}{2E} \quad (= \tfrac{1}{2}kT \text{ by equation (11.32)}),$$

and hence the first result. One can transform this result as follows:

$$p_j^2 = \frac{m_0^2\dot{q}_j^2}{1-v^2/c^2}$$

$$E = c\left(\frac{m_0^2v^2}{1-v^2/c^2} + m_0^2c^2\right)^{\frac{1}{2}}$$

It follows that

$$\frac{c^2p_j^2}{E} = \frac{cm_0^2\dot{q}_j^2}{(1-v^2/c^2)^{\frac{1}{2}}(m_0^2v^2 + m_0^2c^2 - m_0^2v^2)^{\frac{1}{2}}}$$

$$= \frac{m_0\dot{q}_j^2}{(1-v^2/c^2)^{\frac{1}{2}}}$$

$$= m\dot{q}_j^2$$

$$= p_j^2/m.$$

This yields all the alternative forms.

CHAPTER 12.

12.1. For one particle whose energies lie in effect in the range $(0,\infty)$ one has from (12.12)

$$Z_1 = \sum_j \exp\left(-\frac{e_j}{kT}\right) = \int_0^\infty 4\pi v g m h^{-3}(2me)^{\frac{1}{2}}\exp\left(-\frac{e}{kT}\right)de$$

$$= 2\pi v g\left(\frac{2mkT}{h^2}\right)^{3/2}\int_0^\infty \sqrt{x}\,\exp(-x)\,dx \;.$$

The integral is

$$\Gamma\left(\frac{3}{2}\right) = \frac{1}{2}\Gamma\left(\frac{1}{2}\right) = \tfrac{1}{2}\sqrt{\pi}$$

whence writing $\beta = 1/kT$,

$$Z_1 = \left(\frac{2\pi mkT}{h^2}\right)^{3/2} vg = \left(\frac{2\pi m}{\beta h^2}\right)^{3/2} vg \qquad \left(\beta \equiv \frac{1}{kT}\right)$$

12.2. By problem 9.1, the mean energy is

$$U_1 = -\left(\frac{\partial \ln Z_1}{\partial \beta}\right)_v = -\frac{\partial}{\partial\beta}\left(-\frac{3}{2}\ln\beta\right) = \frac{3}{2\beta} = \frac{3}{2}kT \;.$$

In classical statistical mechanics we have by (11.28)

$$U_N = -\left(\frac{\partial \ln Z_N}{\partial\beta}\right)_v = -\frac{\partial}{\partial\beta}[N\ln Z_1 - \ln N!] = NU_1 = \frac{3}{2}NkT.$$

12.3. The free energy is by (9.22)

$$F_N = -kT\ln Z_N = -NkT\ln Z_1 + N\ln N!$$

so that the pressure is by problem 7.1

$$p = -\left(\frac{\partial F_N}{\partial v}\right)_T = NkT\frac{\partial \ln Z_1}{\partial v} = \frac{NkT}{v} \;.$$

This is the 'gas equation' for a classical ideal gas.

12.4. By (11.7) and indicating by a subscript the dependence of $\gamma$ on $N$,

$$F_N = \mu N - A(kT)^{s+2}\Gamma(s+1)e^{\gamma_N}$$

where

$$N = A(kT)^{s+1}\Gamma(s+1)e^{\gamma_N} \;.$$

It follows that

$$F_N = NkT(\gamma_N - 1)$$

and

$$\exp\left(\frac{F_N}{kT}\right) = \exp\{N(\gamma_N - 1)\} \;.$$

12.5. (*a*) We have from equation (11.2)

$$\ln\,\Xi = \frac{pv}{kT} = \frac{G-F}{kT} = \frac{\mu N - F}{kT} = \pm\sum_j \ln(1 \pm t_j) \;.$$

Here $t_j = \exp[\gamma - E_j/kT]$ where $\gamma \equiv \mu/kT$ and

$$E_j = e_j + \varepsilon_0$$

is the total energy; $e_j$ is the energy of a particle above its rest mass. Writing $\eta = \pm 1$ (top signs for fermions), and using (12.16)

$$\ln\,\Xi = \frac{4\pi g v m^2 c_0}{h^3}\eta\int \sqrt{x}\sqrt{(x+2)}(x+1)\ln(1+\eta t_j)\,de$$

$$= \frac{4\pi g v m^3 c_0^3\eta}{h^3}\int_0^\infty \sqrt{x}\sqrt{(x+2)}(x+1)\ln\left\{1+\eta\exp\left(\gamma - \frac{\varepsilon_0}{kT} - \frac{\varepsilon_0 x}{kT}\right)\right\}dx.$$

Write the integrand as $u'w$ with

$$u' \equiv \sqrt{x}\sqrt{(x+2)}(x+1), \qquad w \equiv \ln\left[1+\eta\exp\left(\gamma - \frac{\varepsilon_0}{kT} - \frac{\varepsilon_0}{kT}x\right)\right] \;.$$

Then

$$u = \frac{1}{3}x^{3/2}(x+2)^{3/2}, \qquad w' = -\frac{\eta\varepsilon_0/kT}{\exp[(\varepsilon_0 x/kT + \varepsilon_0/kT - \gamma)] + \eta} \;.$$

The $uw$ term vanishes at both limits and the partial integration yields

$$\Xi = \frac{4\pi g v m^4 c_0^5}{3h^3 kT} \int_0^\infty \frac{x^{3/2}(x+2)^{3/2}}{\exp(\varepsilon_0 x/kT - \gamma') + \eta}\, dx$$

where

$$\gamma' = \frac{\mu - \varepsilon_0}{kT} .$$

(*b*) We multiply the density of states function $dN(e)$ of (12.16) by the mean occupation number $f$ per quantum state, and integrate to obtain the mean total number of particles. The same procedure will work for $U$ except that one must multiply by an extra factor $e = \varepsilon_0 x$. The exponent in $f$ is

$$\frac{E-\mu}{kT} = \frac{1}{kT}(e-\mu') = \frac{e}{\varepsilon_0}\frac{\varepsilon_0}{kT} - \frac{\mu'}{kT} .$$

(*c*) Since $\Xi$ is $\exp(pv/kT)$ we consider

$$kT \ln \Xi/U$$

and this is clearly not a constant as would be necessary for an ideal quantum gas.

12.6. We note first from thermodynamics that

$$\mu N + TS - pv = U' = U + \varepsilon_0 N$$

where $U'$ is the total mean energy including the energy due to the rest mass of the particles. Hence, using the results of problem 12.5 with $f$ denoting the mean occupation number of a quantum state,

$$TS = U + \varepsilon_0 N + pv - \mu N \qquad (pv = kT \ln \Xi)$$

$$\frac{4\pi v g m^4 c_0^5}{3h^3} \int_0^\infty f\{3x(x+1)(x^2+2x)^{\frac{1}{2}} + 3(x+1)(x^2+2x)^{\frac{1}{2}} + (x^2+2x)^{3/2} - \frac{3\mu}{\varepsilon_0}(x+1)(x^2+2x)^{\frac{1}{2}}\}\, dx .$$

Removing $(x^2+2x)^{\frac{1}{2}}$ from the braces, yields the stated



result.

**12.7.** Retain the lowest powers of $x$ for $x \ll 1$ and the highest powers of $x$ for $x \gg 1$, and use the variables

$$y = \frac{e}{kT}, \quad f = \{\exp(y-\gamma') \pm 1\}^{-1}$$

so that $x = (kT/\varepsilon_0)y$. Then we find for $x \ll 1$ and $x \gg 1$

$$N \propto \int \sqrt{(2x)} f\, dx , \qquad N \propto \int x^2 f\, dx$$

$$U \propto \int \sqrt{2}\, x^{3/2} f\, dx , \qquad U \propto \int x^3 f\, dx$$

and one finds the required results.

**12.8.** For a square box the modes of vibration are again specified by integers, but there are only two of them, $n_1$ and $n_2$. The number of modes with wavelengths lying between $\lambda_0$ and $\infty$ is, by analogy with (12.6)

$$N(\lambda_0) = \frac{1}{4}\pi a^2, \quad a \equiv \frac{2D}{\lambda_0}$$

so that

$$N(\lambda_0) = \frac{\pi D^2}{\lambda_0^2}, \qquad dN(\lambda) = -\frac{2\pi D^2}{\lambda^3}\, d\lambda .$$

For a field-free box ($\phi = 0$) we use again the kinetic energy

$$e = \frac{h^2}{2m\lambda^2}, \qquad de = -\frac{h^2}{m\lambda^3}\, d\lambda$$

so that

$$dN(\lambda) = \frac{2\pi D^2 m}{h^2}\, de \qquad (\equiv dN(e)) .$$

The density of states for spin degeneracy $g$ is

$$\frac{dN(e)}{de} = \frac{2\pi m g v}{h^2} \equiv Ae^s .$$

Thus $s = 0$ and the result follows.

12.9. (*a*) Let $e^{\alpha_F} \equiv A$, $e^{\alpha_B} \equiv B$, then we are given that

$$1 = \frac{1}{B} - \frac{1}{A} \quad \text{i.e.} \quad B = \frac{A}{A+1} .$$

The left-hand side of the identity is

$$I \equiv \int_0^\infty \frac{x\,dx}{A^{-1}e^x+1} - \int_0^\infty \frac{x\,dx}{B^{-1}e^x-1}$$

$$= \int_0^\infty \frac{Axe^{-x}dx}{1+Ae^{-x}} - \int_0^\infty \frac{Bxe^{-x}dx}{1-Be^{-x}}$$

Let $s \equiv Ae^{-x}/(1+Ae^{-x})$ in the first integral, and let $t \equiv Be^{-x}$ in the second integral. Then

$$e^x = \frac{A(1-s)}{s}$$

and

$$ds = -\frac{Ae^{-x}}{(1+Ae^{-x})^2}\,dx = -s(1-s)dx .$$

Also

$$e^x = \frac{B}{t} \quad \text{and} \quad dt = -t\,dx.$$

Putting these expressions into $I$,

$$I = \int_0^{A/(1+A)} \frac{\ln\{A(1-s)/s\}}{1-s}\,ds - \int_0^B \frac{\ln(B/t)}{1-t}\,dt .$$

Now put

$$\ln\left\{\frac{A(1-s)}{s}\right\} = \ln\left\{\frac{A}{(A+1)s}\right\} + \ln\{(A+1)(1-s)\}$$

so that the first part of the first integral cancels with the last integral. Hence

$$I = \int_0^{A/(1+A)} \frac{\ln\{(1+A)(1-s)\}}{1-s}\,ds$$



Let $u \equiv (1+A)(1-s)$, $du = -(1+A)ds$, so that

$$I = \int_1^{1+A} \frac{\ln u}{u}\,du = \tfrac{1}{2}\{(\ln u)^2\}_1^{1+A} .$$

This yields the required identity.

(*b*) In two dimensions the density of states function is a constant by problem 12.8, so that the internal energies of the fermion and boson systems are

$$U_F = C\int_0^\infty \frac{x\,dx}{A^{-1}e^x+1}$$

$$U_B = C\int_0^\infty \frac{x\,dx}{B^{-1}e^x-1}$$

where $C$ is the same quantity for both systems and is proportional to $(kT)^2$. Hence

$$U_F - U_B = CI = \tfrac{1}{2}C\{\ln(1+A)\}^2 = \tfrac{1}{2}C[\ln\left(\frac{A}{B}\right)]^2 = \tfrac{1}{2}C\left(\frac{\mu_F-\mu_B}{kT}\right)^2 .$$

Since $C \propto (kT)^2$, the right-hand side is independent of temperature, so that, taking the derivative with respect to $T$,

$$(C_v)_F - (C_v)_B = 0 .$$

CHAPTER 13

13.1. From equation (13.13), noting 1 erg = $10^{-7}$ J,

$$a = \frac{8}{15}\frac{\pi^5k^4}{h^3c_0^3}$$

$$= \frac{8}{15}\pi^5 \frac{(1{\cdot}3806\times10^{-16}\ \text{ergs/K})^4}{(6{\cdot}625\times10^{-27}\ \text{erg s})^3(3{\cdot}000\times10^{10}\ \text{cm s}^{-1})^3}$$

$$= 7{\cdot}654\times10^{-15}\ \text{erg cm}^{-3}\ \text{K}^{-4} .$$

13.2. Consider

$$\Gamma(s+1)I(0,s,+) = \int_0^\infty \frac{x^s\,dx}{\exp x+1}$$

$$= \int_0^\infty x^s \frac{\exp(-x)}{1+\exp(-x)}\,dx$$

$$= \sum_{t=0}^\infty (-1)^t \int_0^\infty x^s \exp\{-(t+1)x\}\,dx$$

$$= \sum_{t=0}^\infty \frac{(-1)^t}{(t+1)^{s+1}} \int_0^\infty y^s \exp(-y)\,dy \ .$$

We have made a series expansion of the denominator and put $y \equiv (t+1)x$. As the integral is $\Gamma(s+1)$, one sees that

$$I(0,s,+) = \sum_{t=0}^\infty \frac{(-1)^t}{(t+1)^{s+1}} = 1 - \frac{1}{2^{s+1}} + \frac{1}{3^{s+1}} - \ldots \ .$$

The Rieman zeta function is defined by

$$\zeta(r) \equiv \sum_{n=1}^\infty n^{-r} \quad (r \geq 1) = 1 + \frac{1}{2^r} + \frac{1}{3^r} + \ldots \ .$$

It follows that

$$2\,\frac{\zeta(r)}{2^r} = 2\sum_{n=1}^\infty (2n)^{-r} = 2\sum_{\substack{n>1\\ n\ \text{even}}} n^{-r} = \zeta(r) - I(0,r-1,+)$$

as may be seen by combining the appropriate series which occur on the right-hand side. Hence we see that

$$I(0,r-1,+) = (1-2^{1-r})\zeta(r) \ .$$

It follows that

$$\Gamma(3)I(0,2,+) = \Gamma(3)(1-\tfrac{1}{4})\zeta(3) = \frac{3}{2}\,\zeta(3) = \frac{3}{2} \times 1{\cdot}202 = 1{\cdot}803$$

$$\Gamma(4)I(0,3,+) = \Gamma(4)(1-\tfrac{1}{8})\zeta(4) = \frac{21}{4}\zeta(4) = \frac{21}{4}\,\frac{\pi^4}{90} = 5{\cdot}681$$

13.3. It was shown in section 13.2 that



$$I(0,s,-) = \zeta(s+1) \ .$$

It follows that

$$\frac{I(0,s,+)}{I(0,s,-)} = 1 - 2^{-s}$$

as required.

13.4. (*a*) For relativistic fermions with high energy $x \gg 1$ in (12.16) and the density of states formula has the first form (12.17). The number of particles and the energy are therefore ($\eta \equiv e/kT$)

$$N = \frac{4\pi gv}{(hc)^3}\int_0^\infty \frac{e^2\,de}{\exp(\eta-\gamma)+1} = 4\pi gv\left(\frac{kT}{hc}\right)^3 \int_0^\infty \frac{\eta^2\,d\eta}{\exp(\eta-\gamma)+1}$$

$$U = 4\pi gvkT\left(\frac{kT}{hc}\right)^3 \int_0^\infty \frac{\eta^3\,d\eta}{\exp(\eta-\gamma)+1} \ .$$

(*b*) As the electron-positron $(e^-,e^+)$ pairs are in equilibrium with black-body radiation ($\nu$) whose chemical potential is zero, it follows from

$$e^- + e^+ = \nu$$

that the equilibrium condition (8.3) is

$$\mu^{(-)} + \mu^{(+)} - \mu^{(\nu)} = 0$$

that is

$$\mu^{(-)} + \mu^{(+)} = 0 \ .$$

If equal numbers of electrons and positrons are created $\mu^{(-)} = \mu^{(+)}$, and $\mu^{(-)} = \mu^{(+)} = 0$ follows. The integrals are given in problem 13.2 and we find

$$N = 4\pi gv\left(\frac{kT}{hc}\right)^3 \frac{3}{2}\,\zeta(3) = 45{\cdot}3v\left(\frac{kT}{hc}\right)^3$$

$$U = 4\pi gv\ kT\left(\frac{kT}{hc}\right)^3 \frac{21}{4}\,\zeta(4) = 142{\cdot}7v\ kT\left(\frac{kT}{hc}\right)^3 \ .$$

(*c*) The first result depends on $I(0,2,+)$ and so should be $\frac{3}{4}$ of the photon number in a cavity of the same volume and at the same temperature by virtue of problem 13.3 . Similarly the energy is $\frac{7}{8}$ of the corresponding energy in black-body radiation.

13.5.(*a*) Energy balance yields $h\nu_1 + h\nu_2 = 2mc^2 + \varepsilon_+ + \varepsilon_-$ since two particles are created, each of rest energy $mc^2$. The forward reaction leads to fermions in final states and this implies by problem 10.4 that the rate has a factor $(1-n_+)(1-n_-)$. The reverse reaction leads to bosons being created by *pair annihilation*, so that bosons appear in the final state with appropriate factors $(1+n_1)(1+n_2)$.

(*b*) We have that the occupation numbers satisfy

$$\frac{n_1}{1+n_1}\,\frac{n_2}{1+n_2}\,\frac{1-n_+}{n_+}\,\frac{1-n_-}{n_-} = \exp\left\{\frac{1}{kT}\left(-h\nu_1-h\nu_2+mc^2+\varepsilon_+-\mu^{(+)}+mc^2+\varepsilon_--\mu^{(-)}\right)\right\}$$

$$= \exp\left\{-\frac{\mu^{(+)}+\mu^{(-)}}{kT}\right\} .$$

Since $\mu^{(+)} = \mu^{(-)}$ for equal numbers of positrons and electrons, and since the left-hand side is unity by part (*a*), it follows that $\mu^{(+)} = \mu^{(-)} = 0$.

(*c*) It was seen in problem 12.7 that the relativistic gas theory, which ought to be used in this problem, leads to formulae in the standard form of our density of states function $Ae^s$, provided the chemical potential is replaced by $\mu - mc^2$. For low enough temperatures, so that the particles move non-relativistically, one can then use equation(11.6) and the fourth line of Table 12.1 to write for the electron and positron concentrations

$$n = g\left(\frac{2\pi m\,kT}{h^2}\right)^{3/2} I\left(\frac{\mu-mc^2}{kT},\tfrac{1}{2},+\right)$$

$$\doteq g\left(\frac{2\pi m\,kT}{h^2}\right)^{3/2} \exp\left(\frac{\mu-mc^2}{kT}\right) .$$



The last form assumes a small concentration of pairs. The required result is found if $g = 2$ and $\mu = 0$.

(*d*) For electrons

$$mc^2/k \sim 5\times 10^9\ \text{K},$$

and for room temperature

$$2\left(\frac{2\pi m\,kT}{h^2}\right)^{3/2} \sim 3\times10^{19}\ \text{cm}^{-3} \sim 3\times10^{25}\ \text{m}^{-3} .$$

Hence

$$n \sim 3\times10^{25}\left(\frac{T}{300}\right)^{3/2}\exp\left(-\frac{5\times10^9}{T}\right)\ \text{m}^{-3} .$$

This yields negligible pair densities up to $T \sim 10^7$ K when there is a rapid increase to $n \sim 10^{36}$ m$^3$ at $T = mc^2/k$ when the mass density is of order $10^3$ g cm$^{-3}$ $\sim 10^6$ kg m$^{-3}$. Above this temperature degeneracy has to be considered. At very high temperatures, when the rest mass is a negligible part of the average electron energy, the thermodynamic laws of radiation and of the gas of pairs are very similar.

**13.6.** Let $B_2$ be the decay product of rest mass $m_2 \neq 0$ and choose an inertial frame in which $B_2$ is at rest. In that frame let the decaying particle A have momentum $p$ and energy $E$. Let $B_1$ be the remaining group of products of the reaction which have zero *or* non-zero rest mass. Then, in an obvious notation, momentum conservation leads to

$$\mathbf{p} = \mathbf{p}_1+\mathbf{p}_2 \quad \text{i.e. } p^2c^2 = p_1^2c^2 , \qquad \text{(i)}$$

since $B_2$ is at rest and therefore $\mathbf{p}_2 = 0$. Since $p$ refers to A and (12.14) states that

$$B^2 = p^2c^2 + (mc^2)^2,$$

(i) implies

$$E^2 = E_1^2 - (m_1c^2)^2 \quad \text{i.e. } E \le E_1. \qquad \text{(ii)}$$

The equality sign is allowed since $B_1$ may have zero rest mass: $m_1 \ge 0$. We now apply energy conservation $E = E_1+E_2$, i.e.

$$E = E_1 + m_2c^2 \quad \text{i.e. } E > E_1 \qquad \text{(iii)}$$

since $B_2$ is at rest. The equality sign is not allowed because $m_2 > 0$. Thus (ii) and (iii) are in contradiction, and the result is established.

13.7. For constant volume

$$S = \int_0^T \frac{dE}{T} .$$

Thus for radiation $E = avT^4$,

$$S = 4av \int T^2 dT = \frac{4av}{3} T^3 .$$

Similarly, if $E = bT^r$, then

$$S = br \int T^{r-2} dT = \frac{br}{r-1} T^{r-1} .$$

If we call this result $cT^s$, then $s = r-1$ and $sc = br$ as required.

13.8. If a smooth surface of separation of media of refractive indices $n_1$, $n_2$ is considered, then Snell's law of refraction enables one to put

$$n_1 \sin\theta_1 = n_2 \sin\theta_2 .$$

It follows that

$$n_1 \cos\theta_1 d\theta_1 = n_2 \cos\theta_2 d\theta_2 .$$



Multiplying,

$$n_1^2 \sin\theta_1 \cos\theta_1 d\theta_1 = n_2^2 \sin\theta_2 \cos\theta_2 d\theta_2 .$$

Since $d\omega_1 = \sin\theta_1 d\theta_1 d\phi_1$, it follows that

$$n_1^2 \cos\theta_1 d\omega_1 = n_2^2 \cos\theta_2 d\omega_2 .$$

Hence the power transmitted through the same area $\delta A$ at the boundary is for unit frequency range

$$K_1 \cos\theta_1 \delta\omega_1 \delta A = K_2 \cos\theta_2 \delta\omega_2 \delta A = K_2 \left(\frac{n_1}{n_2}\right)^2 \cos\theta_1 \delta\omega_1 \delta A$$

where the left-hand side used the data from one side of the area, and the middle term used the data from the other side. It follows that $K/n^2$ *is invariant along a ray* if the refractive index varies in the medium.

13.9. By expression (13.48) the energy emitted by an elemental annular ring of radius $x$ on the disc is, when integrated over frequencies and expressed per unit time,

$$E = \int K \cos\theta \, \delta\omega \delta A .$$

Here the element of area of the disc is $\delta A = 2\pi x \, dx$. The solid angle required so that the emitted light reaches $S$ is

$$\delta\omega = \frac{dA' \cos\theta}{R^2+x^2} .$$

If the surface is Lambertian, $K$ is a constant, and the illuminance is

$$E = 2\pi K \int_0^r \frac{x\cos^2\theta}{R^2+x^2} dx \int_{A'} dA' = 2\pi K A' \int_0^r \frac{R^2 x}{(R^2+x^2)^2} dx$$

since $\cos^2\theta = R^2/(R^2+x^2)$. Hence

$$E = 2\pi K A' R^2 \left| \frac{-1}{2(R^2+x^2)} \right|_0^r = 2\pi K A' R^2 \left\{ \frac{1}{2R^2} - \frac{1}{2(R^2+r^2)} \right\}$$

$$E = \pi K A' \frac{r^2}{R^2+r^2} .$$

For small $r$ the solid angle subtended by the disc at $S$ is $\omega' = \pi r^2/R^2$, so that the illuminance is

$$E = A' K \omega' .$$

13.10. It has been seen from Stefan's constant, p. 231, that the energy density received per unit time per unit area by a black body at temperature $T_\odot$ and of radius $R_\odot$ at a distance $R$ away from it is

$$\sigma T_\odot^4 \frac{4\pi R_\odot^2}{4\pi R^2} .$$

Indeed, if $T_\odot$, $R_\odot$ refer to the sun and $R$ is its distance from the earth this is just the solar constant $f$. For equilibrium with a black body of radius $R_E$ and temperature $T_E$, which simulates the earth,

$$f \pi R_E^2 = 4\pi R_E^2 \sigma T_E^4 .$$

The left-hand side is the energy intercepted by the earth from the sun in unit time. It follows that

$$T_E = \left(\frac{f}{4\sigma}\right)^{\frac{1}{4}} = \left(\frac{1353}{4\times 5\cdot 67\times 10^{-8}}\right)^{\frac{1}{4}} \sim 278 \text{ K}.$$

The equation for $T_E$ can also be written as

$$\frac{T_E}{T_\odot} = \sqrt{\left(\frac{R_\odot}{2R_E}\right)} .$$

13.11. The amended energy per unit frequency range is

$$U_\nu = \frac{8\pi\nu^2 v}{c^3}\left\{1 - \left(\frac{ac}{L\nu}\right)^2\right\} \frac{h\nu}{\exp(h\nu/kT)-1}$$

$$= U_{\nu 0} - \frac{8\pi a^2 L}{c} kT \frac{x}{\exp x - 1} \qquad \left(x \equiv \frac{h\nu}{kT}\right)$$

where $U_{\nu 0}$ is the energy for a large enough volume. Integrating over all frequencies,

SOLUTIONS OF PROBLEMS

$$U = U_0 - \frac{8\pi a^2 L (kT)^2}{ch} \int_0^\infty \frac{x\,dx}{\exp x - 1}$$

$$= U_0 - \frac{4\pi^3}{3} \frac{a^2 L (kT)^2}{ch}$$

$$= \frac{8}{15} \pi^5 k^4 h^{-3} c^{-3} T^4 v \left\{1 - \frac{5}{2}\left(\frac{ach}{\pi kTL}\right)^2\right\}$$

CHAPTER 14

14.1. As in equation (14.3) one has with $x = h\nu/2kT$

$$Z_1 = \sum_{n_1, n_2} \exp\left\{\frac{-(n_1+\frac{1}{2})h\nu - (n_2+\frac{1}{2})h\nu}{kT}\right\}$$

$$= \exp(-2x) \left\{\sum_{n_1=0}^\infty \exp(-2n_1 x)\right\}\left\{\sum_{n_2=0}^\infty \exp(-2n_2 x)\right\}$$

$$= \exp(-2x) \left\{\frac{1}{1-e^{-2x}}\right\}^2$$

$$= \left\{\frac{e^{-x}}{1-e^{-2x}}\right\}^2$$

$$= \left(\frac{1}{2\sinh x}\right)^2$$

$$Z_N = (Z_1)^N = \left(\frac{1}{2\sinh x}\right)^{2N} .$$

Since $E(n_1, n_2) = (1 + n_1 + n_2)h\nu = (1 + j)h\nu$, one needs to vary $n_1$ and $n_2$ subject to $n_1 + n_2 = j$. It is easily seen that the following table solves the problem and shows that the degeneracy is $j+1$.

| | | | | | |
|---|---|---|---|---|---|
| $n_1 =$ | 0 | 1 | 2 | ... | $j$ |
| $n_2 =$ | $j$ | $j-1$ | $j-2$ | ... | 0 |
| $n_1 + n_2 =$ | $j$ | $j$ | $j$ | ... | $j$ |

14.2. In this case one has $w$ sums instead of two sums in the expression for the partition function. Each sum yields the same factor as in problem 14.1. This gives the expression for $Z_N$ given in the problem.

The degeneracy of the level at energy $E_j$ is the number of ways of expressing an integer $j$ as a sum of $w$ integers, zero and repetitions being allowed and order being important. This is the same as placing $j$ indistinguishable balls into $w$ distinguishable boxes, and can be shown to be $(w + j - 1)!/(w - 1)!\, j!$.

14.3. One has with $x \equiv h\nu/2kT$,

$$\ln Z_N = wN \ln \frac{e^{-x}}{1-e^{-2x}} = -wNx - wN \ln(1-e^{-2x})$$

$$F = -kT \ln Z = wNkT\{x + \ln(1-e^{-2x})\}$$

$$S = -\left(\frac{\partial F}{\partial T}\right)_v = wNk\left\{\frac{2x}{e^{2x}-1} - \ln(1-e^{-2x})\right\}$$

$$U = F + TS = \tfrac{1}{2}\, wNh\nu \coth x .$$

14.4. Use equaton (11.30) with $z = h^{-1}$. Then

$$Z = h^{-1}\int_0^\infty \exp\left(-\frac{p^2}{2mkT}\right) dp \int_{-\infty}^{\infty} \exp\left\{-\frac{V(x)}{kT}\right\} dx$$

$$= h^{-1}\,(2\pi mkT)^{\frac{1}{2}}\, I .$$

Here

$$I \equiv \int_{-\infty}^{\infty} \exp\left\{-\frac{V(x)}{kT}\right\} dx = \int_{-\infty}^{\infty} \exp\left(-\frac{ax^2}{2kT}\right) \exp\left(-\frac{bx^3+cx^4}{kT}\right) dx .$$

For small enough $b$ and $c$

$$I = \int_{-\infty}^{\infty} \exp\left(-\frac{ax^2}{2kT}\right)\left\{1 - \frac{bx^3}{kT} - \frac{cx^4}{kT} + \tfrac{1}{2}\left(\frac{bx^3}{kT}\right)^2\right\} dx .$$

The $x^3$-term vanishes since it is odd. For the remaining terms use a formula from p. 436.



One finds

$$I = \left(\frac{2\pi kT}{a}\right)^{\frac{1}{2}}\left(1 - \frac{3ckT}{a^2} + \frac{15\,b^2kT}{a^3}\right)$$

$$Z = A\,kT + B(kT)^2$$

where

$$A = \frac{2\pi}{h}\sqrt{\left(\frac{m}{a}\right)} \qquad B \equiv \frac{6\pi}{a^2h}\sqrt{\left(\frac{m}{a}\right)}\left(\frac{5b^2}{2a} - c\right) .$$

It follows that

$$U = kT^2\,\frac{\partial \ln Z}{\partial T} \doteq kT\left(1 + \frac{B}{A}\,kT\right)$$

$$C_v = k\left(1 + 2\,\frac{B}{A}\,kT\right) .$$

14.5. By (14.20)

$$\frac{C_v}{k} = \frac{9rN}{x_D^3}\int_0^{x_D} \frac{x^4 e^x}{(e^x-1)^2}\, dx .$$

Put $u = x^4$, $v' = e^x/(e^x-1)^2$ and apply a partial integration with $u' = 4x^3$, $v = -(e^x-1)^{-1}$. One finds that the integral becomes

$$-\frac{x_D^4}{e^{x_D}-1} + 4\int_0^{x_D} \frac{x^3 dx}{e^x-1}$$

as required.

14.6. The analogue of (14.5) for one dimension is

$$U = \sum_i \left(\frac{1}{2} + \frac{1}{e^{y_i}-1}\right) h\nu_i \qquad \left(y_i \equiv \frac{h\nu_i}{kT}\right)$$

where the sum is over all $3rN$ degrees of freedom. We use the density of states formula

$$\frac{4\pi v}{h^3}\left(\frac{2}{c_t^3}+\frac{1}{c_\ell^3}\right)(h\nu)^2\,\mathrm{d}(h\nu) = \frac{1}{h^3}\frac{9rN}{\nu_D^3}(h\nu)^2\,\mathrm{d}(h\nu)$$

$$= \frac{9rN}{x_D^3}x^2\,\mathrm{d}x$$

where $x = h\nu/kT$. Converting the sum for $U$ into an integral,

$$U = 3rN\,kT\,\frac{3}{x_D^3}\int_0^{x_D} x^3\left(\tfrac{1}{2}+\frac{1}{\exp x - 1}\right)\mathrm{d}x$$

This is the required expression.

14.7. Note that

$$\mathrm{D}(a) = \tfrac{3}{8}a + \frac{3}{a^3}\int_0^a \frac{y^3}{e^y - 1}\,\mathrm{d}y\ .$$

Hence

$$\mathrm{D}(a) - aD'(a) = \tfrac{3}{8}a + \frac{3}{a^3}\int_0^a \frac{y^3}{e^y-1}\,\mathrm{d}y - \tfrac{3}{8}a + \frac{9}{a^3}\int_0^a \frac{y^3\mathrm{d}y}{e^y-1} - \frac{3a}{e^a-1}$$

and this is exactly $\Delta(a)$.

We also have

$$\frac{1}{k}\frac{\partial U}{\partial T} = 3rN\left\{D(x_D) - TD'(x_D)\,\frac{x_D}{T}\right\} = 3rN\left\{D(x_D) - x_D D'(x_D)\right\}$$

$$= 3rN\,\Delta(x_D)\ .$$

14.8. We are given (14.31)

$$F = \frac{9}{8}\,rNh\nu_D - \frac{3rN}{x_D^3}\,kT\int_0^{x_D}\frac{x^3\mathrm{d}x}{e^x-1} + 3rN\,kT\,\ln\left\{1 - e^{-x_D}\right\}\ .$$

The stated equation for $p$ follows directly by using $p = -(\partial F/\mathrm{d}v)_T$, a relation which can be read off the equation for $\mathrm{d}F$ in problem 7.1. Similarly the relation for the entropy is obtained from $S = -(\partial F/\partial T)_v$. One can calculate $U$ as $F + TS$, and $G = \mu N$ as $F + pv$. These relations imply that



$$\frac{pv}{U} = -\frac{v}{\nu_D}\left(\frac{\partial\nu_D}{\partial v}\right)_T = -\left(\frac{\partial\ln\nu_D}{\partial\ln v}\right)_T$$

$$S = \frac{12rNk}{x_D^3}\int_0^{x_D}\frac{x^3\mathrm{d}x}{e^x-1} - 3rNk\,\ln(1-e^{-x_D})$$

$$\mu = \frac{3}{2}\,r\,h\nu_D + 3kT\,\ln(1-e^{-x_D})\ .$$

14.9. (*a*) From equation (14.3), with $\varepsilon_0 = \frac{1}{2}h\nu$, one has that for a one-dimensional oscillator at temperature $T_s$

$$Z_1 = \frac{e^{-x}}{1-e^{-2x}}\qquad\left(x \equiv \frac{h\nu}{2kT_s}\right)$$

$$= (e^x - e^{-x})^{-1}$$

$$= (2\sinh x)^{-1}\ .$$

For $3N$ distinguishable oscillators, all of the same frequency $\nu$, the free energy is (using (9.22) for example)

$$F_s = -kT\ln Z_{3N} - AN = -3NkT\ln Z_1 - AN\ ,$$

where we have added the interaction term $-AN$, proportional to the number of oscillators. The chemical potential is (using for example the equation for $\mathrm{d}F$ in problem 7.1)

$$\mu_s = \left(\frac{\partial F_s}{\partial N}\right)_{v,T} = -kT\ln\left\{Z_1^3\exp\left(\frac{A}{kT_s}\right)\right\}\ .$$

(*b*) For an ideal quantum gas we have from (11.6)

$$N_g = A_g(kT_g)^{s+1}\,\Gamma(s+1)\,I\left(\frac{\mu_g}{kT},s,\pm\right)$$

where the subscript g indicates that the volume $v_g$ of the gas has to be used. This yields in the classical approximation (11.12)

$$\mu_g = -kT_g \ln\left\{A_g(kT_g)^{s+1}\frac{\Gamma(s+1)}{N_g}\right\} .$$

Equating the two expressions for the chemical potential, as is necessary for equilibrium between solid and gas, i.e. adopting $\mu_g = \mu_s$, $T_g = T_s$

$$\frac{N_g}{v_g} = \frac{A_g}{v_g}(kT_g)^{s+1}\Gamma(s+1)Z_1^{-3}\exp\left(-\frac{A}{kT_g}\right) .$$

(*c*) If the equation of state for the gas phase is $pv_g = N_g kT$, then the left-hand side is $p/kT$. If, furthermore, we adopt from (12.12)

$$A_g = 4\pi g v_g m h^{-3}(2m)^{\frac{1}{2}}, \quad s = \tfrac{1}{2}$$

one finds finally

$$p = \left(\frac{2\pi m k T_g}{h^2}\right)^{3/2} g k T_g (2\sinh x)^3 \exp\left(-\frac{A}{kT_g}\right) .$$

## CHAPTER 15

15.1. Note that, using (15.17')

$$2\frac{\zeta(r)}{2^r} = 2\sum_{n=1}^{\infty}\frac{1}{(2n)^r}$$

$$= 2\sum_{\substack{n>1 \\ (n \text{ even})}}\frac{1}{n^r} = \zeta(r) - C_r .$$

Therefore

$$C_r = \zeta(r)(1 - 2^{1-r}) .$$

15.2. Expanding the denominator in series



$$J_{r-1} = \int_0^{\infty} x^{r-1}\frac{e^{-x}}{1+e^{-x}}\,dx$$

$$= \sum_{s=0}^{\infty}(-1)^s\int_0^{\infty} x^{r-1}\exp\{-(s+1)x\}dx$$

$$= \sum_{s=0}^{\infty}\frac{(-1)^s}{(s+1)}\int_0^{\infty} y^{r-1}e^{-y}\,dy$$

$$= \Gamma(r)\,C_r .$$

The result of problem 15.1 can now be used.

15.3. Putting $z = x - a$,

$$I = \int_{-a}^{\infty}\frac{f(a+z)dz}{1+\exp z}$$

$$= \int_0^a \frac{f(a-y)dy}{1+e^{-y}} + \int_0^{\infty}\frac{f(a+z)dz}{1+e^z} \qquad (y \equiv -z)$$

$$= \int_0^a f(a-y)dy - \int_0^{\infty}\frac{f(a-y)dy}{1+e^y} + \int_0^{\infty}\frac{f(a+z)dz}{1+e^z}$$

$$= \int_0^a f(x)dx + \int_0^{\infty}\frac{f(a+z)-f(a-z)}{1+e^z}\,dz \qquad \begin{pmatrix} x \equiv a-z \\ z \equiv y \end{pmatrix}$$

where we have replaced an upper limit $a$ by $\infty$, which should introduce only an exponentially small error.

15.4. Let $f(x) = x^s/\Gamma(s+1)$ in problem 15.3. The result of that problem then yields

$$I(a,s,+) = \frac{1}{\Gamma(s+1)}\int_0^a x^s\,dx + \frac{\pi^2}{6\Gamma(s+1)}a^{s-1} + \frac{7\pi^4 s(s-1)(s-2)}{360\Gamma(s+1)}a^{s-3} + \dots$$

$$= \frac{a^{s+1}}{\Gamma(s+2)} + \frac{\pi^2 s(s+1)}{6\Gamma(s+2)}a^{s-1} + \frac{7\pi^4(s+1)s(s-1)(s-2)}{360\,\Gamma(s+2)}a^{s-3} + \dots$$

$$= \frac{a^{s+1}}{\Gamma(s+2)}\left\{1 + \frac{s(s+1)}{6}\left(\frac{\pi}{a}\right)^2 + \frac{7(s+1)s(s-1)(s-2)}{360}\left(\frac{\pi}{a}\right)^4 + \dots\right\}$$

as required.

15.5. For zero mass the relativistic result of Table 12.1 yields the density of states function given in the problem. Substitution of this function in (11.6), (11.7), and (11.18) to (11.20) yields the stated results.

15.6. By (11.6) and (11.12) one has for an ideal classical gas

$$N = gv\left(\frac{2\pi mkT}{h^2}\right)^{3/2}\lambda = \frac{pv}{kT} .$$

Now use that at constant temperature and volume $\lambda Z_1/Z_0 = p/p_0$, provided that $Z_1/Z_0$ is independent of pressure. Hence $p_0 = (Z_0 gkT/Z_1)(h^2/2\pi mkT)^{-3/2}$ and

$$\frac{n_t}{N} = \frac{1}{1+Z_0/\lambda Z_1} = \frac{p}{p+p_0} .$$

The discussion of trapping of atoms or molecules at the distinguishable sites of a surface follows in all details section 15.5, provided the symbols are re-interpreted as suggested in the problem.

15.7. The grand partition function is for one adsorption site

$$\Xi_1 = \sum_{j=0}^{M}\lambda^j Z_j = Z_0\sum_{j=0}^{M}\left(\frac{\lambda Z_1}{Z_0}\right)^j = Z_0\,\frac{1-y^{(M+1)}}{1-y} .$$

The mean number of particles trapped at a site is by (15.39) and using $y \equiv \lambda Z_1/Z_0$



$$\lambda\left(\frac{\partial \ln \Xi_1}{\partial\lambda}\right)_{v,T} = \lambda\frac{\partial}{\partial\lambda}\left\{\ln Z_0 + \ln\{1-y^{(M+1)}\} - \ln\{1-y\}\right\} .$$

One finds

$$\lambda\left\{\frac{-(M+1)y^M}{1-y^{M+1}} + \frac{1}{1-y}\right\}\frac{Z_1}{Z_0} = \frac{y}{1-y} - \frac{(M+1)y^{M+1}}{1-y^{M+1}} .$$

For $N$ centres this number of trapped particles must be multiplied by $N$. Hence one finds the result quoted.

This is unlikely to be a good model for electrons trapped at a defect, because the strong Coulomb interaction will make the assumption concerning $Z_j$ inapplicable.

**15.8.** The 'reaction' can be written in the form

$$(r-1) + e = (r) .$$

Consider

$$\frac{n\,N_{r-1}}{N_r} = N_c \exp(\gamma_e-\eta_c)\,\frac{\lambda_D^{r-1}Z_{r-1}}{\lambda_D^r Z_r}$$

$$= \frac{Z_{r-1}}{Z_r}N_c \exp(\gamma_e-\gamma_D) ,$$

and this is the reaction constant. For the valence band and impurities the reaction is

$$(r) + h = (r-1) .$$

One then needs

$$\frac{p\,N_r}{N_{r-1}} = N_v \exp(\eta_v-\gamma_h)\,\frac{\lambda_D^r Z_r}{\lambda^{r-1}Z_{r-1}}$$

$$= \frac{Z_r}{Z_{r-1}}N_v \exp(\gamma_D-\gamma_h) .$$

CHAPTER 16.

16.1. Equation (16.25) is

$$\frac{\sigma_n^2}{\langle n\rangle^2} = \frac{kT}{\langle n\rangle^2}\left(\frac{\partial\langle n\rangle}{\partial\mu}\right)_{v,T}.$$

We have to prove this quantity equal to $-(kT/v^2)(\partial v/\partial p)_T$, i.e. that

$$\left(\frac{\partial\langle n\rangle}{\partial\mu}\right)_{v,T} = -\left(\frac{\langle n\rangle}{v}\right)^2\left(\frac{\partial v}{\partial p}\right)_{\langle n\rangle,T},$$

where the definition of compressibility at fixed $n$ has been used. We thus have to prove a purely thermodynamic result.

First observe that, with $n$ written for $\langle n\rangle$ and the suffix $T$ omitted

$$\begin{aligned}\left(\frac{\partial p}{\partial v}\right)_n &= -\left(\frac{\partial p}{\partial n}\right)_v\left(\frac{\partial n}{\partial v}\right)_p\\ &= +\left(\frac{\partial\mu}{\partial v}\right)_n\left(\frac{\partial n}{\partial v}\right)_p \quad \text{(by a Maxwell relation)}\\ &= -\left\{\left(\frac{\partial\mu}{\partial n}\right)_v\left(\frac{\partial n}{\partial v}\right)_\mu\right\}\left(\frac{\partial n}{\partial v}\right)_p\\ &= -\frac{n^2}{v^2}\left(\frac{\partial\mu}{\partial n}\right)_v \quad \text{(by the result of problem 7.4).}\end{aligned}$$

16.2. From $\langle n\rangle = \frac{1}{\Xi}\sum_{j,n} n\exp\left(\frac{\mu n-E_j}{kT}\right)$,

$$\left(\frac{\partial\langle n\rangle}{\partial T}\right)_{\mu,v} = -\frac{1}{kT^2\Xi}\sum_{j,n} n(\mu n-E_j)\exp\left(\frac{\mu n-E_j}{kT}\right) - \frac{\langle n\rangle}{\Xi}\left(\frac{\partial\Xi}{\partial T}\right)_{\mu,v}$$

$$kT^2\left(\frac{\partial\langle n\rangle}{\partial T}\right)_{\mu,v} = -\mu\langle n^2\rangle + \langle En\rangle + \mu\langle n\rangle^2 - \langle E\rangle\langle n\rangle,$$

where use has been made of

$$-\frac{kT^2\langle n\rangle}{\Xi}\left(\frac{\partial\Xi}{\partial T}\right)_{\mu,v} = -\frac{\langle n\rangle kT^2}{\Xi}\left\{-\frac{1}{kT^2}\sum_{j,n}(\mu n-E_j)\exp\left(\frac{\mu n-E_j}{kT}\right)\right\}$$

$$= \mu\langle n\rangle^2 - \langle n\rangle\langle E\rangle.$$

This is the required result.



16.3. Using the last result established in the preceding problem,

$$kT^2\left(\frac{\partial\langle E\rangle}{\partial T}\right)_{\mu,v} = -\frac{kT^2}{\Xi}\sum_{j,n}\frac{\mu n-E_j}{kT^2}E_j\exp\left(\frac{\mu n-E_j}{kT}\right) - \frac{kT^2\langle E\rangle}{\Xi}\left(\frac{\partial\Xi}{\partial T}\right)_{\mu,v}$$

$$= -\mu\langle nE\rangle + \langle E^2\rangle + \mu\langle n\rangle\langle E\rangle - \langle E\rangle^2$$

as required.

16.4. The results of problems 16.2 and 16.3 have to be added together, and $\sigma_n^2$ has to be replaced by (16.25).

16.5. First note that, writing $E$ for $\langle E\rangle$ and $n$ for $\langle n\rangle$, we have for $dv = 0$

$$dE = \left(\frac{\partial E}{\partial T}\right)_n dT + \left(\frac{\partial E}{\partial n}\right)_T dn = \left(\frac{\partial E}{\partial T}\right)_n dT + \left(\frac{\partial E}{\partial n}\right)_T\left\{\left(\frac{\partial n}{\partial\mu}\right)_T d\mu + \left(\frac{\partial n}{\partial T}\right)_\mu dT\right\}.$$

Hence

$$\left(\frac{\partial E}{\partial T}\right)_{\mu,v} = \left(\frac{\partial E}{\partial T}\right)_{n,v} + \left(\frac{\partial E}{\partial n}\right)_{T,v}\left(\frac{\partial n}{\partial T}\right)_{\mu,v}.$$

The result of problem 16.4 becomes therefore

$$\frac{\sigma_E^2}{kT^2} = \left\{\left(\frac{\partial E}{\partial T}\right)_{n,v} + \left(\frac{\partial E}{\partial n}\right)_{T,v}\left(\frac{\partial n}{\partial T}\right)_{\mu,v}\right\} + \mu\left(\frac{\partial n}{\partial T}\right)_{\mu,v} + \frac{\mu^2}{T}\left(\frac{\partial n}{\partial\mu}\right)_{v,T}.$$

The first term is $C_v$. By problem 7.1

$$d(E-TS) = -SdT - pdv + \mu dn$$

so that

$$\left(\frac{\partial E}{\partial n}\right)_{T,v} = \mu + T\left(\frac{\partial S}{\partial n}\right)_{T,v},$$

and

$$\frac{\sigma_E^2}{kT^2} = C_v + \left\{T\left(\frac{\partial n}{\partial \mu}\right)_{T,v}\right\}^{-1}\left[2\mu T\left(\frac{\partial n}{\partial T}\right)_{\mu,v}\left(\frac{\partial n}{\partial \mu}\right)_{T,v} + T^2\left(\frac{\partial S}{\partial n}\right)_{T,v}\left(\frac{\partial n}{\partial T}\right)_{\mu,v}\left(\frac{\partial n}{\partial \mu}\right)_{T,v} + \left\{\mu\left(\frac{\partial n}{\partial \mu}\right)_{v,T}\right\}^2\right].$$

We know from the Gibbs-Duhem relation that

$$d(pv) = p dv + v dp$$

$$= p dv + S dt + n d\mu .$$

It follows that

$$\left(\frac{\partial S}{\partial \mu}\right)_{T,v} = \left(\frac{\partial n}{\partial T}\right)_{\mu,v} .$$

Using this in the $T^2$-term, it becomes

$$T^2\left(\frac{\partial S}{\partial n}\right)_{T,v}\left(\frac{\partial S}{\partial \mu}\right)_{T,v}\left(\frac{\partial n}{\partial \mu}\right)_{T,v} = T^2\left(\frac{\partial S}{\partial \mu}\right)_{T,v}^2 = T^2\left(\frac{\partial n}{\partial T}\right)_{\mu,v}^2 .$$

This leads at once to the stated equation.

16.6.(*a*) Consider an ensemble with $M_0$ systems having $n_j = 0$ and $M_1$ systems having $n_j = 1$, so that

$$\langle n_j^b \rangle = 0 \times \frac{M_0}{M_0+M_1} + 1 \times \frac{M_1}{M_0+M_1} = \frac{M_1}{M_0+M_1}$$

which is independent of $b$ for $b \neq 0$.

(*b*) In the general case which holds for fermions or bosons (10.2) yields

$$\langle n_j^b \rangle = \frac{\Sigma_{n_1}\Sigma_{n_2}\ldots n_j^b\, t_1^{n_1}\, t_2^{n_2}\ldots}{\Sigma_{n_1}\Sigma_{n_2}\ldots t_1^{n_1}\, t_2^{n_2}\ldots} = \frac{\Sigma_{n_j} n_j^b\, t_j^{n_j}}{\Sigma_{n_j} t_j^{n_j}} \left(\equiv \frac{S_b}{S_0}\right) .$$

Now for bosons,



$$S_0 \equiv \sum_{n_j=0}^{\infty} t_j^{n_j} = \frac{1}{1-t_j}$$

$$S_1 = \sum_{n_j} n_j\, t_j^{n_j} = t_j \frac{dS_0}{dt_j} = \frac{t_j}{(1-t_j)^2}$$

$$S_2 = t_j \frac{dS_1}{dt_j} = \frac{t_j(1+t_j)}{(1-t_j)^3} .$$

Hence

$$\langle n_j^2 \rangle = \frac{S_2}{S_0} = \frac{t_j(1+t_j)}{(1-t_j)^2} = \langle n_j \rangle\, [1+2\langle n_j \rangle].$$

CHAPTER 17

17.1.We have

$$D\frac{\partial \nu}{\partial x} = D'\frac{\partial \nu}{\partial x} + \nu\frac{\partial D'}{\partial x} - \nu U$$

$$= D'\frac{\partial \nu}{\partial x} + \nu\left(\frac{\partial D'}{\partial \gamma}\right)_{T,\tau_0}\frac{\partial \gamma}{\partial x} + \nu\left(\frac{\partial D'}{\partial \tau_0}\right)_{T,\gamma}\frac{\partial \tau_0}{\partial x} - \nu U .$$

At constant temperature $T$ and volume $v$ we know that $\nu$ is just a function of $\gamma$ - see, for example equation (17.17). Hence

$$\left(\frac{\partial \gamma}{\partial x}\right)_{y,z} = \left(\frac{\partial \gamma}{\partial \nu}\right)_{T,v}\left(\frac{\partial \nu}{\partial x}\right)_{y,z} .$$

Equating the coefficients of $\partial\nu/\partial x$, the stated relation results.

17.2.We have

$$\left(\frac{\partial D'}{\partial \gamma}\right)_{T,\tau_0} = \frac{2}{3}\frac{kT}{m}\tau_0(x)\frac{\Gamma(t+5/2)}{\Gamma(3/2)}\frac{I(\gamma,1/2)I(\gamma,t+1/2) - I(\gamma,t+3/2)I(\gamma,-1/2)}{\{I(\gamma,1/2)\}^2} .$$

Now

$$\nu = NI(\gamma,1/2)$$

so that

$$\left(\frac{\partial \nu}{\partial \gamma}\right)_{T,v} = NI(\gamma,-1/2) = \nu\,\frac{I(\gamma,-1/2)}{I(\gamma,1/2)} \; .$$

From a result of problem 17.1 giving $D$ in terms of $D'$

$$D = \frac{2}{3}\frac{kT}{m}\tau_0(x)\,\frac{\Gamma(t+5/2)}{\Gamma(3/2)}\left\{\frac{I(\gamma,t+3/2)}{I(\gamma,1/2)} + \frac{I(\gamma,1/2)I(\gamma,t+1/2)-I(\gamma,t+3/2)I(\gamma,-1/2)}{I(\gamma,1/2)I(\gamma,-1/2)}\right.$$

$$= \frac{2}{3}\frac{kT}{m}\tau_0(x)\,\frac{\Gamma(t+5/2)}{\Gamma(3/2)}\,\frac{I(\gamma,t+1/2)}{I(\gamma,-1/2)} \; .$$

Note that $D$ and $D'$ are different.

Using the Einstein relation in the form of the equation

$$\frac{|q|}{kT}\,\frac{1}{\nu}\left(\frac{\partial \nu}{\partial \gamma}\right)_{T,v} D = \frac{|q|}{kT}\,\frac{I(\gamma,-1/2)}{I(\gamma,1/2)}\,D$$

for the mobility, one finds the required expression.

17.3 and 17.4.

We have

$$\mathbf{J}_j = |q_j|\,v_j\nu_j\,\mathbf{E} - q_jD_j\,\nabla\,\nu_j - q_j\nu_jD_j^T\,\nabla\,T \qquad \text{(i)}$$

$$= -|q_j|\,v_j\nu_j\,\mathrm{grad}\,\phi - q_jD_j\left(\frac{\partial \nu_j}{\partial \mu_j}\right)_{T,v}\nabla\mu_j - q_jD_j\left(\frac{\partial \nu_j}{\partial T}\right)_{\mu,v}\nabla T -$$

$$- q_j\nu_jD_j^T\,\nabla T .$$

By **(17.14)**, $|q_j|\nabla\phi = (\nabla\mu_j - \nabla\overline{\mu}_j)$, so that

$$\mathbf{J}_j = v_j\nu_j\nabla\overline{\mu}_j - \left\{v_j\nu_j + q_jD_j\left(\frac{\partial \nu_j}{\partial \mu_j}\right)_{T,v}\right\}\nabla\mu_j - q_j\left\{D_j\left(\frac{\partial \nu_j}{\partial T}\right)_{\mu,v} + \nu_jD_j^T\right\}\nabla T$$

Equation (17.23) states that

$$q_jD_j = -v_j\nu_j\left(\frac{\partial \mu_j}{\partial \nu_j}\right)_{T,v} \qquad \text{(ii)}$$

so that, using

$$\left(\frac{\partial \mu_j}{\partial \nu_j}\right)_{T,v}\left(\frac{\partial \nu_j}{\partial T}\right)_{\mu,v} = -\left(\frac{\partial \mu_j}{\partial T}\right)_{\nu_j,v},$$

one finds

$$\mathbf{J}_j = v_j\nu_j\,\mathrm{grad}\,\overline{\mu}_j - \left\{v_j\nu_j\left(\frac{\partial \mu_j}{\partial T}\right)_{\nu_j,v} + q_j\,\nu_jD_j^T\right\}\mathrm{grad}\,T \; . \qquad \text{(iii)}$$

If the temperature is regarded as constant, the result of problem 17.3 is obtained.

17.5. Using the equations of the solution of problem 17.4 , three arguments must be given:

A : (i) + (ii) ⇒ (iii)

B : (ii) + (iii) ⇒ (i)

C : (iii) + (i) ⇒ (ii) .

The argument A is a part of problem 17.4 .

Argument B. By (17.14) and (iii), writing $B$ for the term in square braces

$$\mathbf{J}_j = v_j\nu_j\nabla(\mu_j - |q_j|\phi) - B\nabla T$$

$$= v_j\nu_j\nabla\mu_j - |q_j|\,v_j\nu_j\nabla\phi - B\nabla T.$$

From (17.20) and analogous relations for $y$ and $z$ components,

$$\nabla\mu_j = \left(\frac{\partial \mu_j}{\partial \nu_j}\right)_{T,v}\left\{\nabla\nu_j - \left(\frac{\partial \nu_j}{\partial T}\right)_{\mu,v}\nabla T\right\}.$$

Hence

$$\mathbf{J}_j = v_j\nu_j\left(\frac{\partial \mu_j}{\partial \nu_j}\right)_{T,v}\nabla\nu_j + v_j\nu_j\left(\frac{\partial \mu_j}{\partial T}\right)_{\nu_j,v}\nabla T - |q_j|\,v_j\nu_j\nabla\phi - B\nabla T \; . \qquad \text{(iv)}$$

If one now uses (ii), one $\nabla T$-term cancels and (i) is obtained.

Argument C. One can again start with (iii) and take it to (iv) to find

$$\mathbf{J}_j = |q_j|\,v_j\nu_j\,\mathbf{E} + v_j\nu_j\left(\frac{\partial \mu_j}{\partial \nu_j}\right)_{T,v}\nabla\nu_j - q_j\nu_j\,D_j^T\,\nabla T.$$

Comparing this result with (i), the Einstein relation (ii) is found.

CHAPTER 18

18.1. For $2N$ charges and an electrically neutral system

$$\left(\sum_{i=1}^{2N} q_i\right)^2 = \sum_{i=1}^{2N} q_i^2 + 2\sum{}' q_i q_j = 0$$

since $\sum q_i = 0$. Hence

$$\sum{}' q_i q_j = -\frac{1}{2}\sum_1^{2N} q_i^2 = -Nq^2 .$$

It follows that

$$d \equiv \frac{a}{kT}\sum{}' q_i q_j = -\frac{aNq^2}{kT}$$

Hence

$$\tau_2 \equiv -\frac{Td}{2sN} = \frac{aq^2}{2sk}$$

Also $|Q_-| = Q_+ = q$ so that

$$\tau_1 \equiv -\frac{aQ_-Q_+}{sk} = 2\tau_2 .$$

18.2. Since $L_0 = L$, the term $(L_0/L)^d = v^{d/s}/L^d$ cancels out of $Q$ in equation (18.14). It follows that $\ln Q = N\ln v$ + $v$-independent terms so that $p = \rho kT$.

18.3. The number of photons in a range $(\nu, \nu + d\nu)$ incident on a solar cell per unit area per unit time due to black-body radiation at temperature $T_S$ is, by (13.26)

$$f(x)\,dx = \frac{2\pi(kT_S)^3}{h^3 c_0^2}\frac{x^2\,dx}{e^x - 1}$$

where $x \equiv h\nu/kT_S$. The condition (18.18) for an optimum energy gap is therefore

$$x_G \frac{x_G^2}{\exp(x_G) - 1} = \int_{x_G}^{\infty} \frac{x^2\,dx}{e^x - 1} .$$





Some simple numerical work shows that $x_G = 2{\cdot}17$. This can be substituted in (18.17) to yield

$$\eta(x_G = 2{\cdot}17) = 2{\cdot}17\,\frac{\int_{2\cdot17}^{\infty}\{x^2/(e^x-1)\}\,dx}{\int_0^{\infty}\{x^3/(e^x-1)\}\,dx}$$

$$= 2{\cdot}17 \times \frac{1{\cdot}42}{\pi^4/15} = 0{\cdot}47$$

where equation (13.12a) has been used for the denominator.

APPENDIX I

I.1. By (I.5) the equation for $n$ is

$$\frac{\dot{n}}{n\ln n} = s + d.$$

The integration is immediate by putting $n = e^x$ :

$$\int_{n_0}^{n} \frac{dn}{n\ln n} = \int_{x_0}^{x} \frac{e^x\,dx}{xe^x} = \ln(\ln n) - \ln(\ln n_0)$$

$$= (s+d)(t-t_0) .$$

Hence

$$\frac{\ln n}{\ln n_0} = \exp\{(s+d)(t-t_0)\} .$$

I.2. By (I.4)

$$S_i = kD_i \ln N_i = k\ln N_i^{D_i} .$$

If two non-interacting systems are combined, the entropy is added:

$$S = S_1 + S_2$$

that is

$$k\ln N^D = k\ln N_1^{D_1} + k\ln N_2^{D_2} = k\ln(N_1^{D_1} N_2^{D_2}) .$$

Thus with

$$q_i \equiv \frac{\ln N_i}{\ln(N_1 N_2)}$$

$$D = q_1 D_1 + q_2 D_2 \quad (q_1 + q_2 = 1)$$

Thus two non-interacting systems of equal disorder $D_1 = D_2$ yield a system of the same disorder ($q_1 = q_2 = \frac{1}{2}$), even though the entropy increases.

APPENDIX II

II.1. (i) We have for an incremental change

$$\mathrm{d}f = f \sum_{j=1}^{n} \frac{2}{x_j} \, \mathrm{d}x_j \qquad \text{(II.5)}$$

$$\mathrm{d}g = \sum_{j=1}^{n} 2\, x_j \, \mathrm{d}x_j = 0 \,. \qquad \text{(II.6)}$$

For an extremum (II.5) must vanish, subject to the subsidiary condition (II.6). These equations correspond to (II.1) and (II.2) with

$$m \to n, \quad x_j \to \mathrm{d}x_j, \quad A_j \to 2f/x_j, \quad k \to 1, \quad B_{1j} \to 2x_j$$

The required condition (II.4) becomes (changing the sign of $\lambda$ for convenience)

$$\frac{2f}{x_s} - 2\lambda x_s = 0 \quad (s = 1, 2, \ldots, n).$$

It follows that $\lambda x_s^2 = f$ for all $s$ at the extremum. Thus

$$(x_1^2)_{\text{ext.}} = (x_2^2)_{\text{ext.}} = \ldots = \frac{c^2}{n} \,.$$

The last form of writing arises from the constraint $g = c^2$. This extremum corresponds to an extreme value of $f$ given by

$$(f)_{\text{ext.}} = (c^2 n) \,. \qquad \text{(II.7)}$$





This extremum is attained at the points

$$x_1 = \pm\, c/\sqrt{n}, \quad x_2 = \pm\, c/\sqrt{n}, \ \ldots.$$

in an $n$-dimensional space. There are clearly $2^n$ such points, and the extremum is a maximum.

(ii) We have from (II.7) that

$$f = x_1^2\, x_2^2 \ldots x_n^2 \le c^{2n}/n^n$$

whence

$$(x_1^2\, x_2^2 \ldots x_n^2)^{1/n} \le \frac{c^2}{n} = \frac{1}{n} \sum_{1}^{n} x_i^2 \,,$$

as was to be proved.

# PART E: ADDITIONAL MATERIAL

# Summary of key results: thermodynamics

A. FORMALISM: NUMBER OF PARTICLES ($n$) IN THE SYSTEM FIXED

1. Quasi-static process for a closed simple system

$$\mathrm{d}'Q = C_v \mathrm{d}T + l_v \mathrm{d}v = C_p \mathrm{d}T + l_p \mathrm{d}p = m_v \mathrm{d}v + m_p \mathrm{d}p \qquad (3.1a)$$

$$= \mathrm{d}U + p\mathrm{d}v \quad \text{(1st law)} \qquad (2.1),(3.9)$$

$$= T\mathrm{d}S \quad \text{(2nd law for quasi-static processes)} \qquad (5.13)$$

A basic procedure for deriving formulae is to start with this result.

2. Adiabatic process: $\mathrm{d}'Q = 0$.
   Isothermal process: $\mathrm{d}T = 0$ ($\mathrm{d}t = 0$).

3. Expressions exist for $m_v$, $m_p$ in terms of $C_v$, $C_p$, $l_v$ and $l_p$ (3.3)

4. Heat capacity at constant pressure

$$C_p = \left(\frac{\partial U}{\partial T}\right)_p + p\left(\frac{\partial v}{\partial T}\right)_p = \left(\frac{\partial H}{\partial T}\right)_p = T\left(\frac{\partial S}{\partial T}\right)_p . \qquad \text{(section 5.4)} \quad \text{(problem 5.2)}$$

   Heat capacity at constant volume

$$C_v = \left(\frac{\partial U}{\partial T}\right)_v = T\left(\frac{\partial S}{\partial T}\right)_v . \qquad (3.14)$$

B. FORMALISM ALLOWING FOR VARIABLE $n$

5. Gibbs-Duhem equation

$$S\mathrm{d}T - v\mathrm{d}p + n\mathrm{d}\mu = 0. \qquad (7.6)$$

6. Helmholtz free energy

$$F \equiv U - TS = \mu n - pv .$$

Gibbs free energy

$$G \equiv U - TS + pv = \mu n . \qquad (7.4)$$

Enthalpy

$$H \equiv U + pv = \mu n + TS .$$

7. Maxwell relations

$$\left(\frac{\partial T}{\partial v}\right)_{S,n} = -\left(\frac{\partial p}{\partial S}\right)_{v,n} ; \quad \left(\frac{\partial T}{\partial p}\right)_{S,n} = \left(\frac{\partial v}{\partial S}\right)_{p,n}$$

(problem 5.1)

$$\left(\frac{\partial T}{\partial v}\right)_{p,n} = -\left(\frac{\partial p}{\partial S}\right)_{T,n} ; \quad \left(\frac{\partial T}{\partial p}\right)_{v,n} = \left(\frac{\partial v}{\partial S}\right)_{T,n} .$$

8. Second law for non-static processes linking equilibrium states in closed simple systems

$$\delta U - T\delta S + p\delta v = \mu\delta n \leq 0 . \qquad (7.11)$$

9. Intensive variables (section 3.4)

$$T,\ p,\ \mu,\ l_v,\ m_v,\ \frac{v}{U},\ \frac{n}{U},\ \text{etc.}$$

Extensive variables

$$S,\ U,\ v,\ n,\ C_v,\ C_p,\ l_p,\ m_p,\ G,\ F,\ H.$$

C. PHYSICAL AND CHEMICAL SYSTEMS

10. The independent thermodynamic variables for a closed system ($n$=constant) may be any *two* chosen from

$$p,\ v,\ T,\ U(p,v,T),\ S(p,v,T),\ \text{etc.},$$

since we assume an equation of state of the form $f(p,v,T) = 0$. It follows that

$$\left(\frac{\partial p}{\partial v}\right)_T \left(\frac{\partial v}{\partial T}\right)_p \left(\frac{\partial T}{\partial p}\right)_v = -1, \quad \left(\frac{\partial p}{\partial v}\right)_T = \frac{1}{(\partial v/\partial p)_T}, \text{ etc.} \qquad (3.5,6)$$

11. Volume expansion

$$\alpha_p = \frac{1}{v}\left(\frac{\partial v}{\partial T}\right)_{p,n} .$$

Compressibility

$$K_X = -\frac{1}{v}\left(\frac{\partial v}{\partial p}\right)_{X,n}$$

(where $X = t, T$, or $a$).

Grüneisen ratio

$$\Gamma = \frac{\alpha_p v}{K_t C_v} . \qquad (3.10)$$

12. Carnot efficiency $\eta_c \equiv 1 - T_{cold}/T_{hot} < 1$ exceeds all other efficiencies.

(Problem 5.5)

13. Ideal gas

$$pv = NkT \Rightarrow \left(\frac{\partial U}{\partial v}\right)_T = 0 \quad \text{(Joule's law)} \qquad (6.13,14)$$

or

$$pv = At \quad \text{and} \quad \left(\frac{\partial U}{\partial v}\right)_t = 0.$$

Useful identity in this connection

$$T(\partial p/\partial T)_v = p + (\partial U/\partial v)_T. \qquad (6.2)$$

14. Ideal quantum gas $pv = gU$ ($g$ constant); $\Gamma = g$.

15. Van der Waals' equation of state

$$\left(p + \frac{a}{v^2}\right)(v-b) = AT, \ a > 0, \ b > 0, \ A > 0. \qquad (6.9)$$

16. The number of free variables if $\pi$ phases each of $\chi$ components are in equilibrium is

$$f = \chi - \pi + 2 \quad \text{(Gibbs phase rule)} \qquad (7.8)$$

Maxwell equal-area rule:

$$\int v dp = 0 \qquad \text{(section 7.4)}$$

17. The equilibrium condition for the chemical reaction $\Sigma_i \, v_i \, C_i = 0$ is $\Sigma_i \, \mu_i \, \nu_i = 0$. The mass action law for ideal gases is $\Pi_i \, C_i^{\nu_i} = K'(p,T)$ or $\Pi_i \, p_i^{\nu_i} = K(p,T)$. Sums and products are over the participant molecular types of the reaction. (section 8).

# Summary of key results: statistical mechanics

A. GENERAL SYSTEMS

1. *Statistical entropy* (for a system of $n$ states, the $i$th state having probability $p_i$) is $S = -k \sum p_i \ln p_i$, where $\sum p_i = 1$ and $k$ is Boltzmann's constant. The basic procedure is to maximize $S$ subject to subsidiary conditions $\sum p_i f_i = \bar{f}$, that is to maximize (see (2) to (4) below)

$$M = -k \sum p_i \ln p_i + \alpha \sum p_i f_i + \ldots \quad \text{(sections 9.1,2)}$$

2. (a) *Microcanonical distribution (isolated system)*: $p = 1/N$ (section 9.3)

(b) *Canonical distribution or ensemble (closed system)*. Volume fixed and total number of (identical non-interacting) particles fixed. System maintained at temperature $T$ so that average internal energy $\bar{U} = \Sigma\, p_i U_i$ is specified, $U_i$ being the total energy of the system when it is in state $i$. Maximizing

$$M = -k \left(\sum p_k \ln p_k + \beta \sum p_k U_k + \alpha \sum p_k\right)$$

leads to $p_i = Z^{-1} \exp(-\beta\, U_i)$ where $Z = \Sigma \exp(-\beta\, U_i)$ and $S = k \ln Z + k\beta\, \bar{U}$.
Thermodynamic identifications lead to

$$p_i = Z^{-1} \exp -\left(\frac{U_i}{kT}\right) \qquad \text{where} \quad Z^{-1} = \exp\left(\frac{F}{kT}\right)$$

and

$$S = -\frac{F}{T} + \frac{\bar{U}}{T}\,.$$

3. *Grand canonical distribution or ensemble (open system)*. Volume fixed. System maintained at temperature $T$ and chemical potential $\mu$ so that the average internal energy $\bar{U} = \Sigma_{i,j}\, p_{ij} U_i$ and the average number of (identical non-interacting) particles $\bar{n} = \Sigma_{i,j}\, p_{ij} n_j$ are specified. Maximizing

$$M = -k \sum_{ij} (p_{ij} \ln p_{ij} + \beta\, p_{ij}U_i + \gamma\, p_{ij}n_j + \alpha\, p_{ij})$$

leads to $P(n,U_i) = \Xi^{-1} \exp(-\beta\, U_i - \gamma n\,)$

where $\Xi = \Sigma \exp(-\beta\, U_i - \gamma n\,)$

and $S = k \ln \Xi + k\beta\, \overline{U} + k\gamma\overline{n}\;.$

Thermodynamic identifications lead to

$$P(n,U_i) = \Xi^{-1} \exp\left(\frac{\mu n}{kT} - \frac{U_i}{kT}\right)$$

where $\Xi^{-1} = \exp\left(-\frac{pv}{kT}\right)$ and $S = \frac{pv}{T} + \frac{\overline{U}}{T} - \frac{\mu\overline{n}}{T}$ . (9,19,20)

4. *A one-dimensional normal distribution* of zero mean and standard derivation $\sigma$ is given by

$$p(x) = \frac{1}{\surd(2\pi\sigma^2)} \exp\left(\frac{-x^2}{2\sigma^2}\right) \quad (-\infty < x < \infty)$$ .(Problem 9.4) (16.5)

Thus

$$\int_{-\infty}^{\infty} p(x)\,dx = 1 \qquad \int_{-\infty}^{\infty} xp(x)\,dx = 0 \qquad \int_{-\infty}^{\infty} x^2p(x)\,dx = \sigma^2\;.$$

B. PARTICLES WITH POINT INTERACTION

5. *Single-particle theory.* A state of a system of $n$ indistinguishable particles is specified by the occupation numbers $(n_1, n_2, \ldots)$ where $n_1$ particles are in the single-particle energy level $e_1$, $n_2$ particles are in the single-particle energy level $e_2$, etc. Then

$$P(n_1,n_2,\ldots) = \Xi^{-1}\, t_1^{n_1}\, t_2^{n_2} \ldots \quad \text{where } t_i \equiv \exp\left(\frac{\mu - e_i}{kT}\right) \tag{10.1}$$

$$\Xi = \sum_{n_1=0} t_1^{n_1} \sum_{n_2=0} t_2^{n_2} \ldots \tag{10.2}$$

and the mean occupation number is

$$\overline{n}_i = -\frac{\partial \ln \Xi}{\partial \eta_i} \quad \text{where } \eta_i \equiv \frac{e_i}{kT}\;. \tag{10.5}$$

6. *Fermions and bosons*. The assumed range of $n_i$ is (i) fermions, 0, 1; (ii) bosons, 0, 1, 2,...,∞. In all formulae fermions correspond to the *upper* sign and bosons to the lower sign:

$$\Xi = \prod_j (1 \pm t_j)^{\pm 1} \tag{10.4}$$

$$\bar{n}_i = \frac{1}{\exp\{(e_i - \mu)/kT\} \pm 1} = \frac{1}{t_i^{-1} \pm 1} \; . \tag{10.8,9,10}$$

## C. PARTICLES WITH POINT INTERACTIONS IN THE CONTINUOUS SPECTRUM APPROXIMATION

7. *Ideal quantum gas*. Density of states assumption, i.e. approximate number of single-particle quantum states lying in the energy range $(e, e + de)$ is $A_s e^s de$, where $s$ is a constant (e.g. $\frac{1}{2}$,1,2), $A_s$ is independent of $e$ and proportional to $v$. Defining

$$I(\gamma, s, \pm) \equiv \frac{1}{\Gamma(s+1)} \int_0^\infty \frac{x^s \, dx}{e^{x-\gamma} \pm 1} \tag{11.5}$$

then

$$\sum_j \bar{n}_j \to N = A_s (kT)^{s+1} \Gamma(s+1) I \left(\frac{\mu}{kT}, s, \pm\right) \tag{11.6}$$

$$\sum_j \bar{n}_j e_j \to U = A_s (kT)^{s+2} \Gamma(s+2) I \left(\frac{\mu}{kT}, s+1, \pm\right) \quad \text{(Problem 11.1)}$$

$$\frac{pv}{U} = \frac{1}{s+1} = g . \tag{11.21}$$

Note that for $s > -1$

$$\frac{d}{d\gamma} I(\gamma, s+1, \pm) = I(\gamma, s, \pm) \; . \tag{11.13}$$

8. *Black-body radiation*. Photon gas is considered to be an ideal gas of bosons for which

$$\mu = 0 \Leftrightarrow \gamma = 0, \quad s = 2, \quad A = 8\pi v / h^3 \, c_0^3 \tag{13.1}$$

so that

$$U = 3pv = \tfrac{3}{4}\, TS = \tfrac{1}{4}\, TC_v = a\, T^4\, v \quad (a = \text{constant}). \quad (13.10)$$

$$U/v \sim T^4 \text{ (Stefan-Boltzmann law)}, \quad C_v \sim T^3, \quad C_p \sim \infty.$$

The mean photon energy per unit volume per unit frequency range is

$$u(\nu, T) = \frac{8\pi h\, c_0^{-3}\, \nu^3}{\exp(h\nu/kT) - 1} \xrightarrow{\text{classical}} \frac{8\pi\nu^2}{c_0^3}\, kT \qquad (13.15)$$

(Planck's formula) → (Rayleigh-Jeans formula).

9. *Quantum region - low temperature; classical region - high temperature.*

D. OTHER RESULTS.

Mean energy of a three-dimensional oscillator

$$\varepsilon_0 + \sum_{i=1}^{3} \frac{h\nu_i}{\exp(h\nu_i/kT) - 1} \qquad (14.5)$$

Heat capacity per atom of an Einstein solid

$$k \sum_{i=0}^{3} E(x_i), \quad E(x) \equiv \frac{x^2 \exp x}{(\exp x - 1)^2} \qquad (14.10)$$

Energy per particle of low temperature fermions

$$\frac{3}{5}\, \mu(0)\left[1 + \left\{\frac{5}{12}\, \frac{\pi k T}{\mu(0)}\right\}^2\right] \qquad (15.20, 23)$$

Variance of the energy $E$ and of particle number $n$

$$\sigma_E^2 = -\left[\frac{\partial\langle E\rangle}{\partial(1/kT)}\right]_{v,N}, \quad \sigma_n^2 = kT\left(\frac{\partial\langle n\rangle}{\partial\mu}\right)_{v,T} \qquad (16.3, 25)$$

Electrochemical potential ($\phi(\mathbf{r})$ is the electrostatic potential)

$$\bar{\mu}(\mathbf{r}) = \mu(\mathbf{r}) - |q|\phi(\mathbf{r}) \qquad (17.14)$$

Einstein diffusion-mobility ratio

$$\frac{|q|D}{v} = -\frac{|q|}{q}\, \frac{\nu}{(\partial\nu/\partial\mu)_{T,v}} \qquad (17.23)$$

# Useful mathematical and physical data

## USEFUL MATHEMATICAL DATA

$\pi = 3{\cdot}1416 \qquad e = 2{\cdot}7183 \qquad \log_e x = 2{\cdot}3016 \log_{10} x$

$N! \sim (2\pi N)^{\frac{1}{2}} N^N e^{-N}$ (Stirling's formula, $N \gg 1$)

$$\ln N! \sim N \ln N - N + \tfrac{1}{2}\ln(2\pi N)$$

$$\Psi(N) \equiv \frac{\partial \ln N!}{\partial N} \sim \ln(N+\tfrac{1}{2}) \qquad \text{(Gauss's } \Psi\text{-function)}$$

$$\Gamma(u+1) = \int_0^\infty x^u e^{-x} dx \rightarrow u! \text{ if } u \text{ is an integer}$$

$$\Gamma(u+1) = u\Gamma(u), \qquad \Gamma(\tfrac{1}{2}) = \sqrt{\pi}$$

$$\int_0^\infty x^k e^{-\lambda x^2} dx = \frac{1}{2\lambda^{(k+1)/2}} \Gamma\left(\frac{k+1}{2}\right) \qquad \left[\begin{array}{l} k > -1,\ \lambda > 0 \\ \text{See discussion of} \\ \quad \text{equation (16.9)} \end{array}\right]$$

$$\int_0^\infty x^{2r} e^{-\lambda x^2} dx = \frac{1\times 3\ldots(2r-1)}{2^{r+1}} \sqrt{\left(\frac{\pi}{\lambda^{2r+1}}\right)} \qquad \left[\begin{array}{c} r \text{ an integer} \geq 1 \\ \lambda > 0 \end{array}\right]$$

$$\int_0^\infty x^{2r+1} e^{-\lambda x^2} dx = \frac{r!}{2\lambda^{r+1}} \qquad \left[\begin{array}{c} r \text{ an integer} \geq 0 \\ \lambda > 0 \end{array}\right]$$

## USEFUL UNITS

Length:

$$\frac{\hbar^2}{e^2 m} = \text{Bohr radius} = 0{\cdot}529 \times 10^{-10} \text{ m}$$

$$1 \text{ parsec} = 3{\cdot}086 \times 10^{16} \text{ m}$$

$$1\text{Å} = 10^{-10} \text{ m}$$

$$1 \text{ light year} \sim 9{\cdot}46 \times 10^{12} \text{ km}$$

Time:

$$\frac{\hbar^3}{e^4 m} \sim \left\{\begin{matrix}\text{time of description of lowest} \\ \text{Bohr orbit in H-atom}\end{matrix}\right\} \sim 2{\cdot}419 \times 10^{-17} \text{ s}$$

$$1 \text{ year} = 3{\cdot}156 \times 10^{7} \text{ s}$$

$$1 \text{ aeon} = 10^{9} \text{ years}$$

Energy: $$E(\text{eV}) = \frac{1240{\cdot}8}{\lambda(\text{nm})} \quad (1\text{nm} = 10^{-9}\text{m})$$

$$\frac{me^4}{2\hbar^2} \sim \left\{\begin{matrix}\text{ionization energy} \\ \text{of H-atom}\end{matrix}\right\} \sim 13{\cdot}6 \text{ eV}$$

Fundamental constants:

Electron charge $e \sim 4{\cdot}803 \times 10^{-10}$ e.s.u. $\sim 1{\cdot}602 \times 10^{-19}$ C

Electron mass $m \sim 9{\cdot}110 \times 10^{-31}$ kg

Proton mass $M \sim 1{\cdot}673 \times 10^{-27}$ kg

Planck's constant $h \sim 6{\cdot}626 \times 10^{-34}$ Js $\sim 6{\cdot}626 \times 10^{-27}$ erg s

$$\hbar = h/2\pi \sim 1{\cdot}054 \times 10^{-34} \text{ Js}$$
$$= 6{\cdot}582 \times 10^{-16} \text{ eVs}$$

Fine-structure constant $e^2/\hbar c \sim 1/137{\cdot}036$

Speed of light *in vacuo* $c \sim 2{\cdot}9979 \times 10^{8}\text{m s}^{-1}$

Gravitational constant $G \sim 6{\cdot}673 \times 10^{-11}$ N m$^2$ kg$^{-2}$
$\sim 6{\cdot}673 \times 10^{-8}$ dynes cm$^2$ g$^{-2}$

Boltzmann's constant $k \sim 1{\cdot}3806 \times 10^{-23}$ JK$^{-1}$ $\sim 1{\cdot}3806 \times 10^{-16}$ erg k$^{-1}$
$= 0{\cdot}8618 \times 10^{-4}$ eV K$^{-1}$

Energy density of black body radiation $aT^4$

$$a = \frac{\pi^2 k^4}{15 c_0^3 \hbar^3} = 7 \cdot 5641 \times 10^{-16}\ \mathrm{Jm^{-3}\ K^{-4}}$$

Emissive power of a black body $\sigma T^4$, $\sigma$ is the Stefan-Boltzmann constant

$$\sigma = \tfrac{1}{4}\, a c_0 = 5 \cdot 6703 \times 10^{-8}\ \mathrm{Wm^{-2}\ K^{-4}}$$

ASTRONOMY AND COSMOLOGY

Solar radius = $6 \cdot 960 \times 10^8$ m, solar mass = $1 \cdot 989 \times 10^{30}$ kg = $M$

Earth radius = $6 \cdot 371 \times 10^6$ m, earth mass = $5 \cdot 977 \times 10^{24}$ kg

Mean earth-sun distance = $1 \cdot 496 \times 10^{11}$ m

Hubble constant $\sim (18 \times 10^9 \text{ years})^{-1}$

Equivalent mass density of $2 \cdot 7$ K background radiation $\sim 10^{-31}$ kg m$^{-3}$.

Average matter density from galaxy counts $2 \times 10^{-28}$ kg m$^{-3}$.

Planck mass $m_{\mathrm{Pl}} = (\hbar c_0 / G)^{\frac{1}{2}} \sim 2 \cdot 18 \times 10^{-8}$ kg.

Planck length $(\hbar G / c_0^3)^{\frac{1}{2}} \sim 10^{-33}$ cm.

Astronomical orders of magnitudes

| | Earth | Sun | White dwarf | Neutron Star |
|---|---|---|---|---|
| Diameter (km) | $1 \cdot 3 \times 10^4$ | $1 \cdot 4 \times 10^6$ | $2 \times 10^4$ | 13 |
| Number in our galaxy | 1 | $10^{10}$ for spectral class G (to which our sun belongs) | $4 \times 10^9$ | $10^5$ (?) |
| Density (kg m$^{-3}$) | $5 \cdot 5$ | 150 | $10^9$ (at centre) | $10^{17}$ (at centre) |
| Escape velocity at surface | $10^{-5}\ c_0$ | $10^{-3}\ c_0$ | $10^{-2}\ c_0$ | $0 \cdot 7\ c_0$ |
| Temperature (K) (surface) | 300 | 6000 | $10^4$ | $10^7$ |

Ages in $10^9$ years (order of magnitude only):

Oldest rocks 4·3, sun 5·0, material in the galaxy 11-18, oldest stars 10-20, time since the last big bang 20.

ESTIMATES OF RATES OF ENERGY TRANSMISSION

[For 'per capita' statements a population of four thousand million is assumed.]

| | |
|---|---|
| Radiation by the sun | $4 \times 10^{26}$ W |
| Solar radiation intercepted by earth p. cap | $2{\cdot}4 \times 10^{7}$ W |
| Photosynthesis energy yield p.cap | $2{\cdot}5 \times 10^{4}$ W |
| World energy consumption p.cap | $2{\cdot}0 \times 10^{3}$ W |
| World food consumption p.cap | $1{\cdot}0 \times 10^{2}$ W |
| Clear sunlight in summer | $10^{3}$ $Wm^{-2}$ |
| Daylight (overcast) | $10^{2}$ $Wm^{-2}$ |
| Moonlight | $10^{-2}$ $Wm^{-2}$ |
| Human colour perception down to | $10^{-3}$ $Wm^{-2}$ |
| Human black-white perception down to | $10^{-7}$ $Wm^{-2}$ |

# Additional reading

There are very few *small* books which furnish a parallel introduction to *both* thermodynamics *and* statistical mechanics (as does this book). R4 is perhaps the nearest to the **core** of this book in that respect. One or other of the two disciplines is emphasized in the other books mentioned below, and there are so many good books in this category that one cannot know all of them. However, a reasonably representative sample is enumerated below.


R1. I.P.Bazarov (1964). *Thermodynamics*, Pergamon Press, Oxford. 1964. Translated from the Russian; an excellent book; contains no statistical mechanics.

R2. H.B.Callen (1960). *Thermodynamics*, Wiley, New York. A good introduction to thermodynamics, irreversible thermodynamics, and fluctuation theory; contains no statistical mechanics.

R3. P. Dennery (1972). *An introduction to statistical mechanics*, Allen and Unwin, London. Short (117 pp); contains no thermodynamics as such; supplements this book by having a chapter on the imperfect classical gas.

R4. B. Jancovici (1973). *Statistical physics and thermodynamics*, McGraw-Hill, London. Short (147 pp); overlaps with the contents of this book at a more elementary level; crisply written.

R5. L.D.Landau and E.M.Lifshitz (1968). *Statistical physics*, Pergamon Press, Oxford. Large; attempts to answer most questions one is liable to raise in an intuitively appealing way.

R6. P.T.Landsberg, *Thermodynamics with quantum statistical illustrations* (1961). Wiley-Interscience, New York.

R7. P.T.Landsberg (ed). (1970). *International conference on thermodynamics*, Cardiff 1970, Butterworths, London. Problems of cosmology, axiomatics, time, and teaching are discussed in this book.

R8. P.T.Landsberg (ed). (1971). *Problems in thermodynamics and statistical physics*, Pion, London. Has a very wide coverage in the problems-and-solutions format.

R9. A.Münster, (1969). *Statistical thermodynamics*, Vol.I, Springer, Berlin. A specialized treatise; contains no introduction to thermodynamics as such.

R10. R.K.Pathria (1972). *Statistical mechanics*, Pergamon Press, Oxford. A modern treatise; contains no introduction to thermodynamics as such; gives details of interacting systems and phase transitions.

R11. G.S.Rushbrooke (1949). *Introduction to statistical mechanics*, Oxford University Press, London. Clear, not too mathematical; does not contain quantum statistics.

R12. E.B.Stuart, B.Gal-Or, and A.J.Brainard (eds.) (1970). *A critical review of thermodynamics*, Mono Book Corp., Baltimore. A recent book on topics of current interest.

R13. R.C.Tolman (1938). *Principles of statistical mechanics*, Oxford University Press, London. Large and without an introduction to thermodynamics; wordy but lucid. (Dover reprint)

R14. G.Weinreich, (1968). *Fundamental thermodynamics*, Addison-Wesley, Reading, Mass. Imaginative and clear; contains no statistical mechanics.

R15. G.A.P. Wyllie (1970). *Elementary statistical mechanics*, Hutchinson, London. Short (148 pp.) and good; contains little thermodynamics; supplements this book by discussing order-disorder problems and by giving more details of interacting systems.

R16. M.W.Zemansky (1968). *Heat and thermodynamics (5th edn.)*. McGraw-Hill, New York. A large, standard book also containing a short introduction to kinetic theory and statistical mechanics.

# Bibliographical notes

CHAPTERS 1, 2

These chapters are based on R6.

CHAPTER 3

Section 3.3
The reader is warned that a number of 'bogus proofs'that the propagation of sound is adiabatic exist in the literature.

CHAPTER 3

Section 3.6
The zeroth law of thermodynamics has been the subject of a certain amount of recent discussion. The example given in section 3.6 comes from

P.T. Landsberg (1970). Main ideas in the axiomatics of thermodynamics, *Pure Appl. Chem.* **22**, 215.

Other references include:

H.A. Buchdahl (1973). Concerning the absolute temperature function, *Am. J. Phys.* **41**, 98.

R.K. Wangness (1971). Perpetual motion of the zeroth kind, *Am. J. Phys.* **39**, 898.

O. Redlich (1970). The so-called zeroth law of thermodynamics. *J. Chem. Educ.* **47**, 740.

L.A. Turner (1963). Temperature and Carathéodory's treatment of thermodynamics, *J. Chem. Phys.* **38**, 1163.

There are also several more advanced mathematical treatments of the axiomatics of thermodynamics in which the zeroth law is discussed, and these are cited in the references above.

CHAPTER 4

Axiomatic thermodynamics can be handled in a variety of ways. For references to recent work see R7.

CHAPTER 5

Section 5.5
For an introduction to problems concerning entropy and time and recent references see R7, and also the following:

P.C.W. Davies (1974). *The physics of time asymmetry*, Surrey University Press, London.

B. Gal-Or (ed.) (1974). *Modern developments in thermodynamics*, Wiley, New York.

T. Gold and D.L. Schumacher (eds.) (1967). *The nature of time*, Cornell University Press, Ithaca, N.Y.

P.T. Landsberg (1975). *A matter of time*, University of Southampton Inaugural Lecture.

For the inequality on p. 52 see

P.T. Landsberg (1978). A thermodynamic proof of the inequality between arithmetic and geometric means, *Phys. Lett.* **67A**, 42.

*Problem 5.5*
For a discussion of the Carnot cycle with irreversible or non-static elements see, for example, the following.

H.S. Leff and G.L. Jones (1975). Irreversibility, entropy production, and thermal efficiency. *Am. J. Phys.* **43**, 973.

*Problem 5.9*

J.M. Walsh, M.H. Rice, R.G. McQueen, and F.L. Yarger (1957). Shockwave compression of 27 metals. *Phys. Rev.* **108**, 196, equation (15).

*Problem 5.10*

J.F. Ury (1956). A diagram of state for superheated steam, *Appl. sci. Res., Hague A* **6**, 141.

## CHAPTER 6

Section 6.3
The view that it is possible to deduce Nernst's heat theorem from the principle of the unattainability of the absolute zero has been championed for many years, the classical exposition being that of

R.H. Fowler and E.A. Guggenheim (1939). *Statistical thermodynamics*, pp. 223-7, Cambridge University Press, Cambridge.

This view was questioned by

P.T. Landsberg (1956). Foundations of thermodynamics, *Rev. mod. Phys.* **28**, 363

and this matter is expounded in section 6.3. There is little subsequent literature regarding this point and the Fowler and Guggenheim argument still appeared in

E.A. Guggenheim (1967). *Thermodynamics* (5th edn), p. 192, North-Holland, Amsterdam.

Recent discussions of the third law for magnetic systems were given in

E.P. Wohlfarth (1977). The third law of thermodynamics and the theory of spin glasses, *Phys. Lett.* **61A**, 143.

D. Sherrington and J. F. Fernández (1977). Non-analyticity of the entropy and Nernst's law in random systems, *Phys. Lett.* **62A**, 457.

P.W. Anderson, D.J. Thouless and R.G. Palmer (1977). Comment on Wohlfarth's letter, *Phys. Lett.* **62A**, 456.

## CHAPTER 7

Section 7.1
The term *fourth law of thermodynamics* was first used in this context in R6.

Section 7.2
The escaping pressure is explained in this way in R14.

Section 7.4
The connection between the Euler formula and the phase rule is relatively unexplored, and it is not known if it is accidental or of deeper significance. See

D.H. Rouvray (1971). Graph theory in chemistry. *R. Inst. Chem. Rev.* **4**, 173.

Some recent discussion of the phase rule will be found in

*Chem. Br.* **9**, 241, 371, 467 (1973).

Section 7.5
Modern work on phase transitions is introduced very readably in the following

D.L. Goodstein (1975). *States of matter*, Prentice-Hall, Englewood Cliffs, N.J. (Dover reprint)

J.S. Rowlinson (1969). *Liquids and liquid mixtures.* (2nd edn.), Butterworths, London.

Section 7.7

J.T. Lopuszański (1968). On the method of establishing the stability conditions for thermodynamic equilibrium, *Acta Physica Polonica*, **33**, 953.

Sections 7.8, 7.9
For static phenomena near critical points see

L.P. Kadanoff, W. Götze, D. Hamblen, R. Hecht, E.A.S. Lewis, V.V. Palciauskas, M. Rayl, J. Swift, D. Aspnes, and J. Kane (1967). Static phenomena near critical points. *Rev. mod. Phys.*, **39**, 395.

Section 7.10
Chemical reaction studies are reviewed by

G. Nicolis and J. Portnow (1973). Chemical oscillations, *Chem. Rev.* 73, 365.

J. Schnakenberg (1976). Network theory of microscopic and macroscopic behaviour of master equation systems. *Rev. Mod. Phys.* **48**, 571.

H. Haken (1975). Cooperative phenomena in systems far from thermal equilibrium and in non-physical systems. *Rev. Mod. Phys.* **47**, 67.

A semi-conductor application is given in

P.T. Landsberg and A. Pimpale (1976). Recombination-induced non-equilibrium phase transitions in semiconductors. *J. Phys. C* **9**, 1243.

CHAPTER 8

Point (1) is based on

M.L. McGlashan (1977). Amount of substance and the mole. *Phys. Educ.* **12**, 276.

*Problem 8.2* is based on D.A. Lavis and G.M. Bell, Thermodynamic phase changes and catastrophe theory, *Bull. Inst. Maths. and its Appl.* **13**, 34 (1977).

## CHAPTER 9

### Section 9.1

The work of C.E. Shannon was published in

C.E. Shannon (1948). A mathematical theory of communication, *Bell Systems Tech. J.* **27**, 379.

The exposition in this section follows a paper by A.I. Khinchin (1953) published in translation in

A.I. Khinchin (1957). *Mathematical foundations of information theory*, Dover Publications, New York. Translation of a 1953 paper.

### Section 9.5

The classical reference on connection between density matrix and entropy is

J. von Neumann (1932). *Mathematische Grundlagen der Quantenmechanik*, Chap. V, § 2, Springer, Berlin.

The wide variety of views on the causes of entropy increase are illustrated by citations in

P.T. Landsberg (1972). Time in statistical physics and special relativity. In *The study of time* (eds. J.T. Fraser, F.C. Haber, and G. H. Müller) also in Landsberg (1970), *Stud. Gen.* **23**, 1108.

A leisurely exposition of H-theorems is given by

R.C. Tolman (1938). *The principles of statistical mechanics*, Oxford University Press, London. (Dover reprint)

A recent survey is given by

A. Huber (1970). An inequality for traces and its application to extremum principles and related theorems in quantum statistical mechanics, In *Methods and problems of theoretical physics, in honour of R.E. Peierls* (ed. J.E. Bowcock) North-Holland, Amsterdam.

*Problem 9.9* is based on F. Schlögl (1974), Specific heat as general statistical measure, *Z. Phys.* **267**, 77.

*Problem 9.10* is based on B.S. Westcott and P.T. Landsberg (1974), Entropy and $\nu$th-law rectification, *Int. J. Electron* **37**, 219.

## CHAPTER 10

### Section 10.3

The simple n-level example is due to

N.F. Ramsey (1956). Thermodynamics and statistical mechanics at negative absolute temperatures. *Phys. Rev.* **103**, 20.

See also

N. Bloembergen (1973). The concept of temperature in magnetism, *Am. J. Phys.* **41**, 325

for reviews of negative temperatures and magnetic systems. The latter are only touched on in this book (in section 7.9 and sections 10.3 to 10.6).

### Section 10.4

For unusual applications of the theory of sections 10.3 and 10.4 see

E. Callen and D. Shapero (1974). A theory of social imitation, *Phys. Today* July, p. 23.

D.F. Walls (1976). Non-equilibrium phase transitions in sociology. *Collect. Phenom.* **2**, 125.

Section 10.5
Equation (10.25) was given by

A. Abragam and W.G. Proctor (1958). Spin temperature, *Phys. Rev.* **109**, 1441.

The phase space analysis and (10.27) was given by

P.T. Landsberg (1959). Negative temperatures, *Phys. Rev.* **115**, 518

and in R6 where the generalized third law is also discussed.

Section 10.6
Carnot cycles at negative temperatures were discussed by the following and in references given therein.

J.G. Powles (1963). Negative absolute temperatures and rotating temperatures, *Contemp. Phys.* **4**, 338.

R.J. Tykodi (1976). Negative Kelvin temperature. *Am. J. Phys.* **44**, 997.

J. Dunning-Davis (1976). Negative absolute temperatures and Carnot cycles, *J. Phys. A* **9**, 605.

For the formulation of the second law in section 10.6.5 see

P.T. Landsberg (1977). Heat engines and heat pumps at positive and negative absolute temperatures, *J. Phys. A* **10**, 1773.

The example of a coupled heat engine and heat pump was given by

N. Kurti (1976). Thermodynamics and energy, in *Aspects of energy conversion* (eds. I.M. Blair, B.D. Jones, and A.J. van Horn), Pergamon Press, Oxford.

CHAPTER 11

This chapter is based on R6.

Section 11.6
This work is discussed fully from a classical point of view in a number of books (for example R11, p. 44).

Section 11.7
The rather general equipartition theorem (11.32) is due to R.C. Tolman (1914) and it is applied to special advantage in problem 11.8.

CHAPTER 12

This work is based on R6. It is extended in the problems to give an introduction to the relativistic quantum gas. For further details of these systems see the following and references cited therein:

W.D. Bauer, E. Hilf, K. Koebke, and F. Schmitz (1974). The perfect quantum gases. In *Modern developments in thermodynamics* (ed. B. Gal-Or), Wiley, New York.

H. Arzeliès (1968). *Thermodynamique relativiste et quantique*, Gauthier-Villars, Paris.

A. Lanlignel (1968). Theorie quantique relativiste des gaz parfaits de Fermi et de Bose, *Acta Phys. Polon.* **34**, 895.

J.L. Anderson and H.R. Witting (1974). Relativistic quantum transport coefficients, *Physica* **74**, 489.

The expansion of an ideal relativistic gas and its Bose condensation are discussed in the following:

P.T. Landsberg, W.C. Saslaw, and A.J. Haggett (1971). On the adiabatic expansion of relativistic gas, *Mon. Not. R. Astron. Soc.* **154**, 7P.

P.T. Landsberg and J. Dunning-Davies (1965). Ideal relativistic Bose condensation, *Phys. Rev. A*, **138**, 1049.

The result of problem 12.9(b) was given by

R.M. May (1964). Quantum statistics ofideal gases in two dimensions, *Phys. Rev. A* **135**, 1515.

## CHAPTER 13

### Section 13.4

For reviews of cosmology see for example

E.R. Harrison (1968). The early universe, *Phys. Today*, June, p. 21.

H. Bondi (1952). *Cosmology*, Cambridge University Press, Cambridge.

S. Weinberg (1972). *Gravitation and cosmology*, Wiley, New York.

S. Weinberg (1977). *The first three minutes*, André Deutsch, London.

D.W. Sciama (1971). *Modern cosmology*, Cambridge University Press, Cambridge.

M. Berry (1976). *Principles of cosmology and gravitation*, Cambridge University Press, Cambridge.

M. Rowan-Robinson (1977). *Cosmology*, Oxford University Press, Oxford.

P.T. Landsberg and D.A. Evans (1977). *Mathematical cosmology: An introduction*, Oxford University Press, Oxford.

Discussions with Dr. J.G. McEwen, Physics Department, Southampton University, are gratefully acknowledged in connection with the construction of Table 13.1.

The 3K cosmic black-body background radiation is much discussed in the cosmological literature as it furnishes a strong argument for the big-bang hypothesis. It is isotropic to an accuracy of order 1 in $10^{-4}$ (when allowance is made for a speed of order 600 km $s^{-1}$ of our galaxy through it). This represents probably the most accurate of the important measurements in cosmology. For a general recent discussion see M. Rowan-Robinson (1977). Aether drift detected at last, *Nature* **270**, 9.

### Section 13.6

The problem of negative heat capacities has been considered by

D. Lynden-Bell and R.M. Lynden-Bell (1977). On the negative specific heat paradox, *Mon. Not. R. astr. Soc.* **181**, 405.

W. Thirring (1970). Systems with negative specific heat, *Z. Phys.* **235**, 339.

P. Hertel and W. Thirring (1971). A soluble model for a system with negative specific heat, *Ann. Phys.* **63**, 520.

G. Horwitz and T. Katz (1977). Steepest descent technique and stellar equilibrium statistical mechanics, *Astrophys. J.* **33**, 251.

Section 13.7

C.C.O. Ezeilo (1977). Effective radiative disc radius of the sun, *Solar Energy* **19**, 387.

Section 13.8

G.E. Nicodemus (1963). Radiance, *Am. J. Phys.* **31**, 368.

R.T. Ross, Jr. (1966). Thermodynamic limitations on the conversion of radiant energy into work, *J.Chem.Phys.* **45**, 1.

P.T. Landsberg and D.A. Evans (1968). Thermodynamic limits for some light-producing devices, *Phys. Rev.* **166**, 242.

P.T. Landsberg (1977). A note on the thermodynamics of energy conversion in plants, *Photochemistry and photobiology* **26**, 313.

Optical refrigeration was found experimentally by

T. Kushida and J.E. Geusic (1968). Optical refrigeration in Nd-doped yttrium aluminium garnet, *Phys. Rev. Lett.* **21**, 1172.

The thermodynamic inequality for the efficiency of light emission was tested experimentally by

W.J. Bitter, Intradev, and F. Williams (1969). Electroluminescence of ZnS single crystals, *J. Phys. Chem. Solids* **30**, 503.

See also

O. Kafri and R.D. Levine (1974). Thermodynamics of adiabatic laser processes: optical heaters and refrigerators, *Optics Commun.* **12**, 118.

For recent reviews of photosynthesis see

R.S. Knox (1977). Photosynthetic efficiency and exciton transfer and trapping. In *Primary Processes of Photosynthesis* (Ed. J. Barber). Elsevier/North Holland, Amsterdam.

G. Porter and M.D. Archer (1976). In vitro photosynthesis, *Interdisciplinary Science Reviews* **1**, 119.

Section 13.9

P.T. Landsberg and J.R. Mallinson (1976). Thermodynamic constraints, effective temperatures and solar cells, In *Electricité solaire (colloque internationals), Toulouse, 1976*. Nationale d'Etudes Spatiales, Toulouse.

*Problem (13.9)*
The temperature distribution in the lower levels of the atmosphere lies between 220 and 260 K and is discussed from the theoretical and experimental point of view by, for example,

W.L. Smith (1970). Iterative solution of the radiative transfer equation for the temperature and absorbing gas profile of an atmosphere, *Appl. Optics* **9**, 1993.

*Problem (13.11)*
is based on

H.P. Baltes (1974). Comment on black-body radiation in small cavities, *Am. J. Phys.* 42, 505. See also H.P. Baltes (1976), Planck's radiation law for finite cavities and related problems, *Infrared Phys.* **16**, 1.

## CHAPTER 14

Fig. 14.3 is based on D.G. Henshaw and A.D.B. Woods (1961). Modes of atomic motions in liquid helium by inelastic scattering of neutrons, *Phys. Rev.* **121**, 1266.

## CHAPTER 15

### Sections 15.1, 15.2, 15.3

These sections give an outline of the Sommerfeld theory of metals (1927); among recent developments note

> Y. Weissman (1976). Quantum spatial distribution of a non-degenerate free charge carrier gas in semiconductors subject to high external electric fields at low temperatures, *J. Phys. C* **9**, 2353.

A temperature-proportional heat capacity has also been found recently for certain amorphous materials such as $SiO_2$, $GeO_2$, and Se by

> R.C. Zeller and R.O. Pohl (1971). Thermal conductivity and specific heat of noncrystalline solids, *Phys. Rev.* **B4**, 2029.

A theoretical discussion of this point is given by

> P.W. Anderson, B.I. Halperin, and C.M. Verma (1972). Anomalous low-temperature thermal properties of glasses and spin glasses, *Phil. Mag.* **25**, 1.

### Sections 15.4 and 15.5

Section 15.4 gives A.H. Wilson's theory of semiconductors (1931). The statistics of defects given in section 15.5 is based on

> P.T. Landsberg (1952). A note on the theory of semiconductors, *Proc. Phys. Soc. A* **65**, 604.

> P.T. Landsberg (1956). Defects with several trapping levels in semiconductors, *Proc. Phys. Soc. B* **69**, 1056.

Section 15.7 is developed from

> J. Orear and E.E. Salpeter (1973). Black holes and pulsars in the introductory physics course, *Am. J. Phys.* **41**, 1131.

For further introductory material see also

> P.F. Browne (1975). Aspects of stellar evolution reappraised, *Contemp. Phys.* **16**, 51.

> S. Chandrasekhar (1974). The black hole in astrophysics: The origin of the concept and its roles.

> I. Asimov (1977). *The collapsing universe*, Hutchinson, London.

For a semi-popular discussion of white holes, see

> J. Gribbin (1977). *White holes*. Paladin, St. Albans.

### Section 15.8

Black hole thermodynamics is discussed in the following

> S.W. Hawking (1976). Black holes and thermodynamics, *Phys. Rev. D* **13**, 191.

> J.D. Bekenstein (1975). Statistical black-hole thermodynamics, *Phys. Rev. D* **12**, 3077.

P.C.W. Davies (1977). The thermodynamic theory of black holes, *Proc. Roy. Soc. A* **353**, 499.

D.W. Sciama (1976). Black holes and their thermodynamics, *Vistas Astron.* **19**, 385.

V.P. Frolov (1976). Black holes and quantum processes in them, *Sov. Phys.-Uspekhi* **19**, 244.

Section 15.9
The discussion follows

M.J. Adams and P.T. Landsberg (1969). The theory of the injection laser, in *Semiconductor Lasers* (ed. C.H. Gooch), Wiley, London.

Equation (15.59) was derived by

M.G.A. Bernard and B. Duraffourg (1961). Laser conditions in semiconductors, *Phys. Status Solidi* **1**, 699.

Equation (15.60) was derived by

P.T. Landsberg (1967). The condition for negative absorption in semiconductors, *Phys. Status Solidi* **19**, 777.

*Problem (15.5)*
was developed from

B. Kuchowicz (1974). Neutrino statistics: Elementary problems and some applications, *Fortschritte d. Phys.* **22**, 525.

*Problem (15.8)*
is based on

P.T. Landsberg (1960). In *Solid state physics in electronics and telecommunication* (ed. M. Désirant) **1**, 436, Academic Press, London.

CHAPTER 16

Section 16.4
The section is based on

A.A. Kokolov and G.V. Skrotsky (1972). Fluctuations of energy of the thermal radiation, *Lett. Nuovo Cim.* **4**, 637.

P.T. Landsberg (1975). The radiation law and the systematics of ideal fluids. In *Proceedings of the International Research Symposium on Statistical Physics, 1974*. S. Bose Institute of Physical Sciences, Calcutta.

Section 16.6
There is a good literature on Bose condensation and its connection with liquid helium. For example, see R10 and

J.G.M. Armitage and I.E. Farquhar (eds.) (1975). *The helium liquids*, Academic Press, London.

D. Baeriswyl and W. Czaja (1975). Bose-Einstein condensation in exciton systems and liquid helium, *RCA Rev.* **36**, 5.

For photon bunching mentioned on p. 314 see for example

J.P. Dudeja and S. Chopra (1977). Temperature-dependence of photon bunching in thermal radiation, *Phys. Lett.* **64A**, 271.

Section 16.7
For extension of this work, see

G. Nicolis and I. Prigogine (1971). Fluctuations in nonequilibrium systems, *Proc. Nat. Acad. Sci. (U.S.A.)* **68**, 2102.

## CHAPTER 17

The difference between the generalized Einstein relation (17.24), which allows for carrier degeneracy, and the simple non-degenerative form appears to explain certain aspects of semi-conductor device characteristics; see for instance

P.T. Landsberg (1952). On the diffusion theory of rectification, *Proc. R. Soc. A* **213**, 226.

K.A. Shore and M.J. Adams (1976). The effect of carrier degeneracy on transport properties of the double heterostructure injection laser, *Appl. Phys.* **9**, 161.

*Problem (17.2)*
References for problem 17.2 are

R. Stratton (1972). Semiconductor current-flow equations, *I.E.E.E. Trans. ED* **19**, 1288.

P.T. Landsberg and S.A. Hope (1976). Diffusion currents in semiconductors, *Solid-state Electronics*, **19**, 173.

## CHAPTER 18

Section 18.2
The virial theorem for Coulomb interactions in $d$ dimensions has recently been noted by

G. Toulouse (1975). Le gaz d'électrons et la transition métal-isolant a $d$ dimensions, *J. Phys. (Paris)* **36**, 1137.

M. Parrinello and N.H. March (1976). Thermodynamics of Wigner crystallization, *J. Phys. C* **9**, L147.

Section 18.3
The model of this section has been discussed in the following

R.M. May (1967). Exact equation of state for a 2-dimensional plasma, *Phys. Lett.* **25A**, 282.

G. Knorr (1968). The partition function of a two-dimensional plasma, *Phys. Lett.* **28A**, 166.

E. H. Hauge and P.C. Hemmer (1971). *Phys. Norv.* **5**, 209.

J.J. Gonzãlez and P.C. Hemmer (1972). On the existence of a two-dimensional classical plasma, *Phys. Lett.* **42A**, 263.

Section 18.4
For a general introduction to the topic of this section see

T.S. Moss, G.J. Burrell, and B. Ellis (1973). *Semiconductor optielectronics*, Butterworths, London.

P.T. Landsberg (1975). An introduction to the theory of photovoltaic cells, *Solid State Electron.* **18**, 1043.

Section 18.5
The discussion of special relativistic thermodynamics in recent years was initiated by

H. Ott (1963). Lorentz Transformation der Wärme, *Z. Phys.* **175**, 70.

This was a posthumous publication suggesting that a moving body appears hot. See also the bibliographical notes for Chapter 12 and Appendix IV.

Section 18.6
The Carnot cycle in general relativity has been discussed by

R. Ebert and R. Göbel (1973). Carnot cycles in general relativity, *Gen. Relativity Gravitation* **4**, 375.

The whole subject of general relativistic problems in thermodynamics and cosmology is discussed by

R.C. Tolman (1934). *Relativity, thermodynamics and cosmology*, Clarendon Press, Oxford. (Dover Reprint)

For more recent references see

P.T. Landsberg and D. Park (1975). Entropy in an oscillating universe, *Proc. R. Soc.* **346**, 485.

Section 18.7
For further information see

R. Hagedorn (1965). Statistical thermodynamics of strong interactions at high energies, *Suppl. Nuov. Ciment.* **3**, 147.

H. Satz (1977). Statistical concepts in hadron physics, *13th I.U.P.A.P. Conference on Statistical Physics, Haifa, August 1977.*

Section 18.8
This section is influenced by

H.B. Callen (1974). Thermodynamics as a science of symmetry, *Found. Phys.* **4**, 423.

Section 18.9
For a popular exposition of the order-from-fluctuations concept see

I. Prigogine, G. Nicolis, and A. Babloyantz (1972). Thermodynamics of evolution, *Phys. Today*, November and December.

A detailed account is given in

G. Nicolis and I. Prigogine (1977). *Self-organisation in nonequilibrium systems*, New York.

For the order-from-growth concept see the reference Landsberg (1972) for section 9.5.

APPENDIX I

Section 1.5
See the reference Landsberg (1972) under the bibliographical notes for section 9.5.

APPENDIX IV

Section IV.1
This section is based on

P.T. Landsberg (1970). Special relativistic thermodynamics - a review. In *A critical review of thermodynamics* (eds. E.B. Stuart, B. Gal-Or, and A.J. Brainard), Mono Book Corporation, Baltimore.

See also

D. Eimerl (1975). On relativistic thermodynamics, *Ann. Phys.* **91**, 481

and references given there.

The erroneous argument noted at the end of section IV.2 appears in

E.A. Guggenheim (1967). *Thermodynamics* (5th edn.) North-Holland, Amsterdam.

E.A. Guggenheim (1967). *Elements and formulae of special relativity*, chap. 6, Pergamon Press, Oxford.

The data on energy transmission on p.438 are based on an Appendix in

L.O. Björn, *Light and life* (London: Hodder and Stoughton, 1976).

# Author index

# Subject index

Page numbers in italic refer to Problems (Solutions of Problems are not indexed).

A CATALOG OF SELECTED

# DOVER BOOKS

IN SCIENCE AND MATHEMATICS

# A CATALOG OF SELECTED
# DOVER BOOKS
## IN SCIENCE AND MATHEMATICS

QUALITATIVE THEORY OF DIFFERENTIAL EQUATIONS, V.V. Nemytskii and V.V. Stepanov. Classic graduate-level text by two prominent Soviet mathematicians covers classical differential equations as well as topological dynamics and erqodic theory. Bibliographies. 523pp. 5⅜ × 8½. 65954-2 Pa. $10.95

MATRICES AND LINEAR ALGEBRA, Hans Schneider and George Phillip Barker. Basic textbook covers theory of matrices and its applications to systems of linear equations and related topics such as determinants, eigenvalues and differential equations. Numerous exercises. 432pp. 5⅜ × 8½. 66014-1 Pa. $8.95

QUANTUM THEORY, David Bohm. This advanced undergraduate-level text presents the quantum theory in terms of qualitative and imaginative concepts, followed by specific applications worked out in mathematical detail. Preface. Index. 655pp. 5⅜ × 8½. 65969-0 Pa. $10.95

ATOMIC PHYSICS (8th edition), Max Born. Nobel laureate's lucid treatment of kinetic theory of gases, elementary particles, nuclear atom, wave-corpuscles, atomic structure and spectral lines, much more. Over 40 appendices, bibliography. 495pp. 5⅜ × 8½. 65984-4 Pa. $11.95

ELECTRONIC STRUCTURE AND THE PROPERTIES OF SOLIDS: The Physics of the Chemical Bond, Walter A. Harrison. Innovative text offers basic understanding of the electronic structure of covalent and ionic solids, simple metals, transition metals and their compounds. Problems. 1980 edition. 582pp. 6⅛ × 9¼. 66021-4 Pa. $14.95

BOUNDARY VALUE PROBLEMS OF HEAT CONDUCTION, M. Necati Özisik. Systematic, comprehensive treatment of modern mathematical methods of solving problems in heat conduction and diffusion. Numerous examples and problems. Selected references. Appendices. 505pp. 5⅜ × 8½. 65990-9 Pa. $11.95

A SHORT HISTORY OF CHEMISTRY (3rd edition), J.R. Partington. Classic exposition explores origins of chemistry, alchemy, early medical chemistry, nature of atmosphere, theory of valency, laws and structure of atomic theory, much more. 428pp. 5⅜ × 8½. (Available in U.S. only) 65977-1 Pa. $10.95

A HISTORY OF ASTRONOMY, A. Pannekoek. Well-balanced, carefully reasoned study covers such topics as Ptolemaic theory, work of Copernicus, Kepler, Newton, Eddington's work on stars, much more. Illustrated. References. 521pp. 5⅜ × 8½. 65994-1 Pa. $11.95

PRINCIPLES OF METEOROLOGICAL ANALYSIS, Walter J. Saucier. Highly respected, abundantly illustrated classic reviews atmospheric variables, hydrostatics, static stability, various analyses (scalar, cross-section, isobaric, isentropic, more). For intermediate meteorology students. 454pp. 6⅛ × 9¼. 65979-8 Pa. $12.95